PRACTICAL TRANSFORMER DESIGN HANDBOOK

2ND EDITION

ERIC LOWDON

TAB Professional and Reference Books

Division of TAB BOOKS Inc.
Blue Ridge Summit, PA

To Isobel
For an endless supply of tea, coffee, patience, and encouragement.

SECOND EDITION
FIRST PRINTING

Copyright © 1989 by Eric Lowdon.
First edition copyright © 1980 by Eric Lowdon.
Printed in the United States of America

Library of Congress Cataloging in Publication Data

Lowdon, Eric.
    Practical transformer design handbook / by Eric Lowdon. — 2nd ed.
       p.   cm.
    Includes index.
    ISBN 0-8306-3212-3
    1. Electric transformers—Design and construction—Handbooks, manuals, etc. I. Title.
TX2551.L68   1988
621.31′4—dc19   88-7782
                                                            CIP

TAB BOOKS Inc. offers software for sale. For information and a catalog, please contact TAB Software Department, Blue Ridge Summit, PA 17294-0850.

Questions regarding the content of this book should be addressed to:

    Reader Inquiry Branch
    TAB BOOKS Inc.
    Blue Ridge Summit, PA 17294-0214

Contents

Transformer Resistance—Losses and Guesstimates—The Geometry of Losses—More on Regulation—Losses and Temperature Rise

Preface to the Second Edition

In the past decade or so since this book was first written, little has changed in transformer design basics. Therefore the content of this edition is also little changed. The slant and style, too, which were appreciated in the first edition, remain as before.

There are, however, *some* changes in the book. The main justification for calling it, rather grandly, the *Second Edition* is that its physical format has been greatly improved by the new publisher. Thicker than before, it is otherwise dimensionally smaller and thus easier to use and easier to store in the normal bookshelf spacing.

A list of reference material has been added. Some of these references are rather old, but nevertheless, they are among the best at the required technical level. Moreover, in some cases they are the *only* ones available on a subject that has received scant attention in recent decades. Some references might not be easy to find; check out the big public libraries and the engineering libraries of colleges and universities. The opportunity has been taken, as well, to improve the clarity of the text by cleaning up some minor errors and omissions.

A new chapter, Chapter 17, has been added. Here, in the light of the recent furor in scientific circles surrounding new discoveries in *superconductivity*, I thought it would be instructive and even entertaining to speculate on the possible effects of these events, among other things, on the future of low-frequency power transformer design.

Finally, I extend sincere thanks to the good people of Allied Chemical Company for data supplied.

Introduction

The fact that this book is in your hands indicates that you have at least a passing interest in transformers. The title caught your eye, you glanced at the table of contents, skimmed through the index, briefly ruffled through the pages, and then you returned to the introduction. Now you want to know if it was written with your interests in mind.

Even though we haven't met, the author's confident reply is "Quite likely." Surely you are a circuit designer, laboratory technician, experimenter, amateur, hobbyist, electrical science student, or teacher—or none of these things, but you want to know something about transformers for one reason or another. Whatever the case, the fact that you have an interest means that there is probably something here for you.

This might seem like a wide range of levels of interest to cater for in one small book, and so it is. Although the book does contain material at professional and academic levels, it is essentially a practical how-to-design-it book, constructed on a substantial but simple how-it-works frame. After all, even professionals don't really object if things are made easy, do they?

However, this is not a project book. There are no "fasten A to B using C mount on D" kinds of instructions. Rather, it deals with the *methods* used in designing transformers to meet one's own specific needs. Construction data is dealt with in broad, general terms.

This text pays a great deal of attention to the mechanical as well as the electrical aspects of design, emphasizing the need to ensure that the design will in fact physically fit together like the proverbial Swiss watch. This consideration is especially important to the casual experimenter who can't afford to disassemble his or her work and start afresh just because allowances made were insufficient for the the various fits.

There is mathematics, but no calculus—only elementary algebra and arithmetic. This is all reviewed in Chapter 1, which you have the option to read. Many readers, however, find the tutorial exercises in Chapter 1 and the general theory in Chapter 2 useful as a short refresher course before getting down to the transformer specifics beginning in Chapter 3.

I have taken the liberty of developing a line of thought along simple mathematical lines when it seems useful or interesting to do so. The reader is not obliged to take part in these excursions. The final answer—the one that counts—is always on the bottom line. In addition, most of the bottom lines—the working formulas, in other words—are gathered together for convenience in Chapter 7. This makes it easy to select the equation most comfortable for one's style of working or the one most appropriate for the problem on hand.

Thus, my view of potential readers is that you work or play at most levels of electrical/electronic crafts. You probably know what transformers look like and what they do. You have possibly worked with transformers and perhaps even know something of design parameters. Your technical education includes a "grounding" in ac theory (even if it is a little rusty at the present), and ohms, amperes, volts, and watts are familiar terms.

There are several respects in which the book departs from professional interests in favor of the casual experimenter. For example, I address the subject of money from time to time and the pecuniary advantages to be gained from rummaging through junk heaps and cannibalizing old TV and radio sets in search of cores and other materials. These discussions might be viewed with distaste by some, but you should remember that even in the highest echelons of engineering, cost-cutting is an important design factor. At these rarefied heights, however, the odor of money grubbing is disguised by perfumed phrases such as "cost-performance tradeoffs" and "sales profit margins." Industry rarely considers the city dump as a viable source of cheap cores, but the major reason for not doing so is that the cost of separating and classifying the materials would make it too expensive. The independent experimenter, however, works within a different arena and can usually explore such sources with advantage. Certainly he or she may follow the old pioneering axiom of "waste not, want not," which today is making a strong comeback in many fields.

In the days of my childhood, kids saw little merit in the mere acquisition of knowledge. The standard putdown hurled at those who attempted to inflict unsolicited information on my gang was the contemptuous accusation "you read that in a book," implying that *anyone* can do *that*. Nevertheless, very little of the material in here sprang full-fledged from the author's mind. Much of it was indeed culled from books, but it was also tempered through the years in the crucible of personal experience. Some of it was acquired at the feet of masters, and some might even appear as small fragments of original research. It is accordingly not just knowledge, then, but knowledge distilled from studying and working with transformers for a long time.

Finally, thanks are due to two courteous companies, Magnetic Metals and Indiana General, for their kind permission to use some of their material. Specifically, the author would like to thank the marketing managers of these companies, Brian J. Scully and John P. Breikner, for their cooperation. Thanks are also due to D. Douglas Ward, General Manager of Allen-Bradley—Magnetics Division. Allen-Bradley purchased the Indiana General assets and rights to manufacture the Indiana General soft ferrite lines in 1985.

ONE

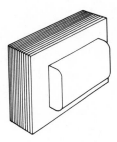

Symbols, Abbreviations, and Other References

Mathematical symbols are used to simplify writing equations that contain more than two or three factors. Ideally, they would consist of the initial letter(s) of the word(s) they represent, but in electrical work, this is only partially possible because of the vast number of different quantities and properties involved and because of the acute shortage of available letters.

ELECTRICAL AND ELECTRONIC SYMBOLS

For example, in the standard lexicography of electrical mathematics, the letter R immediately brings to mind the word "resistance." But then which symbols are used for reactance, remanence, resistivity, revolutions, reluctance, and so on? The only solution is to use different letter forms and to assign nonassociative symbols. For the examples just quoted, the commonly used symbols are X, B_r, p, η, and $\mathcal{R}$, respectively. See Table 1-1.

This kind of mind-bending memory stress continues with the use of symbols like I, E, Z, H, B, λ, ω, and Φ, to give just a small sample. These are the standard symbols, respectively, for current, voltage (or emf), impedance, magnetizing force, flux density, wavelength, angular velocity (of a vector), and total flux. The last three symbols are from the Greek alphabet. They are lambda (λ), omega (ω), and uppercase phi (Φ), respectively.

**Table 1-1. Electrical and Magnetic Symbols
and Abbreviations**

Property	Sym.	Unit	Abr.
Angular velocity	ω	radians per second	
Capacitance	C	farad	F
Capacitive reactance	X_C	ohms	Ω
Current density	J,S	amperes per unit area	
Current	I	ampere	A
Coercivity	H_C	oersted	Oe
Efficiency	η	-	
Emf, voltage	E, V	volts	V
Flux	Φ	maxwell or line	
Flux density	B	Gauss; Lines or max well per unit area	
Frequency	f	hertz	Hz
Impedance	Z	ohms	Ω
Inductance	L	henrys	H
Inductive reactance	X_L	OHMS	Ω
Magnetizing force	H	oersteds	Oe
Magnetomotive force	mmf	ampere-turns	
Permeability	μ		
Power, real	P	watts	W
Power, apparent	VA	volt-amperes	VA
Ratio	r, n		
Resistance	R	ohms	Ω
Reluctance	$\mathcal{R}$		
Remanence	B_r	gauss	
Specific resistance	p	ohms/unit length	

However, it is evident that even by adding the Greek alphabet to the ABCs of the Roman alphabet, only a very limited initial-letter identification is possible, and duplication is rife. In the context of the present work, the problem is compounded by the fact that the subject matter of this book deals simultaneously with mechanical, electrical, and magnetic properties. Quantities such as the length, breadth, and area of a core window are likely to occur together in an equation containing the width, thickness, and area of the core center limb, plus voltage, flux density, current density, and other factors. The symbols selected for Table 1-1, then, are not necessarily standard symbols. They have been chosen for maximum convenience in the light of the problems of duplication and letter identification.

Dimension Symbols

Table 1-2 lists the symbols and abbreviations for some physical dimensions. Here, again, there are a few duplications, or near duplications, but there should be no confusion when these signs are used in practice.

Table 1-2. Mechanical Symbols and Abbreviations

Property	Sym.	Unit	Abr.	Property	Sym.	Unit	Abr.
General							
Diameter	D	centimeter, inch	cm, in	**Windings**			
Radius	r	centimeter, inch	cm, in	Winding length	l_W	centimeter, inch	cm, in
Core				Length mean turn	l_r	centimeter, inch	cm, in
Window height	F	centimeter, inch	cm, in	Cooling surface area	C_s	square centimeter,	cm^2, in^2
Window length	G	centimeter, inch	cm, in			square inch	
Core depth, center limb	D	centimeter, inch	cm, in	Margin	M	centimeter, inch	cm, in
				Bobbin thickness	b	centimeter, inch	cm, in
Core width, center limb	E	centimeter, inch	cm, in	Wire area (single strand)	A	square centimeter, square inch	cm^2, in^2
Window area	W	square centimeter, square inch	cm^2, in^2			*or*	
						circular mils	CM
Core area	a	square centimeter, square inch	cm^2, in^2	Winding area	w	square inch, circular mils	in^2, CM
Magnetic path length	l_M	centimeter, inch	cm, in	Winding space factor		-	-
Core gap	g	centimeter, inch	cm, in	(w/W)	K		
Stacking factor	s	-	-	Turns ratio	r		-
Volume	u *or* v	cubic centimeter, cubic inch	cm^3, in^3	Wire length	l	inch	in
				Temperature	T, t	degrees Celsius	°C
Weight	Wt	gram, pound	g, lb	Winding	N	number of turns	-

Prefix Symbols

Certain letters are used in front of unit symbols to denote multiples and submultiples of the units. Again, it is necessary to dip into the common inventory of letters supplied by the two available alphabets to obtain these letters. Obviously, there are bound to be further duplications. Table 1-3 gives most of the common prefixes (although only a few of these prefixes are needed in transformer design).

Table 1-3. Prefixes

Prefix	Symbol	Value	Number
pico	p	one-trillionth	1×10^{-12}
nano	n	one-billionth	1×10^{-9}
micro	μ	one-millionth	1×10^{-6}
milli	m	one-thousandth	1×10^{-3}
centi	c	one-hundredth	1×10^{-2}
deci	d	one-tenth	1×10^{-1}
deca	da	ten	1×10^{1}
hecto	h	one hundred	1×10^{2}
kilo	k	one thousand	1×10^{3}
mega	M	one million	1×10^{6}
giga	G	one billion	1×10^{9}
tera	T	one trillion	1×10^{12}

While on the subject of duplication, note that the combination of "mu" (μ) and other symbols can lead to problems. For instance, μH means *microhenry* (or one-millionth of a henry). Even though this is a common unit, it could also be construed as μ *times* H or "permeability times magnetizing force" (see Chapter 3). When the quantity is taken in context, there should not be any confusion, but it is possible.

The Greek Alphabet

Table 1-4 is self-explanatory. Both the uppercase and lowercase forms are given with the pronunciations. However, only a few of these symbols are used in this book. The familiar "pi" (π), capital omega (Ω), lowercase omega (ω), eta (η), and, of course, mu (μ) should be readily recognized.

Abbreviations

It is essential in all this confusion to distinguish between the letter symbols for properties, such as I for current, and the abbreviation for the unit of the property, such as A for ampere (the unit of current). This is further shown by the use of P as the symbol for power and W as the abbreviation for watt, the unit of power. Strictly speaking, the abbreviations for units should never be used in equations as substitutes for the properties, although it is frequently done. And, let's face it, it is sometimes clearer to do so.

Letter		Name
Small	**Capital**	
α	A	Alpha
β	B	Beta
γ	Γ	Gamma
δ	Δ	Delta
ϵ	E	Epsilon
ζ	Z	Zeta
η	H	Eta
θ	Θ	Theta
ι	I	Iota
κ	K	Kappa
λ	Λ	Lambda
μ	M	Mu
ν	N	Nu
ξ	Ξ	Xi
o	O	Omicron
π	Π	Pi
ϱ	P	Rho
σ	Σ	Sigma
τ	T	Tau
υ	Υ	Upsilon
ϕ	Φ	Phi
χ	X	Chi
ψ	Ψ	Psi
ω	Ω	Omega

Table 1-4. Greek Alphabet

Therefore, a pox on pedantry if defiance of the convention will make things easier. But watch it—carelessness in this respect can lead to confusion. Observe that letter symbols are usually italicized, e.g., *I*, *f*, whereas abbreviations of units are not, e.g., A, Hz.

It should be noted that the matter of associative symbols is much better with regards to units than it is to properties. In every case that is listed, the abbreviation is in fact the initial letter of the word (except for ohms, but omega (Ω) is sufficiently close to O to qualify as associative).

ANATOMY OF A TRANSFORMER

The sketches given in Fig. 1-1 provide a reference to the terms and dimensions that are used in connection with transformers. The various points illustrated in the sketches are all discussed in due course in the text. However, it should be helpful to anyone who is unfamiliar with the inner structure of a transformer to study the drawings and get their bearing, as it were, before launching further into the book.

CONVERSION FACTORS

Some readers may be less than enthralled to learn that many of the units used in engineering are related to the length of the arm of England's King Henry I. This is the alleged original standard for the yard in the English system of units, while the 1/10,000,000 part of the quadrant of the meridian that passes through France from Dunkirk to Formentara is the presumed length of the meter in the continental or metric system. Nevertheless, a number of unit systems are in common usage, including the familiar fps (foot, pound, second), cgs (centimeter, gram, second), and the less familiar SI (Systeme Internationale) systems. All these systems stem from one or the other of the original two standard measures, among other things.

The point being stressed here is that there are a number of systems in everyday usage, and one of the quirks of transformer design is that units from several or all of them invariably crop up simultaneously in even the simplest equations. This means that conversion factors must be used to make the values compatible. Moreover, because different systems of units and different presentations are frequently applied to given equations in various literature, the equations seem to change form between one treatise and another in a distressingly vague fashion. Examples of this are discussed later.

In the meantime, values that cover the conversions used in this book are provided in Table 1-5. The quantity to be converted is listed in the left-hand column, and the conversion factor for the unit into which it is to be changed is found in the appropriate column to the right. For example, what is 6.0 inches in centimeters? Find the inch unit in the left-hand column and then move across the chart to the centimeter column, where you read 2.54. Thus, 6.0 inches is $6 \times 2.54 = 15.24$ centimeters. With the possible exception of that useful but curious measurement, the *circular mil*, these conversion factors may possibly be remembered from your school days and need no further explanation. The circular mil, however, may not be so well known, and we will now discuss it.

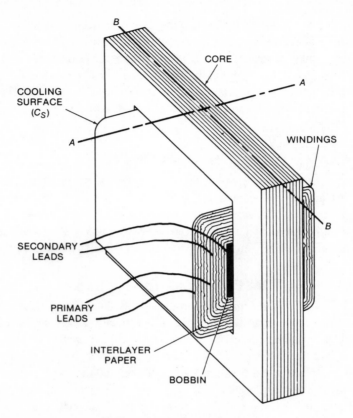

(A) Various parts of a transformer.

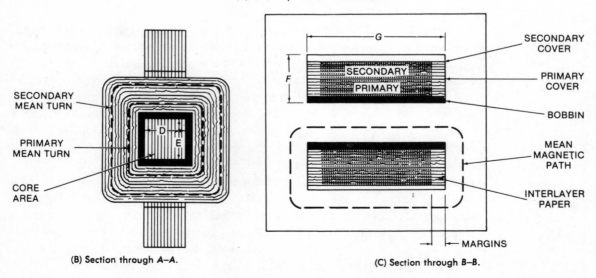

(B) Section through *A–A*.

(C) Section through *B–B*.

Fig. 1-1. Anatomy of a transformer.

Table 1-5. Conversion Factors

Length

	mil	in	ft	yd	mm	cm	m
mil	1	0.001			0.0254		
inch	1000	1	0.083	0.0278	25.4	2.54	
foot		12	1	0.33	304.8	30.48	0.305
yard		36	3	1	914.4	91.44	0.914
millimeter	39.37	0.03937			1	0.1	0.001
centimeter		0.3937	0.0328		10	1	0.01
meter		39.37	3.28	1.094	1000	100	1

Area

	cir mil	sq mil	sq in	sq ft	sq yd	sq cm	sq m
circular mil	1	0.7854					
square mil	1.274	1	10^{-6}				
square inch	1274000	10^6	1	0.00694		6.451	
square foot			144	1	0.1111	929.03	0.0929
square millimeter	1973	1550	0.00155				
square centimeter	197300		0.155	0.00108			

Volume / **Weight**

	cu in	cu cm			oz	lb	gram
cubic inch	1	16.39		ounce	1	0.0625	28.35
cubic centimeter	0.061	1		pound	16	1	453.6

Current Density

$$\text{Circular mils per ampere} = \frac{1273240}{\text{amperes per square inch}}$$

or

$$\text{Amperes per square inch} = \frac{1273240}{\text{circular mils per ampere}}$$

Temperature

$$\text{Temperature (°F)} = \frac{9 \times °C}{5} + 32$$

or

$$\text{Temperature (°C)} = \frac{5(°F - 32)}{9}$$

THE CIRCULAR MIL

The current-carrying capacity of a conductor and its resistance are directly proportional to the cross-sectional area of the conductor. The area, in turn, is proportional to the square of the diameter of the conductor; this can be seen from the familiar formula for the area of a circle:

$$A = \pi \left(\frac{D}{2}\right)^2$$
$$= \pi r^2$$

where

A is the area of the circle,
π is pi (3.14),
D is the diameter,
r is the radius (D/2).

In the course of a design, wire gauges are often chosen on the basis of a stated current density, which can be in terms of amperes per unit area or area per unit current— for instance, in amperes per square inch of conductor area, or in square inches of conductor area per ampere. Often the conductor resistance is a factor. Perhaps a conductor is required that has half the resistance of one that was previously chosen; thus, look for a conductor that has double the area (for a given length). Because the area of a circle is proportional to its diameter squared, if the relative diameters of two circles are known, it is easy to determine just how much greater or less the area of one is than the other without knowing the actual areas. For example, if the ratio of two circles, A and B, is

$$\frac{\text{diameter of } A}{\text{diameter of } B} = 2.0$$

then the area of circle A is $2^2 \times B$, or $4B$. If the diameter of circle A is five times that of circle B, then the area of circle A is 5^2B, or $25B$.

However, using the formula $\pi(D/2)^2$ is something of a chore, so a cleaner, uncluttered unit of circular area was introduced, based on the area being proportional to the diameter squared. This unit of measurement is called the *circular mil* (CM) and is expressed simply as

$$A = D^2$$

where
   A is the area in circular mils,
   D is the diameter expressed in straight mils.
A mil is one-thousandth of an inch (0.001 inch), so a circle that is 1 mil in diameter has an area of

$$A = 1^2 = 1 \text{ circular mil}$$

A circle 5 mils in diameter has an area of

$$A = 5^5 = 25 \text{ circular mils}$$

This, in fact, is the diameter and area of No. 36 AWG wire, as shown in the wire table given in Chapter 6 (Table 6-1).

The choice of wire gauges for a given current density is then conveniently made in terms of circular mils per ampere. If a wire is to carry 5 amperes and a current density of 1000 circular mils per ampere is desired, then select a conductor with an area of 1000 $\times$ 5 = 5000 circular mils or the closest one to it.

Circular mils are not confined to wire area or even to circles. In working with circular cores (see Toroids in Chapters 5 and 6), it is often convenient to express the area of the center window of the core in circular mils. If a toroidal core has an inside diameter of 1.5 inches, the diameter in mils is 1.5 $\times$ 1000 = 1500 mils, and its area is

$$A = D^2 = 1500^2 = 2,250,000 \text{ CM}$$

A square area can also be expressed in circular mils just as a circular area can be expressed in ordinary square measure. Figure 1-2 shows an area of 1 circular mil inside

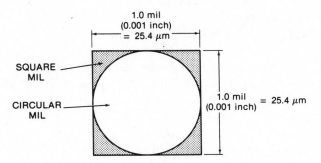

Fig. 1-2. Circular mil drawn inside a square mil (not to scale).

an area of 1 square mil. It will be observed that the area of the circle is expressed in circular mils; thus, $A = 1^2 = 1$ CM. This gives the same number value as that obtained for the area of the square when it is expressed in square mils; thus, $A = 1 \times 1 = 1$ square mil.

The area of the circle expressed in square measure is

$$A = \pi \left(\frac{D^2}{2} \right)$$

$$= 3.14 \left(\frac{1.0^2}{2} \right)$$

$$= 0.785 \text{ square mil}$$

The conversion factor from CM to mil^2 and, vice versa, is

$$\text{mil}^2 = \text{CM} \times 0.785$$

$$\text{CM} = \frac{\text{mil}^2}{0.785}$$

or

$$\text{CM} = \frac{\text{inch}^2 \times 10^{-6}}{0.785} = 1.274 \times 10^{-6} \text{ inch}^2$$

Another use for the circular mil is found in the equation

$$R = \frac{k \times l}{A}$$

where

    R is the resistance of the wire in ohms,
    l is the length of the wire in feet,

A is the area in circular mils,

k is the resistance of 1 mil-foot of wire in ohms.

The value of 1 mil-foot is for one foot of wire that is 1.0 mil in diameter, 1.0 CM in area. If *k* for annealed copper is taken as 10.4 at 20°C, the resistance of a No. 20 AWG copper wire (1020 CM) that is 30 feet long is

$$R = \frac{k \times l}{A}$$
$$= \frac{10.4 \times 30}{1020}$$
$$= 0.306 \text{ ohm}$$

This gives a good degree of accuracy if a resistance figure is not available but the area or diameter of the wire is known.

Finally, do not confuse circular mils with millimeters. Among other important differences, the former owes its existence to King Henry I, while the latter is a metric unit with French ancestors.

WORKING WITH EQUATIONS

The most common and useful trick in working with equations is being able to separate any given factor and express its value in terms of all the others. This usually is not difficult but simply a matter of applying a few simple rules. Many people feel a lot happier with a formula when they know how its form was changed. However, the accomplishment really pays off when it comes to interpreting the data one finds in literature, including that of the manufacturers. Here, the forms of basic equations are often so altered by the inclusion of conversion factors and the manipulation of exponents, that at first sight, they resemble mystical incantations from Merlin's blue book of magic rather than the dolled-up versions of commonplace formulas that they really are.

In this section, the reasons for and the results of various equation manipulations is explained. The equations used as examples are "real" and are used throughout the book. In the meantime, their significance is ignored. At present, our interest lies only in the methods of handling equations in general.

Balancing the Equation

One well-known version of the Ohm's law relationship is the following formula:

$$I = \frac{V}{R}$$

where

I is the current in amperes,

V is the potential difference in volts,

R is the resistance in ohms.

This is a statement of equality—an equation, in other words—from which the symbolic or numerical relationships of all the quantities can be easily deduced. Because $I = V/R$, it can be stated with certainty that

$$V = I \times R$$

and

$$R = \frac{V}{I}$$

These two expressions are derived from the first by "rearranging" the symbols in such a way that equality is always maintained. In the same way that the two pans of a weighing scale can be kept in perfect balance by adding, subtracting, dividing, or multiplying the material in each pan by exactly the same amount, so, too, can the equation be kept in perfect balance by performing identical operations to each side of the equation.

For example, the second arrangement ($V = I \times R$) is obtained as follows. Both sides of the equation are multiplied by R; thus

$$I \times R = \frac{V \times R}{R}$$

On the right-hand side of the equation, the quantity R divides into R, giving a 1. In other words, R is said to "cancel out." The equation then becomes

$$I \times R = V \quad \text{or} \quad V = I \times R$$

If you need to find the quantity R in terms of the other two factors, as shown in the third version, the same principle is applied. Starting again with the first arrangement, divide both sides by V to isolate R to get

$$\frac{I}{V} = \frac{V}{V \times R}$$

In this case, the symbol V cancels out on the right-hand side, leaving

$$\frac{I}{V} = \frac{1}{R}$$

To obtain R (rather than $1/R$) simply invert both sides (sticking to the principle that what is done to one side must be done to the other). Thus,

$$\frac{V}{I} = \frac{R}{1}$$

$$\frac{V}{I} = R$$

$$R = \frac{V}{I}$$

No matter how many symbols are involved, that is all there is to it. Whatever is done to one side of the equation must be done to the other side.

Now consider the following equation. It looks complicated because it has more symbols, but in reality, it's just as easily dealt with as Ohm's law.

$$V = 4FfaNB \times 10^{-8}$$

What is N in terms of the other quantities? Divide both sides by $4FfaB \times 10^{-8}$ to isolate N to get

$$\frac{V}{4FfaB \times 10^{-8}} = \frac{4FfaNB \times 10^{-8}}{4FfaB \times 10^{-8}}$$

On the right-hand side of the equation, $4FfaB \times 10^{-8}$ cancels out, leaving N. Thus,

$$N = \frac{V}{4FfaB \times 10^{-8}}$$

In a slightly more complicated maneuver, what would the steps be to obtain N_1/N_2 from the following equation:

$$R_T = R_S + R_P \left(\frac{N_1}{N_2}\right)^2$$

The object is to isolate N_1/N_2. Subtract R_S from both sides. The equation is then

$$R_T - R_S = R_S - R_S + R_P \left(\frac{N_1}{N_2}\right)^2$$

On the right, $R_S - R_S$ is obviously zero. Therefore, the equation becomes

$$R_T - R_S = R_P \left(\frac{N_1}{N_2}\right)^2$$

Divide both sides by the quantity R_P.

$$\frac{R_T - R_S}{R_P} = \frac{R_P}{R_P}\left(\frac{N_1}{N_2}\right)^2$$

On the right side of the equation, $R_P/R_P = 1$; therefore,

$$\frac{R_T - R_S}{R_P} = \left(\frac{N_1}{N_2}\right)^2$$

To eliminate the brackets of the right side of the equation, take the square root on each side of the equation. The square root of $(N_1 N_2)^2$ is, of course, N_1/N_2, and the square root of the left-hand side of the equation is $\sqrt{[(R_T - R_S)/R_P]}$. The equation then becomes

$$\frac{N_1}{N_2} = \sqrt{\frac{R_T - R_S}{R_P}}$$

The small numbers at the lower right of N (subscript $_1$ and subscript $_2$) have no arithmetical significance. Like the small letter shown at the lower right of some symbols, they are simply part of the identification.

Watch the Units

It is essential to define not only the quantities represented by the symbols but also the units in which they are stated. It is not enough to say that I equals current. The unit of current—amperes, milliamperes, or whatever—must be stated, and the equation must correctly reflect the units used. For example, to find the voltage V when $I = 500$ milliamperes and $R = 100$ ohms, use Ohm's law. If you say that $V = 500 \times 100 = 50,000$ volts, the answer is wrong. The 500 milliamperes should have been expressed as 0.5 ampere. Therefore, $V = 0.5 \times 100 = 50$ volts. However, the equation could have been arranged to reflect the use of milliamperes. For example, if it is desirable to plug the number 500 directly into the equation, the equation must be written with a conversion factor of 1000. Thus

$$V = \frac{I}{1000} \times R$$

Because a milliampere is one-thousandth of an ampere, dividing the quantity I by 1000 allows the milliampere figure to be used directly. Thus,

$$V = \frac{500}{1000} \times R$$
$$= 0.5 \times 100$$
$$= 50 \text{ volts}$$

To use another example, suppose you want to deal directly in milliamperes and kilovolts. Then, for I in milliamperes, $I/1000$ is the number of amperes, and with V in kilovolts, $V \times 1000$ is the number of volts. Therefore, the equation required for these units is

$$V \times 1000 = \frac{I}{1000} \times R$$

or

$$V = \frac{I}{1,000,000} \times R$$

This is easily rearranged to give R, in ohms, by using the principles already discussed. This gives

$$R = \frac{V \times 1{,}000{,}000}{I}$$

where

    I is the current in milliamperes,
    V is the volts in kilovolts,
    R is the resistance in ohms.

If V is 2 kilovolts and I is 200 milliamperes, then

$$R = \frac{2 \times 1{,}000{,}000}{200} = 10{,}000 \text{ ohms}$$

is the resistance.

EXPONENTS

The small number seen to the right and slightly above a quantity, as shown in the *Number* column of Table 1-3 and elsewhere, is called an *exponent*. It is shown, for example, in 10^2. In this case, it simply means 10×10. The number 10^3 means $10 \times 10 \times 10$, or 10 raised to the power of 3. The number 10 is called the *base*; the system using 10 as a base with exponents is sometimes referred to as the *Standard System of Scientific Notation*. If the last equation in the preceding section had been written using this system, it would have appeared as

$$R = \frac{2 \times 10^6}{2 \times 10^2} = 10^4 \text{ ohms}$$

This illustrates why it is used. The system is neat, tidy, and easy to work with.

A simple way to treat exponents to the base 10 is to visualize the number 1 as an unwritten but ever present adjunct. For example, 10^6 may be visualized as 1×10^6. Then 10^6 is the same as 1.0 with the decimal point removed six times to the right thus, 1 0 0 0 0 0 0, or 1,000,000. Using the same idea with a *negative* exponent, the decimal point is moved to the left. For example, 10^{-8} may be visualized as 1×10^{-8}, which is 1.0 with the decimal point moved eight times to the left; thus, . 0 0 0 0 0 0 0 1, or 0.00000001. In the same way, $10^3 = 1000$, and $10^{-3} = 0.001$. When no exponent is shown, as in the number 10, then an exponent of 1 is assumed. Therefore, 10 is really 10^1 or 1×10^1, or 1.0 with the decimal point removed once to the right, thus, 10. = 10. It follows, then, that 10^0, or 1.0×10^0 as we now visualize it, is simply 1.0, or 1.

Exponents can be manipulated in various ways to make counting easier. To multiply 10^3 by 10^3, just *add* the exponents, to obtain 10^6. Thus $10^3 \times 10^3 = 10^6 = 1{,}000{,}000$. To divide 10^5 by 10^3, *subtract* the exponents, to obtain 10^2. Thus $10^5 \div 10^3 = 10^2 = 100$. If the exponents are all negative, the process is the same, e.g. $10^{-5} \times 10^{-2} = 10^{-7}$ and $10^{-8} \div 10^{-3} = 10^{-5}$.

   When there is a mixture of positive and negative exponents, remember that addition and subtraction is done algebraically. For instance, $10^{-6} \times 10^2$ is 10^{-4}. If 10^{-6} is divided by 10^2, that is to say $10^{-6} \div 10^2$, then, subtracting, we get for the exponents $-6 - (+2) = -8$, for a result of 10^{-8}.

   A frequent convenient dodge is to move a term with an exponent from the bottom line of an expression to the top line and vice versa by changing the sign of the exponent. For instance, the equation

$$N = \frac{V}{4FfaB \times 10^{-8}}$$

can be written as

$$N = \frac{V \times 10^8}{4FfaB}$$

This operation does not change the statement. This is easily proved by substituting numbers in both versions of the equation and solving the equation. Examples of this type of numbering are given below:

10^0	= 1	10^{-1}	= 0.1
10^1	= 10	10^{-2}	= 0.01
10^2	= 100	10^{-3}	= 0.001
10^3	= 1000	10^{-4}	= 0.0001
10^4	= 10,000	10^{-5}	= 0.00001
10^5	= 100,000	10^{-6}	= 0.000001
10^6	= 1,000,000	10^{-7}	= 0.0000001

Roots

   Exponents may also be used to denote roots. Thus the square root of x can be stated as

$$x^{1/2} \quad \text{or} \quad \sqrt{x} \quad \text{or} \quad \sqrt[2]{x}$$

while the cube root is shown as

$$x^{1/3} \quad \text{or} \quad \sqrt[3]{x}$$

The fractional exponent denotes a root. The bottom number of the fraction denotes the degree of the root—square root, cube root, or whatever. The top number denotes the power to which the root is raised. In the examples shown, the square root and the cube root of x was raised to the power of 1, or in other words, simply the square root and the cube root. If a number is written as $x^{2/3}$, it means the square of the cube root of x, or

$$x^{2/3} = \sqrt[3]{x} \times \sqrt[3]{x} \quad \text{or} \quad x^{1/3} \times x^{1/3}$$

As an example of the above, solve the following problem. What is $27^{2/3}$? The solution follows.

$$27^{2/3} = \sqrt[3]{27} \times \sqrt[3]{27}$$

or

$$27^{2/3} = 27^{1/3} \times 27^{1/3}$$
$$= 3 \times 3$$
$$= 9$$

PLUGGING IN THE NUMBERS

While all this might be great fun for those fascinated by algebraic and arithmetical manipulation, our real aim is to eventually plug in, or substitute, numbers in place of the symbols to get answers to our problems.

The Basic Transformer Equation

Consider the following equation. This is usually referred to as the basic transformer equation; we saw it earlier in a slightly different form.

$$V = 4FfaBN \times 10^{-8}$$

This equation appears time and again throughout the book, so it's a good subject on which to demonstrate number substitution. At the same time, we can touch on some aspects of transformer design. Some idea of what the symbols in the above equation refer to is given in Figs. 1-3 and 1-4, although it is not important that they be fully understood at this stage. They will be discussed in detail later.

Suppose that for certain symbols, the values are known. For example, f stands for *frequency* in hertz; in designing power transformers for most domestic supplies, f will be 60 hertz. Symbol F represents something called the *form factor* (which is explained later); it is always 1.11 for a sine wave. The letter B is the *flux density* and the number assigned to it can be anything you like within certain limits that are imposed by the kind of material used for the core and the design requirements. For this factor, a very conservative figure might be 75,000 lines per square inch of core area. The core area referred to is the cross-sectional area a shown in Fig. 1-3. The value needed for a is dependent on various factors. For power transformers, the output power is a determining factor, and for designs to work from a 60-hertz source, a is sometimes related to power (P) by the equation

$$a = 0.16 \sqrt{P}$$

where a is the cross-sectional area of the core in square inches.

The units used in the main equation and the symbol identifications are:

V is the voltage across a considered winding in volts,

F is the form factor (normally 1.11),

f is the input frequency in hertz,

a is the cross-sectional area of the core in square inches,

B is the flux density in lines (or maxwells) per square inch,

N is the number of turns on a considered winding.

The number of turns on the winding is often the unknown number.

Plugging the known numbers into the equation of $V = 4FfaBN \times 10^{-8}$, yields

$$V = 4 \times 1.11 \times 60 \times a \times 75 \times 10^3 \times N \times 10^{-8}$$

Multiplying gives

$$V = 19,980 \times 10^{-5} \times N \times a$$

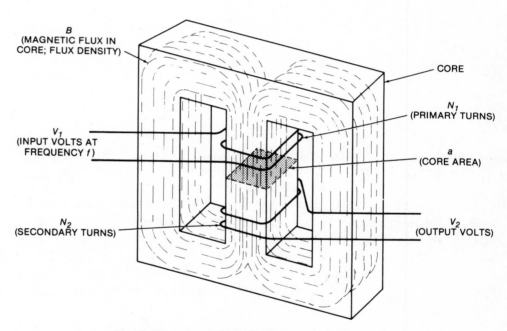

Fig. 1-3. Illustration of the "factors" given in a basic equation.

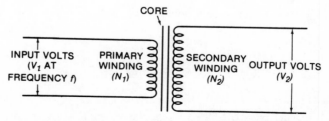

Fig. 1-4. A standard schematic for transformers like those depicted in Figs. 1-1 and 1-3.

Rearranging this to give N in terms of the other factors gives

$$N = \frac{V}{19{,}980 \times 10^{-5} \times a}$$

$$= \frac{V \times 10^5}{19{,}980 \times a}$$

and dividing out leaves

$$N = \frac{V \times 5}{a}$$

as the number of turns.

Turns Per Volt

It is often more convenient to arrange the above result into the form of

$$\frac{V}{N} = \frac{5}{a}$$

The term N/V is known as the *turns-per-volt* figure for a transformer—that is to say, the number of turns on the windings for each volt across them. This ratio is the same for each winding on a transformer (ignoring, for the moment, the effect of losses).

Thus, by substituting known values into the relatively complex equation that we started with, it is reduced to a very simple and easily handled statement that is the exact equivalent of the original equation, so long as the values assigned to F, f, and B remain valid. All that is needed to use this abbreviated equation is to estimate the core area a, and then plug in the value. For example, suppose a transformer is required to supply 8.0 volts at 5.0 amperes from a 110-volt domestic (residential) supply line. The output power P is $8 \times 5 = 40$ watts. From the simple equation given earlier,

$$\begin{aligned} a &= 0.16 \sqrt{P} \\ &= 0.16 \sqrt{40} \\ &= 0.16 \times 6.325 \\ &= 1.0 \text{ square inch} \end{aligned}$$

Plugging this value into the preceding equation gives

$$\begin{aligned} \frac{N}{V} &= \frac{5}{1.0} \\ &= 5 \text{ turns per volt} \end{aligned}$$

The primary voltage is 110 volts; therefore the turns required on the primary are $110 \times 5 = 550$ turns. The secondary voltage is 8 volts. Therefore, the turns needed for the secondary winding are $8 \times 5 = 40$ turns. It is as easy as that. Because you now know how the equation was derived, equally simple equations can be made to suit other conditions of frequency, flux density, and so on.

GRAPHS

Another method of keeping things simple is to convert equations into graphs. As a simple example, the abbreviated equation just derived is graphed. First, a few values are selected for area *a* within the expected range. Suppose six values are chosen ranging from 0.5 square inch to 3.0 square inches. When the value of *a* is 0.5 square inch, the turns-per-volt figure is

$$\frac{N}{V} = \frac{5}{0.5}$$

$$= \ 10 \ \text{turns per volt}$$

In the same way, other values can be found for N/V when *a* is 1.0, 1.5, 2.0, 2.5, and 3.0. The values work out at 5, 3.3, 2.5, 2.0, and 1.66, respectively. A graph may now be drawn of N/V against *a*, as shown in Fig. 1-5. To illustrate the use of the graph, suppose a core has a cross-sectional area of 1.5 inch^2. How many turns per volt are needed for the windings? Find 1.5 on the bottom axis, draw a line to meet the curve as shown in Fig. 1-5 by the dotted line, and read the value of $N/V = 3.3$.

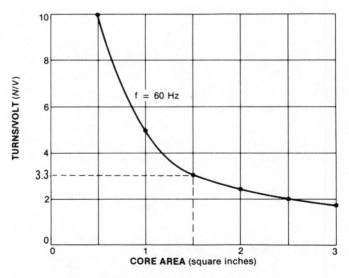

Fig. 1-5. Graph of turns per volt versus core area.

The scope of a graph like this can be further extended by determining, from the basic equation, new values for the numerator in the abbreviated formula for several values of frequency or for different values of flux density. To illustrate the problem for different frequencies, suppose values of 25, 400, and 1000 hertz are chosen. Then, from the basic equation of $V = 4FfaBN \times 10^{-8}$ (where B and F are the same as before), rearrange the equation and substitute the known values to get

$$\frac{N}{V} = \frac{10^8}{4 \times 1.11 \times f \times a \times 75 \times 10^3}$$

$$= \frac{10^5}{333 \times f \times a}$$

Plugging the value for the frequency ($f = 25$ Hz) into the equation gives

$$\frac{N}{V} = \frac{10^5}{333 \times 25 \times a}$$

$$= \frac{12}{a}$$

A curve may now be derived from this by substituting the various values for area a, as was done previously for the value of $f = 60$ Hz. Repeating the above procedure for 400 and 1000 hertz gives the values for two more curves. The values obtained at the various frequencies and cross-sectional areas are shown in Table 1-6. They are graphed as three curves in Fig. 1-6, along with the curve that was plotted in Fig. 1-5.

Observe that because the values for N/V become so small at the higher frequencies of 400 and 1000 hertz, it is necessary to draw these curves to a different scale in order to keep them readable (see scale B). Scale A is used for the curves for 25 and 60 hertz. Two further examples have been drawn into the graph. For $a = 0.9$ inch^2, N/V is 0.33 turns per volt at 1000 hertz. In the second, N/V is 10.9 turns per volt for a core area of 1.1 square inches at 25 hertz. Note the vast difference in the required number of turns per volt at these two frequencies.

The shape of these curves, however, is not ideal for practical use, because the slope is too shallow at the bottom end of the curve for any changes in N/V (with a change in a) to be read easily, and at the top end of the curve the slope is too steep for any

**Table 1-6. Values for Plotting the Curves
in Fig. 1-6**

Core Area (a)	Turns/Volt (N/V)			
	25 Hz	**60 Hz**	**400 Hz**	**1000 Hz**
0.5	24	10	1.5	0.6
1.0	12	5	0.75	0.3
1.5	8	3.33	0.5	0.2
2.0	6	2.5	0.375	0.15
2.5	4.8	2	0.3	0.12
3.0	4	1.66	0.25	0.1

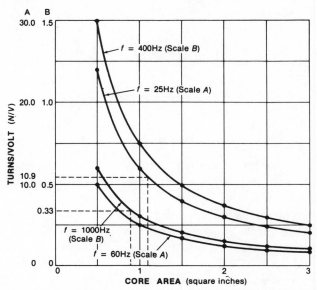

Fig. 1-6. Graph of turns per volt versus core area at various frequencies.

changes in a (with a change in N/V) to be read accurately. This situation is addressed later after a brief look at the question of slope.

Slope of Graph Curves

An important and sometimes puzzling feature of curves is the slope of the curve line. Many curves are straight lines, such as that shown in Fig. 1-7. This curve is the

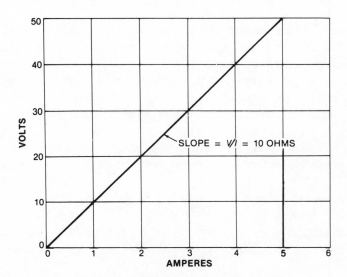

Fig. 1-7. Volts versus amperes—Ohm's law graphed.

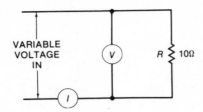

Fig. 1-8. Circuit for determining and plotting the voltage against the current.

result of plotting the voltage against the current as they were measured in the circuit shown in Fig. 1-8. Here the voltage across resistor R is increased incrementally from 0 to 50 volts and the current through the resistor is recorded and plotted. The slope of this line is given by the equation V/I that, from Ohm's law, also gives the value of the resistance. Wherever the slope of the line is measured on this particular graph, the ratio of V/I is the same and is equal to 10 ohms. For example, at $I = 5$ amperes and $V = 50$ volts, the slope is $50/5 = 10$ ohms.

When the lines curve as in the N/V versus a graph in Fig. 1-6, the slope is not the same at all parts of the curve, as is obvious from visual inspection. It should also be clear that any third quantity derived from the slope, as was done in the example above for resistance, is not the same at all parts of the curve.

An important example of this curve concept is in a graph known as the B-H curve in which flux density (B) is plotted against magnetizing force (H). A typical B-H curve is shown in Fig. 1-9. A third quantity (permeability in this case) is derived from the slope and given by the formula

$$\mu = \frac{B}{H}$$

But the factor B/H is not constant. It is different at various points on the slope because the slope varies. Therefore, the permeability can be defined only at given points on the curve. The different slopes are indicated by tangents drawn on the curve shown in Fig. 1-9. Two examples of slope measurement are shown by the small triangles that are drawn at two points on the slope given in Fig. 1-10. Here, a small change in B for a small change in H is being measured; that is to say, the length of the vertical line of the triangle is plotted against the length of its base. The smaller the triangle, the more accurate the measurement. This can be done by using calculus, which deals in infinitely small changes, but we are more concerned at this point with the principle rather than with the mathematics.

To restate the problem, a third quantity that is derived from a curve of this kind is not constant; it varies throughout the curve. In the case of the B-H curve, the permeability is highest where the curve is steepest and it falls to near zero where the curve levels out at the top and bottom.

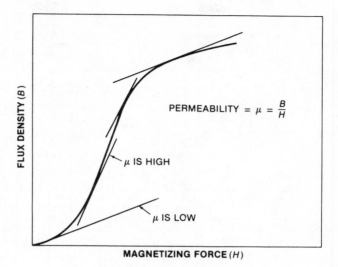

Fig. 1-9. A typical **B-H** curve.

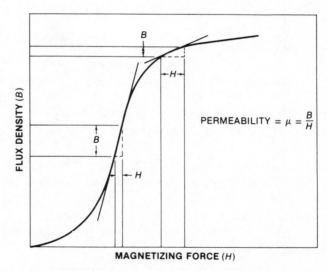

Fig. 1-10. Factors **B** and **H** vary, causing **M** to vary.

Linear and Logarithmic Graphs

The graphs that were drawn earlier in Figs. 1-5 and 1-6 were constructed on linear graph paper, meaning paper on which the values along the x and y axis are evenly spaced on the paper. Many curves, however, such as those shown in Fig. 1-11, are drawn on paper that is constructed according to logarithmic law. Note that the numerical values

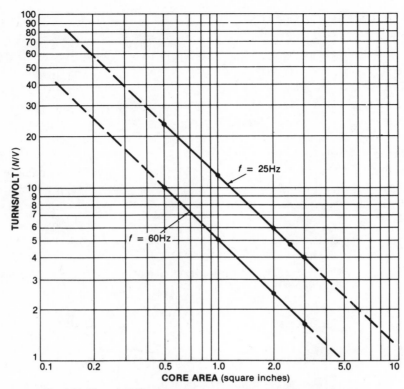

Fig. 1-11. The values listed in Table 1-6 plotted on logarithmic scale paper.

are spaced logarithmically along both axes. This particular example of scale is known as *logarithmic* (2 cycle by 2 cycle), because two cycles of the values 1 to 10 appear on each axis.

The interesting thing about this type of graph paper is that the curves drawn on it are identical in value to those drawn in Fig. 1-6 and were constructed directly from the plots recorded in Table 1-6. This can be proved by simply comparing them. The dead-straight lines appearing in Fig. 1-11 make reading the curves much easier than before. The straightness of the lines is due, of course, to the fact that the paper "matches" the law obeyed by the curves.

KINDS OF VOLTAGE AND CURRENT

Up to now, voltage and current have been discussed without trying to define them. What is there to define? There are the familiar dc and ac types, but apart from that, a volt is a volt and an ampere is an ampere—or are they? The fact is that values of alternating voltage and current can be stated in several different ways, each of which has a special significance for the transformer designer.

Peak or Maximum Values

In Fig. 1-12, the peak value of the sine wave voltage is unmistakably 155.6 volts in the positive direction and the same in the negative direction. This sine wave can be described as 155.6 volts *peak* or *maximum*. Because the insulation needed in a transformer depends (among other things) on the *peak* voltage, this figure is obviously a practical consideration for the designer. There are circuit situations that create voltage stresses in the windings that are several times greater than the peak voltage. Similarly, peak currents create peak flux values, and this, too, is of interest to the designer. Sometimes, alternating voltage values are stated as peak-to-peak values; in Fig. 1-12, this is equal to 311.2 volts, or double the peak value. It should be noted here that although a sine wave is depicted in Fig. 1-12, the peak or peak-to-peak values can be used to describe any waveshape.

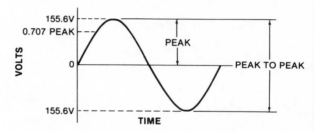

Fig. 1-12. Sine wave that illustrates peak, peak-to-peak, and rms voltage.

Instantaneous Values

Although this value is not used much in practical design, it is necessary to know what it means in order to understand the important definition for rms values. The *instantaneous* value is simply the value present or measured at any given instant in time. For example, in Fig. 1-13, the instantaneous value at time *t* is 10 volts.

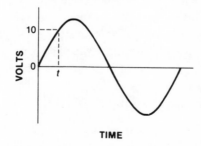

Fig. 1-13. Instantaneous value of voltage at time *t*.

RMS or Effective Values

In a dc circuit the power consumed is equal to $V \times I$. In an ac circuit with a resistive load, the power is also equal to $V \times I$, but which V and which I? It can be shown that the power (for example, heat) produced by an ac current in a resistive load is the same as that produced by a dc current that is equal to 0.707 of the peak value of the current sine wave. In other words, the *effective* value of a sine wave waveform is equal to 0.707 of its peak value. A peak sine-wave current of 1.0 ampere produces the same power as does a direct current of 0.707 ampere (assuming equal time periods).

Figure 1-14 is, in fact, a sketch of the waveform of a standard 110-volt ac supply. It has a peak value of 155.6 volts but an effective or rms value of 155.6 × 0.707 = 110 volts. An rms value is the most commonly used way of expressing ac current and voltage magnitudes. On the data plate of a TV set, an electric kettle, or whatever, the voltage stated is an rms value, although it might not say so. The letters stand for "root mean square." The number 0.707 is equal to $1/\sqrt{2}$. For the record, it is the square root of the mean value of the sum of the squares of the instantaneous values taken over one cycle of voltage. Let's take another look at Fig. 1-14.

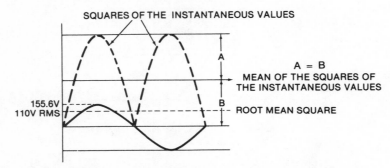

Fig. 1-14. Derivation of the rms value. (The curves are not drawn to scale.)

The mean value of a sine wave must be zero because it is perfectly symmetrical in the positive and negative directions. This does not mean that it cannot produce power. The heating of a stove element, for example, does not depend on the polarity of the current passing through it. However, a mean value can be found by first squaring all the instantaneous values. In this case, the negative values all become positive because the square of a negative is always positive; for instance, −2 × −2 = +4. When the squared values are plotted, they look like the dashed-line curve shown in Fig. 1-14. (This line is not drawn to scale.) A mean value of the squares can now be determined as shown, and the square root of this value gives the rms value. The rms value of a sine wave is, thus, 0.707 of the peak value.

Average Value

The average value is not often used except in relation to rectifier circuits. However, because rectifier circuits represent one of the most common arrangements used in conjunction with transformers, it (average value) deserves at least a passing mention here. It also appears in the basic transformer equation as a ratio factor called the *form factor*.

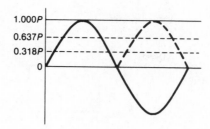

Fig. 1-15. Derivation of average value.

The average value of a full-cycle sine wave is twice the value of the average of a half cycle. This is the only way in which the average value can be stated, because as mentioned earlier, the true average (or mean) value of a sine wave is zero. Mathematically, the average value is $(2 \times \text{peak})/\pi$, or 0.637 peak, which is quite close to the rms value. The average value of a half cycle is, of course, peak/π or 0.318 peak. These relative values are shown in Fig. 1-15.

Form Factor

The ratio of the rms value to the average value yields a number called the *form factor*. Thus

$$F = \frac{\text{rms}}{\text{average}}$$

For a sine wave

$$F = \frac{0.707}{0.637} = 1.11$$

is the form factor.

This is a frequently recurring number in transformer calculations, although it is usually hidden in the arithmetical manipulation. The value of 1.11 is true only for sine waves (or other waveforms where the proportions are deliberately arranged to simulate a sine wave in this respect. For instance, a square-type wave can be arranged to do this, as will be explained in Chapter 10). Other waveshapes have other values because their rms and average values are different. In a true square wave, for instance, the form factor is 1.0.

Various Values in Relation to Square Waves

As will be seen later, certain design problems and analysis center around square waves. The rms and average values of a square wave are the same as its peak value. This is easy to see. Take the rms value, for instance. The instantaneous values of a square wave are obviously the same as the peak values at all instants of time t. See Fig. 1-16. Therefore, the mean of the sum of the squares of the instantaneous values

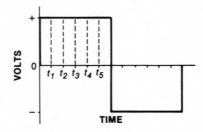

**Fig. 1-16. At any specific point in time, the instantaneous values of a
square wave equal the peak value.**

is P^2, and the rms value (the root of the mean of the sum of the squares of the
instantaneous values) is $\sqrt{P^2}$, which equals P.

Similarly, the average of the full cycle squared must also be equal to P. It follows,
then, that the form factor for a square wave is 1.0. For a half-cycle square wave, the
average must be $P/2$; but the rms value for a half cycle is $P/\sqrt{2}$. The latter result stems
from the following. The mean of the sum of the squares of the instantaneous values of
a half cycle is $P^2/2$. The root of this equation is $\sqrt{(P^2/2)}$ or $P/\sqrt{2}$.

Summary of Peak, Average, and RMS Relationships

The references given in Table 1-7 are frequently useful when studying a circuit for
voltage relationships.

Table 1-7. Summary of Average and RMS Values

Value	Sinusoidal Waves	Square Waves
RMS	Full cycle $= \dfrac{\text{Peak}}{\sqrt{2}}$	Full cycle $=$ Peak
Average	Full cycle $= \dfrac{2 \times \text{Peak}}{\pi}$	Full cycle $=$ Peak
RMS	Half cycle $= \dfrac{\text{Peak} \times \sqrt{2}}{4}$	Half cycle $= \dfrac{\text{Peak}}{\sqrt{2}}$
Average	Half cycle $= \dfrac{\text{Peak}}{\pi}$	Half cycle $= \dfrac{\text{Peak}}{2}$

TWO

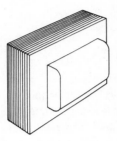

Elementary
Electromagnetics

DEVELOPMENT OF TRANSFORMERS_____

The transformer, in common with a host of other devices, owes its existence to the pioneering work of Danish physicist Hans Christian Oersted, who discovered electromagnetism in 1820, and to Michael Faraday of England and to Joseph Henry of the United States. The latter two men, working independently, discovered electromagnetic induction (more or less concurrently) in 1830-1831. These momentous events laid the foundation for the vast electrical/electronic field that governs so much of our living style and culture today. But the transformer, more than most other devices, is the direct descendant of these early experiments. Mind you, it is perhaps a little unfair to single out only these names for credit, because they were led to their discoveries by the people who preceded them, just as the pioneers who came after them had the advantage of *their* work.

Although Oersted is generally credited with the discovery of electromagnetism, it was a man called Arago who first performed the classic iron-filings experiment illustrated in Fig. 2-1. A wire is inserted perpendicularly into a sheet or card, and a strong dc current is passed through the wire. Iron filings are sprinkled onto the card and, when the card is gently tapped, the filings arrange themselves into concentric lines around the wire, thus making visible the presence of a magnetic field. This magnetic field is due to the current through the wire.

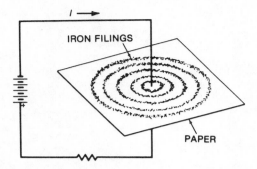

Fig. 2-1. Iron filings demonstrate the presence of a magnetic field.

Oersted, however, demonstrated the dependence of the field direction on the direction of the current by conducting an experiment similar to that shown in Fig. 2-2. Instead of iron filings, a compass was placed in the vicinity of the wire. When a current was passed through the wire in one direction as shown in Fig. 2-2A, the needle of a compass pointed in one direction. When the current was reversed, the needle also reversed as shown in Fig. 2-2B. It was shown that if the current was repeatedly reversed, the magnetic field would follow the current reversals.

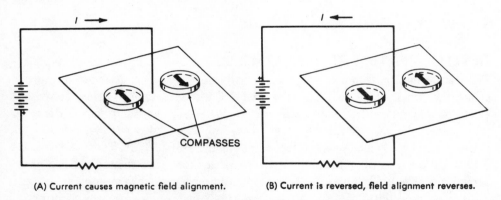

(A) Current causes magnetic field alignment.   (B) Current is reversed, field alignment reverses.

Fig. 2-2. The dependence of field direction on the current direction is demonstrated by use of compasses.

The next person on the scene was Andre M. Ampere, a French physicist, who showed that if a wire is wound into the form of a coil, its magnetic effect is greatly intensified. The coil, in fact, exhibits a magnetic field exactly like that of a bar magnet. This is shown in Fig. 2-3. The coil has a north pole and a south pole and a neutral equatorial region just like a bar magnet. Moreover, the polarity can be reversed by reversing the current through the coil. This, again, demonstrates the dependence of the field direction on the current direction.

Both Arago and Ampere demonstrated that hard steel needles could be made into permanent magnets by placing them inside a coil carrying dc current; when the current was switched off, the needles retained their magnetism. Then, in 1825, Sturgeon

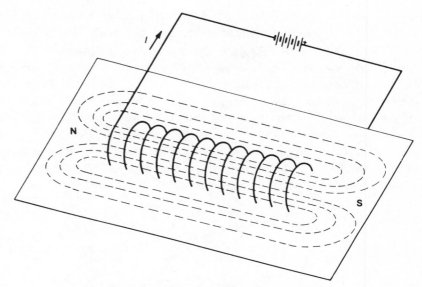

Fig. 2-3. A coil with current in it acts as a magnet.

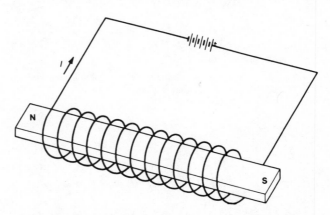

Fig. 2-4. Placing a soft iron bar inside a coil makes an electromagnet.

demonstrated that if *soft* iron is inserted into a coil carrying dc current, the iron is magnetized but the magnetization vanishes when the current is switched off. This important combination of a soft iron bar inside a current-carrying coil was called an *electromagnet* (Fig. 2-4). It is the basis of transformer action.

Note that the current direction in these experiments (Figs. 2-3 and 2-4) is assumed to be that of conventional current, namely from positive to negative outside the source. This is the kind of current that the early pioneers dealt with.

Michael Faraday, an English chemist and physicist, then entered the scene. Noting that magnetic effects can be produced by a current, as demonstrated by Oersted and the others, he felt that the reverse should be possible—that is to say, it should be possible somehow to produce current by means of magnetic effects. Seven years later, he performed an experiment like the one shown in Fig. 2-5. In August, 1831, he connected a solenoid (a long coil) to a galvanometer, and by moving a permanent magnet in and out of the coil, he obtained a corresponding deflection of the galvanometer needle while the magnet was in motion relative to the coil. In other words, he proved that whenever the magnetic lines of force pass through or "cut" the windings of a coil, a current is produced in the circuit. If the permanent magnet is replaced by Sturgeon's electromagnet and is moved in and out of the solenoid, as illustrated in Fig. 2-6, the effect is precisely the same. It was also noted that the magnitude of the induced current is proportional to the speed at which the magnet is moved. Thus, the needle deflection is greatest when the relative motion is fastest and is zero when there is zero motion.

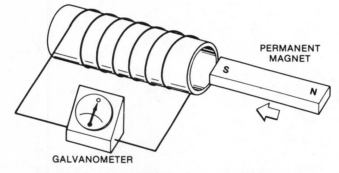

Fig 2-5. A magnet moved into and out of a coil will generate an electric current.

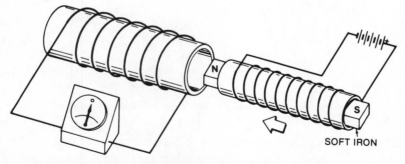

Fig. 2-6. An electromagnet moved into and out of a coil generates current in the coil.

Taking the experiment a step further, suppose the electromagnet (call it the primary coil) is inserted into the solenoid as shown in Fig. 2-7A (call the solenoid the secondary coil). Then, instead of moving the primary coil, the current is simply switched on and off, thus creating a movement of the magnetic field. Current is again induced in the secondary coil. This induced current occurs when the primary current is switched on and, again, when it is switched off. While the dc current in the primary is steady, no current flows in the secondary. This phenomenon is known as *induction*.

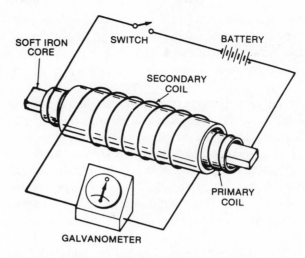

(A) A two-coil device wound on an iron core.

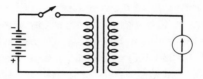

(B) Schematic diagram of circuit.

Fig. 2-7. Switching of battery current in primary winding circuit generates current in secondary winding circuit.

What we now have is a two-coil device wound around a soft-iron core. It looks suspiciously like a transformer and is depicted schematically in Fig. 2-7B. However, it has a dc input and does nothing at all except at the instants of switching on and switching off, at which times the magnetic field is either building up or collapsing.

It is worth noting, for future reference, that the effect is much more pronounced when the circuit is broken than when it is switched on. The reason for this is that the time constant of the circuit (that is to say, the time needed for the field to build up or collapse) depends on the inductance of the coil and on the resistance in the circuit in accordance with the following simple formula:

$$\tau = \frac{L}{R}$$

where

τ is the time constant,
L is the inductance,
R is the resistance.

(The time constant is the time needed for the current to reach 63.2 percent of its full value.) When the switch is "made," the only resistance is that of the coil. Because this is usually small, the time constant is large, and the buildup of the field is relatively slow. This slow movement of the field creates a low current. At the "break" time, however, the circuit resistance suddenly becomes exceedingly high (because the contacts are open), resulting in a low time constant, a very rapid collapse of the field, and a high induced current.

So far, the phenomenon of induction has been discussed in terms of *induced current*, but as smoke is relative to fire, when there is current, there is voltage. The current is caused by an induced voltage across the secondary windings. The magnitude of the induced emf (electromotive force), according to a law stated by Faraday, is proportional to the rate of change of the flux, or lines of force. There is also an emf that is developed by *self-induction* across the primary itself. This self-induced emf opposes the applied emf (this is called Lenz's law). These statements lead directly to the basic transformer equation that was mentioned in Chapter 1 and is discussed in more detail later in this chapter.

If a current is repeatedly and rapidly switched on and off, the input to the primary is essentially a series of squarish dc pulses as shown in Fig. 2-8A. The circuit shown in Fig 2-8B includes an interrupter of the trembler type (the electric bell principle), in order to perform that function. This circuit is the basis for the well-known induction coil of the early days of electricity. It is sometimes called the *Ruhmkorff coil*, for the Parisian scientific instrument maker (1803-1877) who added improvements to earlier models. The usual purpose of an induction coil was to transform low-voltage direct current into a very high secondary-pulsed voltage for the purpose of creating sparks across a spark gap. The primary consisted of a relatively few turns of wire and the secondary of many turns, perhaps thousands, in order to provide a high secondary-to-primary turns ratio. In those

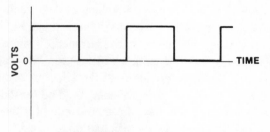

(A) An idealized input waveform.

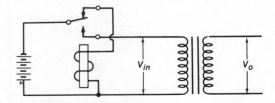

(B) An induction-type coil circuit with an interrupted dc input.

Fig. 2-8. A circuit for turning the current rapidly on and off.

days, the device was used mainly for physiological purposes. Then, later, it was used in radiotelegraphy as the heart of the venerable spark transmitter and, also, for other purposes.

In modern times, it is best known as the automobile ignition coil. This system was patented by Kettering in 1908, and with some modern refinements, is still going strong. Not surprisingly, most of the improvements to the induction-coil circuit since its inception have centered around the method of interrupting current. Even today, this is a principal preoccupation of designers. The trembler was discarded early in the game. In automobiles, it was replaced by the engine-driven points, which in turn are gradually giving way to transistorized circuitry. The principle will also be recognized by those familiar with the subject as that of the dc-to-dc power converter and the dc-to-ac power inverter. This is discussed in greater detail in a later chapter.

CONVENTIONAL TRANSFORMERS

It is now evident that with the development of the induction coil with its interrupted dc input, we are but a step away from the conventional transformer with its continuously varying ac input. Until the latter part of the 19th Century, however, there was no need to hitch primaries to ac power, because the power in use was almost universally direct current. It wasn't that ac power was not available; it was just considered to be a kind of synthetic electricity. The only ''proper'' kind of electricity was direct current, the kind that came from batteries or was synthesized by generators. These generators were fitted with commutators to convert the alternating current inherent to the machine to direct current at its output. Direct-current motors then needed commutators to reverse the process and convert the direct current to current producing a varying field.

At the Polytechnic Institute in Gratz, Austria, Nichola Tesla, a 19-year-old student, brashly suggested to his professor that the dc Gramme motor being demonstrated was impressive but that it sparked a lot around the commutator brushes, so why not dispense with them by using alternating current as the source of power instead of direct current. The good professor, it is reported, devoted an entire lecture period to destroying the suggestion point by point, and he proved the impossibility of running motors on alternating current. He compared the idea of an ac motor to the naive dreams of those who sought to design perpetual motion machines.

It took Tesla seven years to find the answer. It came to him suddenly while he strolled in the park one February afternoon. In a magnificent vision, he saw clearly not only the phased coils necessary to make an ac motor work but also the circuitry, the materials, and even the tolerances to which it had to be made. In one blinding flash, he had it. Then, without putting pen to paper, so he claims, he designed, constructed, and tested in his mind single-phase, two-phase, three-phase, and polyphase power systems to feed his motors—all of which worked perfectly when he finally got around to constructing the real thing. And, one of the many ancillary devices he designed for his system was, of course, the transformer.

The scientific contributions of some of the engineers and physicists named in this thumbnail history are honored today with the names of electrical units—the oersted, the farad, the henry, and the tesla. Arago has not made the list yet, but Lenz has his law, and Charles Kettering, along with other automobile pioneers, has every rusting automobile junkyard in the world as a transient monument.

IMAGERY OF ELECTROMAGNETICS

In the classic iron filings experiment, it was demonstrated that a magnetic field or flux exists in the space around a conductor. Here, the field has the appearance of spaced circular lines, but, of course, this is simply the effect it has on iron filings sprinkled on paper. In reality, the field is totally pervasive, filling all the space around the wire at every point along its length. It dwindles in intensity at greater distances from the wire until finally it tapers away to virtually nothing. There is no sharply defined limit to the field nor does it exist in lines.

Nevertheless, the concept of lines of flux or force is entrenched in the physics of magnetism. It helps us to visualize an abstract phenomenon and permits pictures to be drawn to reinforce explanations. For example, we speak of total flux in terms of the total number of lines of flux, and of flux density in terms of the number of lines of flux occupying a given area, like blades of grass on a lawn.

Other, perhaps arguable notions, such as the idea of ''flow'' of magnetic flux and electrical analogies to it, are traditionally employed when analyzing magnetic effects. All of these appear in the following discussion, but they must be recognized for what they are—simply aids to achieving a useful image of a mysterious and invisible force that is evident only by its effects.

THE MAGNETIC CIRCUIT

The equations and simplified analysis used in this section are not often employed directly in practical transformer design. They are really part of the theoretical basis for data, put out by manufacturers and in various literature in the practical form of graphs and tables of various kinds. However, at least a nodding acquaintance with magnetic terms is highly desirable if available data is to be fully understood and utilized. Moreover, a brief study is valuable in providing a proper perspective and in the understanding of design in general.

As we have seen, the field around a coil that is carrying current may be viewed conveniently as a ''flow'' of flux in the medium surrounding the coil. In this context, anything in the space around the coil, even a vacuum, is viewed as a medium. The magnitude of the flux flow is determined by the product of the current (I) and the number (N) of turns in the coil plus the magnetic conductivity of the medium. The property of flux conduction is called *permeability* (μ) while the force $N \times I$ required to create the flow is called the *magnetomotive force* (mmf).

The magnetic circuit is, therefore, the space in which the flux flows around the coil, regardless of what the space is filled with. Most materials are poor conductors of magnetic

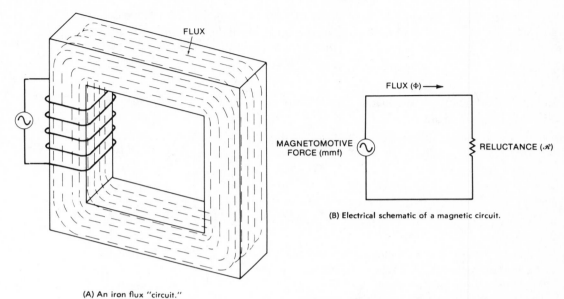

FLUX

FLUX (Φ) ⟶

MAGNETOMOTIVE
FORCE (mmf)

RELUCTANCE (ℛ)

(B) Electrical schematic of a magnetic circuit.

(A) An iron flux "circuit."

Fig. 2-9. A magnetic circuit and its electrical counterpart.

flux; in other words, they have low permeability. A vacuum has a permeability of 1.0, and nonmagnetic materials like brass, air, paper, copper, and so on, have permeabilities of the same order. By contrast, a few materials, including iron, nickel, cobalt, and their alloys, have high permeabilities sometimes ranging into the hundreds of thousands; these form the basis for practical core materials. In practice, the term "magnetic circuit" is also generally applied to the ferromagnetic material that comprises the core.

The operation of the magnetic circuit can be compared to that of the electrical circuit. The ampere-turns $(I \times N)$, which we have called the magnetomotive force (mmf), is analogous to electromotive force (voltage). A poor conductor of flux has a high magnetic resistance, appropriately named *reluctance* (ℛ). The greater the reluctance, the higher the magnetomotive force required to obtain a given flow of flux. A magnetic circuit, together with its electrical counterpart, is shown in Fig. 2-9.

ELECTROMAGNETIC "OHM'S LAW"

Flux (Φ), magnetomotive force (mmf), and reluctance (ℛ) are related by an equation that has the same form as Ohm's law $(I = V/R)$. Thus,

$$\Phi = \frac{mmf}{ℛ} \tag{2-1}$$

Reluctance (ℛ)

The reluctance of a core depends on the composition of the metal and its physical dimensions in exactly the same way that the electrical resistance of a conductor is related to its length (l), cross-sectional area (a), and its specific resistance p (the resistance per unit length). In the case of the electrical conductor, the resistance is

$$R = \frac{pl}{a}$$

In the case of the magnetic equation, $1/\mu$ is analogous to p (it is called the *reluctivity*). The reluctance is given by

$$\mathcal{R} = \frac{1}{\mu} \times \frac{l}{a}$$
$$= \frac{l}{\mu a}$$

(2-2)

where
  l is the length of the magnetic path (in effect, the core length),
  a is the cross-sectional area of the core,
  μ is the permeability.

These factors are depicted in Fig. 2-10. Thus, a high-permeability material has a low reluctance for a given l and a.

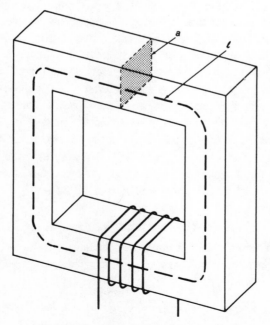

Fig. 2-10. Factors *a* and *l* in an iron core.

If the core is composed of two materials in series, like iron with a gap in it as is shown in Fig. 2-11, the total reluctance of the core is the sum of the iron reluctance

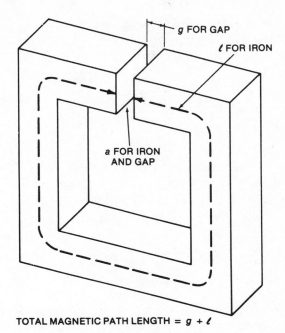

TOTAL MAGNETIC PATH LENGTH = $g + l$

Fig. 2-11. An iron core with an air gap.

and the air gap reluctance in the same way that two series resistances are added. The reluctance of the gap is given also by Equation 2-2; in this case, the symbol g is used for the length of the gap, and because the permeability (μ) of air is 1.0, the term $1/\mu$ disappears. The reluctance of an air gap is, then,

$$\mathcal{R} = \frac{1}{\mu} \times \frac{l}{a}$$

$$= \frac{1}{1} \times \frac{g}{a} \qquad (2\text{-}3)$$

$$= \frac{g}{a}$$

The total reluctance, $\mathcal{R}_T$, of the core pictured in Fig. 2-11 is, then,

$$\mathcal{R}_T = \frac{l}{\mu a} + \frac{g}{a}$$

$$= \frac{(l/\mu) + g}{a} \qquad (2\text{-}4)$$

Because the permeability of the gap is much lower than that of even very low grade iron, its reluctance is very much higher and the total reluctance of the circuit is likely to depend more on the gap than on the iron. Suppose, for example, that l = 8.0 inches, a = 1.0 square inch, μ = 5000 for the iron, and g = 0.005 inch; the total reluctance is, therefore,

$$\mathcal{R}_T = \frac{8}{(5000)(1.0)} + \frac{0.005}{1.0}$$
$$= 0.0016 + 0.005$$
$$= 0.0066$$

Thus, even with a low-permeability iron, an air gap only 0.005-inch wide has an obviously devastating effect on the reluctance of the total core. In this case, the reluctance has increased by a factor of 4. In the case of high-permeability iron, where μ may be on the order of more than 100,000 the same small gap results in

$$\mathcal{R}_T = \frac{8}{1.0 \times 10^5} + 0.005$$
$$= 0.00508$$

Here the reluctance is due almost entirely to the gap. It is about 64 times greater than for the iron alone. Because the total flux in the core, as given by Equation 2-1, is

$$\Phi = \frac{mmf}{\mathcal{R}}$$

it is substantially reduced by the presence of the gap. Obviously, the effects of even a minute air gap (or brass, or any other low-permeability substance) in the core cannot be ignored.

SUMMARY

In the preceding, it has been assumed that the area of the iron is the same as the area of the air gap, and it looks that way in the diagram of Fig. 2-11. In fact, however, the area of the gap is greater than that of the core due to the *fringe effect* in which the flux spreads out as it bridges the gap. This is shown in Fig. 2-12. For most practical purposes, however, the iron area may be taken as the gap area. Air gaps are not always bad. It will be seen shortly that when both dc and ac currents are present in the windings, a gap can be an advantage.

Magnetomotive Force

The effect of magnetomotive force is given by the equation

$$mmf = 1.257NI \qquad\qquad (2\text{-}5)$$

where
   N is the number of turns,
   I is the current in amperes.

REAL GAP AREA ON a_2

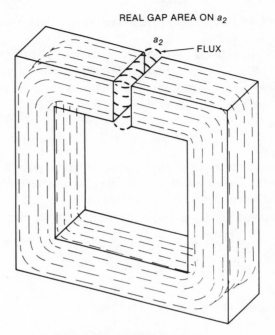

Fig. 2-12. The fringe effect and the gap area.

Therefore, *NI* is the ampere-turns factor. Thus, the greater the number of turns on the coil and the greater the current passing through the turns, the greater the mmf and the greater the flux. Substituting for mmf and $\mathcal{R}$ (Equations 2-2 and 2-5), Equation 2-1 can now be restated as

$$\Phi = \frac{mmf}{\mathcal{R}}$$

$$= \frac{1.257NI}{1/\mu a} \tag{2-6}$$

$$= \frac{1.257NI\mu a}{l}$$

This, of course, assumes no gap in the core. The expression given here for mmf is in units called *gilberts*. The unit of flux is in *maxwells* or *lines*; a maxwell is one line of magnetic flux.

Dividing the mmf by the length of the magnetic circuit *l* gives the magnetizing force in gilberts per unit length; if the unit length is stated in centimeters, the magnetizing force (*H*) will be in oersteds. Thus,

$$H = \frac{1.257NI}{l} \text{ oersteds} \tag{2-7}$$

But, sometimes mmf is stated simply in gilberts per centimeter. (Do you suppose some-one is trying to confuse us?)

Magnetomotive force should not be confused with magnetizing force. The two are related as cause and effect. The magnetizing force is the intensity of magnetization created by the magnetomotive force. If the flux (Φ) is divided by the core area (a), we get flux density (B) in lines or maxwells per unit area. Thus,

$$B = \frac{\Phi}{a} \text{ lines (or maxwells) per unit area} \qquad (2\text{-}8)$$

If the area is stated in square inches, then B is in lines per square inch. If the core area is stated in square centimeters, then B is in lines per square centimeter or *gauss*; a gauss is one line or maxwell per square centimeter (cm^2).

Finally, the permeability (μ) is given by the ratio of the flux density (B) to the magnetizing force (H). Thus,

$$\mu = \frac{B}{H} \qquad (2\text{-}9)$$

There is more about permeability in the next chapter.

Words of Warning

It is only fair to warn you that should you extend your studies of electromagnetism, you will encounter in various literature a somewhat confusing variety of terms and definitions, such as has been hinted at here. This stems from the fact that a number of systems and subsystems of units are in common use, each with its own different definitions—sometimes using the same symbols and sometimes not.

We will not delve further into the subject here—there are excellent textbooks available on the subject. However, as a token of what to expect and what to prepare for, consider the terms one is likely to encounter. The following terms are used to describe flux density—maxwells/inch^2, maxwells/cm^2, gauss, lines/inch,^{2}, lines/cm^2, webers/meter^2, teslas, and so on. So far, the property of inductance has not been mentioned by name, although induction, induced voltage, and induced current, which are faces of the same coin, have been treated briefly. A chapter has been reserved later on for the subject of inductance.

THREE

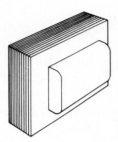

Properties
of Transformers

Figure 3-1 shows a simple iron-core transformer. It consists of a primary winding (N_1), sometimes called "primary" for short, and a secondary winding (N_2), sometimes called "secondary" for short, each mounted on a separate limb of a rectangular core *(C)*. This is the so-called core configuration. Although it is not as common as the "shell" type of Fig. 1-1 in Chapter 1, it is completely practical and illustrates clearly the operation of the transformer. An alternator *(G)* supplies ac voltage V_1, which drives current I_M through the primary winding, creating an alternating flux in the core, as indicated by the broken lines. The flux "links" or passes through the secondary and induces in it a secondary voltage (V_2). The current (I_M) in the primary winding of the unloaded transformer establishes the flux and is called the *magnetizing* current. In practical transformers, the magnetizing current is usually very small, and for the discussion that follows, it will be considered so small as to be negligible. When "current" is referred to, then, it means load current only.

KINDS OF TRANSFORMATIONS

Voltage Transformation

An important relationship exists between the voltages across the primary and secondary windings and the numbers of turns (N_1 and N_2) of the windings. It is

$$\frac{V_1}{V_2} = \frac{N_1}{N_2} \tag{3-1}$$

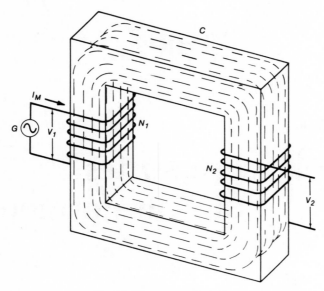

Fig. 3-1. Core type configuration.

That is to say, the ratio of the primary and secondary voltages (V_1 and V_2) is equal to the ratio of the number of turns in the primary and secondary windings. A rearrangement of the equation gives

$$\frac{N_1}{V_1} = \frac{N_2}{V_2} \tag{3-2}$$

and

$$V_2 = \frac{N_2}{N_1} V_1 \tag{3-3}$$

In words, the number of turns per volt on the primary winding is the same as the number of turns per volt on the secondary winding. If this ratio is known, the number of turns required for a given voltage on any winding can be easily calculated.

Example 1—Suppose that N/V is 8. If the primary voltage is 110, then 110×8 = 880 turns of wire are needed for the primary winding (N_1). If V_2 is to be 25 volts, the turns required for the secondary winding (N_2) must be 25×8 = 200. Or, using Equation 3-3,

$$V_2 = \frac{200}{880} \times 110 = 25 \text{ volts}$$

is the secondary voltage.

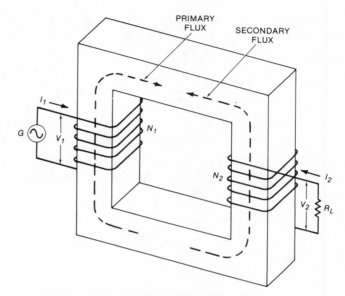

Fig. 3-2. On load the primary and secondary fluxes are in opposition.

Current Transformation

When a load is connected to the secondary winding (Fig. 3-2), a current (I_2) flows through the secondary winding and creates a magnetic flux in the core in *opposition* to that established by the magnetizing current. This flux would reduce the effective flux, except that a reduction in flux in the core causes an increase in the primary winding current, which restores the flux to its former value. In this way, a balance of the interacting forces takes place any time the load current changes; the original value of flux is always maintained. The primary winding load current increases along with the secondary winding load current in the following relationship:

$$\frac{I_2}{I_1} = \frac{N_1}{N_2} \tag{3-4}$$

This states that the ratio of the secondary winding load current to the primary winding load current is equal to the turns ratio. Note once more that I_1 and I_2 are due to the load only and do not include the magnetizing current. A rearrangement of the equation gives

$$I_1 N_1 = I_2 N_2 \tag{3-5}$$

The terms $I_1 N_1$ and $I_2 N_2$ are, for obvious reasons, called the primary winding and secondary winding ampere-turns.

Example 2—Using the same numbers as in the first example for volts and turns with R_L as 10 ohms in Fig. 3-2, the secondary winding current is, from Ohm's law, $I_2 = V_2/R_L = 25/10 = 2.5$ amperes. Rearranging Equation 3-5 to find I_1 gives

$$I_1 = \frac{I_2N_2}{N_1} = \frac{2.5 \times 200}{880} = 0.568 \text{ ampere} \tag{3-6}$$

as the primary winding current.

Equation 3-5 says that the primary winding ampere-turns equal the secondary winding ampere-turns. If there were more than one secondary winding, as in Fig. 3-3, the primary winding ampere-turns would equal the sum of the ampere-turns of the secondaries. Thus

$$I_1N_1 = I_2N_2 + I_3N_3 + I_4N_4 + \ldots + I_nN_n \tag{3-7}$$

Quantities that are equal to the same quantity are equal to one another. So it follows from Equations 3-1 and 3-4 that

$$\frac{V_1}{V_2} = \frac{I_2}{I_1} \tag{3-8}$$

and a rearrangement of Equation 3-8 reveals another important relationship:

$$V_1I_1 = V_2I_2 \tag{3-9}$$

This says that the primary winding volt-amperes equal the secondary winding volt-amperes. As with the ampere-turns, the sum of the volt-amperes of the secondaries (if there is more than one secondary) is equal to the primary winding volt-amperes. For the moment, the effects of losses are not being considered. (Losses add to the primary volt-amperes, making the primary volt-amperes (VA) greater than the secondary volt-amperes in practice.)

Example 3—From the last example, there is 25 volts and 2.5 amperes in the secondary winding, giving 62.5 volt-amperes for the secondary. For the primary winding, there is 110 volts at 0.568 amperes, giving 62.5 volt-amperes.

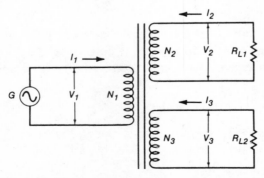

Fig. 3-3. Here $N_1I_1 = N_2I_2 + N_3I_3$ and $I_1V_1 = I_2V_2 + I_3V_3$.

Example 4—If a second winding were added to the transformer of, say, 10 volts at 2.0 amperes, or 20 volt-amperes, the total secondary volt-amperes would be 82.5 volt-amperes. The primary winding volt-amperes would also be 82.5 volt-amperes and the primary winding current would be 82.5/110 = 0.75 ampere.

Impedance Transformation

Consider now what happens when Equation 3-1 is multiplied by Equation 3-4.

$$\frac{V_1}{V_2} = \frac{N_1}{N_2} \tag{3-1}$$

multiplied by

$$\frac{I_2}{I_1} = \frac{N_1}{N_2} \tag{3-4}$$

gives

$$\frac{V_1 I_2}{V_2 I_1} = \left(\frac{N_1}{N_2}\right)^2$$

This is the same as

$$\frac{V_1}{I_1} \times \frac{I_2}{V_2} = \left(\frac{N_1}{N_2}\right)^2$$

which is identical with

$$\frac{V_1}{I_1} \div \frac{V_2}{I_2} = \left(\frac{N_1}{N_2}\right)^2$$

From Ohm's law the secondary winding load resistance R_L is V_2/I_2. Therefore V_1/I_1 is a kind of "reflection" of R_L in the primary winding that can be designated R_L'. Substituting R_L and R_L' in the equation gives

$$\frac{R_L'}{R_L} = \left(\frac{N_1}{N_2}\right)^2 \tag{3-10}$$

Rearranging the equation to express R_L' in terms of the other factors gives

$$R_L' = R_L \left(\frac{N_1}{N_2}\right)^2 \tag{3-11}$$

where

R_L is the actual load resistance,

R_L' is the resistance presented by the primary winding to the supply source.

In other words, the primary winding acts like a resistance equal to the load resistance across the secondary multiplied by the square of the turns ratio (see Fig. 3-4).

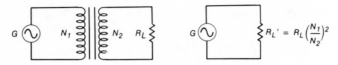

(A) The transformer plus R_L appears to the generator as R_L' (on the right).

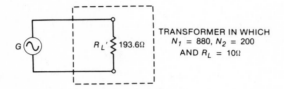

(B) A transformer with specific values for N_1, N_2, and R_L as it appears to the generator.

Fig. 3-4. Illustrating the resistance presented by the primary winding to the supply source.

Example 5—Using the numbers in the previous examples, we have

$$R_L' = 10 \left(\frac{880}{200} \right)^2 = 193.6 \text{ ohms}$$

for the primary winding load resistance.

A frequently used variant of this formula is the following:

$$\frac{N_1}{N_2} = \sqrt{\frac{R_L'}{R_L}} \tag{3-12}$$

In words, the turns ratio is equal to the square root of the impedance ratio.

TAPPED WINDINGS

The simple circuit shown in Fig. 3-5A contains hidden magic. Suitably arranged, it will not only step up and step down voltage but provide an incredibly high operational efficiency in doing so, often close to 100 percent (efficiency is measured as the circuit output power divided by the input power).

Voltage Transformation

Fig. 3-5A illustrates a version of the step-down case. Here V_1 is an ac input. The

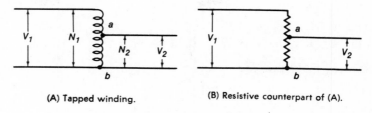

(A) Tapped winding.   (B) Resistive counterpart of (A).

Fig. 3-5. Use of a tapped transformer winding.

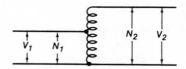

Fig. 3-6. Tapped winding as a step-up transformer.

circuit is rather obviously a reactive voltage divider similar in action to its dc counterpart shown in Fig. 3-5B. It is not surprising, then, that it steps the voltage down. When it is examined a little more closely, however, it is seen to exhibit the properties of a transformer rather than a simple voltage divider. Current-voltage-turns relationships are the same as those found in the two-winding arrangement discussed earlier.

In Fig. 3-5A, the number of turns N_2 between any two points a and b divided by the voltage V_2 between these points is equal to the total turns N_1 divided by the total voltage V_1 across the coil. In symbols, the ratio reads $N_2/V_2 = N_1/V_1$. This is already familiar as Equation 3-2, and N/V is the turns-per-volt figure for the coil.

Moreover, the circuit is reversible, as is the standard two-winding arrangement. If an input V_1 is applied to the short portion of the coil as in Fig. 3-6, a stepped-up voltage V_2 is obtained across the entire winding N_2. The turns-per-volt relationship still holds good, and the output voltage is given as in the standard case by a rearrangement of Equation 3-2

$$V_2 = \frac{N_2}{N_1} \ V_1$$

Example 6—If, in Fig. 3-6, $V_1 = 117$ Vac, $N_1 = 300$ turns, and $N_2 = 350$ turns, what is V_2? From the preceding equation,

$$V_2 = \frac{350}{300} \times 117 = 136.5 \text{ volts}$$

is the secondary voltage.

Current Transformation

Referring to Fig. 3-7, the current-turns relationship is the same as for the two-winding arrangement as given by Equation 3-4. However, I_1 and I_2 flow through the portion N_1

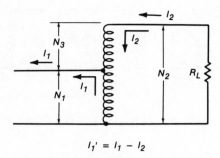

$$I_1' = I_1 - I_2$$

Fig. 3-7. Directions of currents in windings.

in opposite directions, and therefore the resultant current I_1' in N_1 is the difference between the two.

Example 7—Using the same voltage and turns figure as in Example 6 and making R_L = 27.3 ohms, then I_2 is

$$I_2 = \frac{136.5}{27.3} = 5 \text{ amperes}$$

Now I_1 is given by Equation 3-4.

$$I_1 = \frac{I_2 N_2}{N_1}$$

$$\frac{5 \times 350}{300}$$

$$= 5.833 \text{ amperes}$$

But I_1 and I_2 are in opposition in N_1. Therefore the current I_1' is

$$I_1' = I_1 - I_2$$
$$= 5.833 - 5.0$$
$$= 0.833 \text{ ampere}$$

through winding N_1.

Volt-Amperes (VA)

As in any transformer, the secondary volt-amperes, or VA, must equal the input VA as stated by Equation 3.5. But herein lies the special magic referred to earlier. It is easily illustrated by an example.

Example 8—Using the same quantities as before, the input VA is $V_1 I_1$ = 117 × 5.833 = 682.5 VA and the secondary VA is $V_2 I_2$ = 136.5 × 5.0 = 682.5 VA. Note, however, that because the resultant primary winding current is only 0.833 ampere, the VA actually handled by the transformer is only 117 × 0.833 = 97.5 volt-amperes.

In other words, a very small transformer seems to handle an output seven times larger than its rating. This magic is, of course, illusory. The transformer actually delivers the power put out by the portion of the winding N_3. The number of turns in N_3 is $350 - 300 = 50$ turns. The voltage across N_3, according to the normal transformer relationship, is

$$V_3 = \frac{N_3}{N_1} \ V_1 = \frac{50}{300} \times 117 = 19.5 \text{ volts}$$

and the VA delivered by this portion of the winding is $V_3 \times I_2 = 19.5 \times 5.0 = 97.5$ volt-amperes.

The percent of losses incurred by the transformer will be the same as for any other, but because it is so small in relation to the total output, the actual loss will be a much smaller percent of the total output, leading to a high circuit efficiency. The magic, then, might be illusory, but the benefits are very real.

KINDS OF TRANSFORMERS

Each of the relationships indicated by Equations 3-1 to 3-12 form a basis for designing different kinds of transformers.

Voltage Transformer

Equation 3-1 is the basis for the so-called voltage (sometimes called ''potential'') transformer. The voltage transformer is designed to achieve an accurate voltage ratio constant within fine limits over its load range. It is commonly used to supply instruments such as voltmeters, voltage coils on wattmeters, protective relays, and the like. It finds uses for stepping-up very low voltages for ease of measurement and stepping-down very high voltages to low levels for safety in measurement. The method of connection is shown in Fig. 3-8.

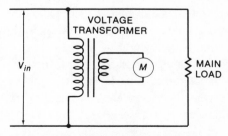

Fig. 3-8. Connection for voltage transformer.

Because it is of somewhat limited interest for the experimenter and in any case is similar to the next type to be discussed, its design is not specifically dealt with in this book. Instead, its close relative, the common power transformer, will receive a great deal of attention.

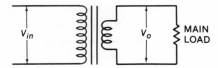

Fig. 3-9. Connection for power transformer.

Power Transformer

Essentially, the power transformer is a voltage transformer based on Equation 3-1. It, too, is required to achieve an accurate voltage ratio that is substantially constant over its range of loads. In all transformers, the secondary winding voltage falls off as the load current increases. The difference between the full-load and no-load voltages is usually expressed as a percentage and is called the *regulation*. In most cases, the regulation is expected to be reasonably good in a power transformer (Fig. 3-9)—not as good as in a voltage transformer, but good. Regulation is expressed as

$$\% \text{ Regulation} = \frac{100 \text{ (No-load voltage)}}{\text{Full-load voltage}} \qquad (3\text{-}13)$$

that is, as a percent.

The power transformer is the most widely used of all transformers and ranges in size from the mighty high-power components used in distribution systems to the miniscule types found in charger adapters for pocket calculators. This range includes transformers in radios, TV's, toys, and equipment of all kinds in the home and industry. Figure 3-10 shows a few uses of power transformers. Chapter 8 focuses on power transformers.

Current Transformer

Not quite so well-known is experimenter circles is the current transformer, the design of which is based on Equation 3-5. Just as the object of the voltage transformer is to achieve an accurate voltage transformation, so the purpose of the current transformer is to achieve a highly accurate and constant transformation ratio. It is commonly used to supply wattmeters and protective overcurrent relays. As will be discussed later, though, it has some interest for the experimenter as an instrument transformer.

This type of transformer is usually placed in series with a high-current circuit, as shown in Fig. 3-11, and steps the current down to drive a low-current device of some kind, frequently a low-reading current meter. Because of this connection mode, it is sometimes referred to as a *series transformer*.

Impedance Transformer

Equations 3-10, 3-11, and 3-12 are used in the design of a common class of transformers variously referred to as *impedance, matching, speaker, input, output,* or *coupling transformers,* depending on the specific application. This type of transformer is designed to achieve an accurate impedance transformation, particularly in low-frequency

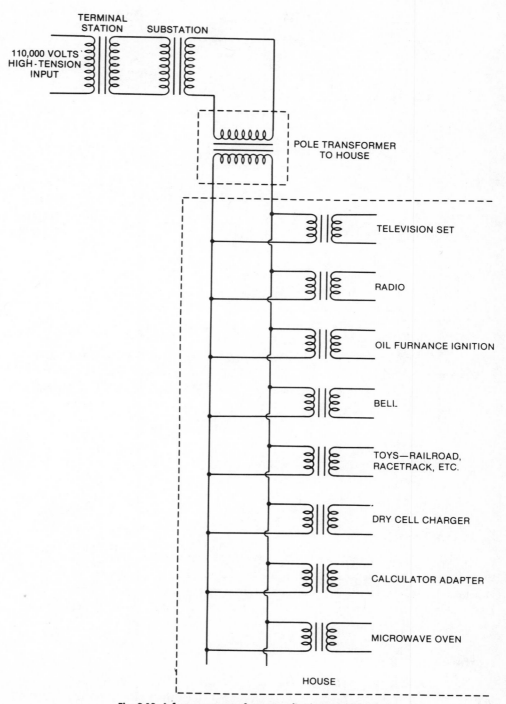

Fig. 3-10. A few power-transformer applications in the home.

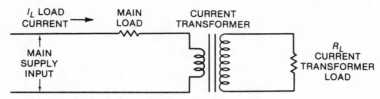

Fig. 3-11. Current-transformer connection.

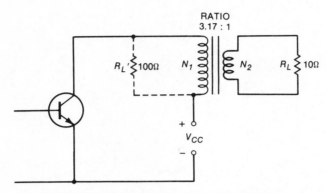

Fig. 3-12. Common impedance-transformer application.

amplifiers. Figure 3-12 shows a practical example of one form of circuit in which it is required to transform a 10-ohm load to suit a 100-ohm transistor input. In this case, the ratio (r) required for the windings is given by Equation 3-12 with $r = N_2/N_1$:

$$r = \sqrt{\frac{R_L'}{R_L}} = \sqrt{\frac{100}{10}} = \sqrt{10} = 3.17\!:\!1$$

where

R_L' is the primary load resistance,
R_L is the actual load resistance.

Isolation Transformer

A glance at any transformer diagram will reveal the important feature of isolation: there is no direct connection between the windings. They are connected only by the intangible lines of magnetic flux in the core. In some types of circuits, the transformer might have a 1:1 turns ratio—that is to say, no step-up or step-down action; the only reason for the transformer is to isolate the circuit from the supply.

Autotransformer

The autotransformer is based on the characteristics of the tapped winding depicted in Figs. 3-5, 3-6, and 3-7. This device is really more in the nature of a circuit arrangement than a distinct breed of component, except in some special cases. Electrically, it is almost identical with the standard two-winding transformer connected as shown in Fig. 3-13. Here the primary is connected across the supply in the normal way, but the secondary is connected in series with the supply so that its voltage adds to the supply voltage when the output is taken across both windings. The circuit can be readily analyzed as a two-winding type.

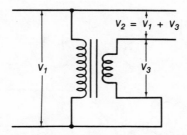

$$V_2 = V_1 + V_3$$

Fig. 3-13. Standard two-winding transformer connected as autotransformer.

A comparison of the standard and autotransformer winding configurations is made in Figs. 3-14A and 3-14B. The same inputs and outputs are involved. The standard configuration requires two full-current windings, while in the autotransformer only the low-voltage portion carries the full current. This not only saves on copper but also on iron, because less core is needed to enclose the smaller windings.

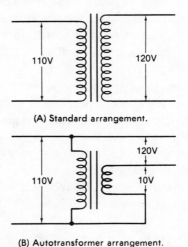

(A) Standard arrangement.

(B) Autotransformer arrangement.

Fig. 3-14. Comparison of autotransformer and standard transformer with identical outputs.

Unfortunately, the valuable isolation feature of the standard transformer is lost. Autotransformers are generally used to obtain small increments of voltage above (or below) the input voltage, and the closer the output-to-input voltages ratio comes to 1.0, the better the performance. However, it still offers substantial advantage at ratios of 2:1 (or 1:2).

The well-known Variac and similar types of transformers specially designed for adjusting voltages employ a continuously variable wiper contact similar to that shown in Fig. 3-15.

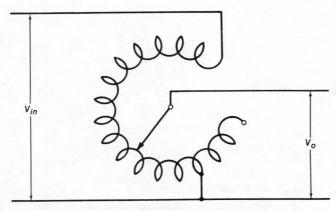

Fig. 3-15. Continuously variable autotransformer.

At Heart They Are All the Same

No matter what they are called, *all* loaded transformers transmit power and transform voltage, current, and impedance. The factor that makes one transformer different from another is simply that in its design, one aspect of transformation has been selected for emphasis over others. The other aspects are still there, however, though they may be largely ignored.

For example, it is quite possible to use a power transformer as an impedance transformer if the turns ratio and other features are right for the job on hand. To take another example, a current transformer when connected normally may have, say, 1 volt across its secondary winding and a small fraction of a volt across its primary winding. If the secondary load is disconnected while current is passing through the primary, several hundreds of volts might suddenly appear across the secondary winding. In addition to proving that the current transformer transforms voltage, this new voltage can cause some quickening if one's fingers happen to be what completes the secondary circuit.

TRANSFORMER RATINGS

Transformer ratings are nearly always stated in volt-amperes rather than watts although both are given by multiplying volts and amperes. But volt-amperes and watts are not the same thing.

Watts Versus Volt-Amperes

If the load is a pure resistance, then the current in the load is in phase with the voltage across it, and the volts dropped multiplied by current in the load is watts. This power is often referred to as *true power*. If, however, the load is a pure inductance or a pure capacitance, the voltage across it is not in phase with the current through it. In the case of an inductance, plotting voltage and current against time produces two curves as shown in Fig. 3-16. If the power curve is developed from this by multiplying the instantaneous values of voltage and current and plotted, the curve shown as the dotted line is obtained. The dotted line shows that power is consumed in the first quarter cycle of current to build up the magnetic field, and then the power returned to the circuit in the following quarter cycle. Because the power curve swings positive and negative in one half of the current cycle and then repeats this pattern in the second half, the net power consumed is zero.

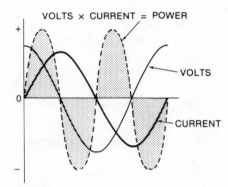

Fig. 3-16. The net power consumed in a pure inductance is zero.

A similar situation is obtained with a capacitor. Only if there is resistance in the circuit is there a net or true power consumption. A transformer always has resistance in its windings, and although a reactive current might not be in phase with the voltage across the reactive load, it is always in phase with the voltage drop that it creates across the winding resistance. This phase agreement causes power to be consumed in the windings that in turn creates undesirable heat in the copper. This power loss is a major limiting factor in transformer ratings, and because it can occur even if the load is not consuming watts but rather only out-of-phase volt-amperes, the transformer must be rated in volt-amperes.

There is a tendency in transformer design to use the terms "watts" and "volt-amperes" interchangeably, even though it is not strictly accurate. But the distinction between the terms should be kept in mind if highly reactive loads are involved.

The ratio of watts to volt-amperes, that is to say, of true power to apparent power in a load, is known as the *power factor* (pf). Because the volt-amperes can never be less than the watts, the power factor must always be 1.0 or less. When the power factor is 1.0, the load is totally resistive and consumes watts only. If the power factor is less than 1.0, the load is partly reactive and volt-amperes as well as watts must be considered.

TWO BASIC DESIGN EQUATIONS

Although there are many equations associated with transformer design, all of them fundamental, there are two in particular that are usually considered as the basic design equations.

The first one, which we have already encountered, is sometimes referred to as the *voltage equation*, and the second is the *power capability equation*. They are, respectively,

$$V = 4FfaNB \times 10^{-8} \qquad (3-14)$$

where

V is the voltage across the considered winding in volts rms,

F is the form factor,

f is the frequency in hertz,

a is the core cross-sectional area,

N is the number of turns on the considered winding,

B is the flux density in maxwells per unit area.

and

$$P = 0.707JfWaB \times 10^{-8} \qquad (3-15)$$

where

J is the current density of the transformer in amperes per square centimeter,

f is the frequency in hertz,

W is the area of core window in square centimeters,

a is the cross-sectional area of core in square centimeters,

B is the flux density in gauss (maxwells per square centimeter).

In Equations 3-14 and 3-15, if a is in square centimeters, then B must be in gauss, or if a is in square inches, then B must be in lines or maxwells per square inch. In short, the units must be compatible. Note that the symbols W and a are as defined in previous chapters. Usually in practice, the unit area is square centimeters.

The Voltage Equation

One essential equation appears in so many guises (some of them not easily recognizable as arrangements of Equation 3-14) that it is well to be familiar with the various forms of this equation and the reasons why it varies.

The arrangement of Equation 3-14 is the one commonly encountered in textbooks together with the rearrangements below:

$$\frac{N}{V} = \frac{10^8}{4FfaB} \qquad (3-16)$$

and

$$\frac{N}{V} = \frac{1}{4FfaB \times 10^{-8}}$$ (3-17)

Sometimes T (or some other letter) is made to represent N/V, giving

$$T = \frac{10^8}{4FfaB}$$ (3-18)

Frequently it is assumed that the equation will be used with sinusoidal inputs, so F is immediately assigned the value of 1.11. This value, combined with 4, gives 4.44 in place of $4F$:

$$\frac{N}{V} = \frac{10^8}{4.44faB}$$ (3-19)

Of course, the preceding equations are simply rearrangements of the basic equation. It is usually more practical for design purposes, however, to add conversions to them that enable a to be expressed directly in inches and B in gauss. In manufacturers' data, for example, flux density is usually in gauss and most people are more comfortable with an inch scale in their hands than a centimeter scale (on the North American continent at any rate). This conversion is accomplished by including the factor 6.45 in the bottom line (there are 6.45 sq cm to the square inch). Applying this to Equation 3-19,

$$\frac{N}{V} = \frac{10^8}{4.44 \times 6.45faB} = \frac{10^8}{28.64 \, faB}$$ (3-20)

or, dividing,

$$\frac{N}{V} = \frac{3.49 \times 10^6}{faB}$$ (3-21)

Applying the factor 6.45 to Equation 3-18 gives

$$T = \frac{N}{V} = \frac{10^8}{25.8FfaB}$$ (3-22)

or, dividing out,

$$T = \frac{N}{V} = \frac{3.876 \times 10^6}{FfaB}$$ (3-23)

is the equation.

Often the equation is encountered with a *stacking factor* included. The purpose of the stacking factor is to compensate for the fact that the measured area of the core is always bigger than the actual area of iron. (This is explained in detail in Chapter 5.) The symbol for the stacking factor used in this book is lower case s. The stacking factor is usually somewhere around 0.85 and 0.95.

If the stacking factor is included as a symbol, it is of course easily spotted; it will appear in the denominator of Equations 3-22 or 3-23. But if it is included as a number (which it might well be), it is not so easy to recognize, and this can lead to errors due to duplication. For example, would you see that the following equation includes a conversion factor for a in square inches and B in gauss, a form factor of 1.11 for a sine wave, and a stacking factor of 0.9?

$$\frac{N}{V} = \frac{10^8}{25.77faB} \tag{3-24}$$

or

$$\frac{N}{V} = \frac{3.88 \times 10^6}{faB} \tag{3-25}$$

Surely, some experience is necessary with the equations.

Incidentally, the reason for using the form factor is as follows. When the equation is derived from first principles, the *average* rate of flux in the core is used to derive the average voltage V_{avg} as

$$V_{avg} = 4faNB \times 10^{-8} \tag{3-26}$$

But usually our interest is in volts rms. It was shown in Chapter 1 that form factor F = rms/average, or rms = $F \times$ average. Therefore

$$V_{rms} = 4FfaNB \times 10^{-8}$$

which is Equation 3-14.

This leads to one of the more subtle forms of the equation that you watch for. In converter-inverter design (Chapter 10), the equation is presented as

$$V = 4faNB \times 10^{-8} \tag{3-27}$$

As can be seen, the F appears to be absent. In truth, though, it is not missing. It has simply been assigned the value of 1.0 because the circuits in Chapter 10 work with *square* waves, not sine waves. Recall from Chapter 1 that the form factor of a square wave is 1.0.

In design practice, the equation is most often used to solve for N/V as in most of the forms given here. But it is also used frequently to solve for other factors. Thus, based on Equation 3-22,

$$B = \frac{V \times 10^8}{25.8FfaN} \tag{3-28}$$

$$f = \frac{V \times 10^8}{25.8FaNB} \tag{3-29}$$

or even for a,

$$a = \frac{V \times 10^8}{25.8FfNB} \tag{3-30}$$

Finally, consider the following variant of Equation 3-14:

$$V = 4FfN\Phi \times 10^{-8} \tag{3-31}$$

If this is compared to the basic equation (Equation 3-14), you can see that the symbol Φ is now being used to replace $a \times B$. The reason this can be done is in Chapter 2. Equation 2-8 states that

$$B = \frac{\Phi}{a}$$

where

B is the flux density,
Φ is the total flux in the core,
a is the cross-sectional area of the core.

The preceding equation can be restated as $\Phi = a \times B$. Equation 3-31 is commonly used in the form

$$\Phi = \frac{V \times 10^8}{4FfN} \tag{3-32}$$

that expresses the total flux in the core.

When working with these equations where conversion factors have been included, a small trap awaits the unwary. Consider one form, such as Equation 3-22. It is not valid to rearrange this equation as

$$aB = \frac{V \times 10^8}{25.8FfN}$$

The answer will be wrong because this equation includes the factors needed to allow a to be stated in square inches and B in gauss. In this case, $a \times B$ is "multiplying oranges and apples." To keep matters straight, when moving a from the right to the left side of the equation, its conversion factor must be moved with it; that is to say

$$6.45aB = \frac{V \times 10^8}{4FfN} \tag{3-33}$$

Otherwise stated, to match with Equation 3-22,

$$6.45aB = \Phi = \frac{V \times 10^8}{4FfN} \tag{3-34}$$

where
  a is in square inches,
  B is in gauss.

Also,

$$aB = \Phi = \frac{V \times 10^8}{4FfN} \tag{3-35}$$

where
  a is in square centimeters,
  B is in gauss.

The Power Capability Equation

As with the voltage equation, the arrangement of Equation 3-15 is frequently changed to suit specific requirements, and conversion factors are often included. Again it is most convenient when dimensions can be stated in inches and flux density in gauss. There is a choice also for the current density; it is often stated in amperes per square inch but might be more convenient in circular mils per ampere.

Equation 3-15 is expressed as follows:

$$P = 4.55JfWaB \times 10^{-8} \tag{3-36}$$

where
  J is the current density in amperes per square inch,
  W is the area of the core window in square inches,
  a is the cross-sectional area of the core in square inches,
  B is the flux density in gauss,
  f is the frequency in hertz,
  P is the power in volt-amperes.

A useful arrangement of this expression is

$$Wa = \frac{P \times 10^8}{4.55JBf} \tag{3-37}$$

The term Wa is called the Wa product, a number that is valuable in core selection.

The current density can be stated in circular mils per ampere with the dimensions in inches and flux density in gauss. Using S to denote this quantity, the equation becomes

$$P = \frac{fBWa}{17.26S} \tag{3-38}$$

or

$$Wa = \frac{17.26SP}{fB} \qquad (3\text{-}39)$$

is the Wa product.

Another version of this product is to state a and W in terms of their linear dimensions. Referring back to Fig. 1-1 in Chapter 1, $W = FG$ and $a = DE$. Therefore

$$P = \frac{fB \times DE \times FG}{17.26 \times S} \qquad (3\text{-}40)$$

Thus, any given dimension of the core area or window area can be extracted by suitable rearrangement. For example,

$$D = \frac{17.26PS}{fB \times EFG} \qquad (3\text{-}41)$$

is the horizontal dimension of the core area.

Note that when current density is stated in circular mils per ampere, the smaller the number, the greater the density is. From Equation 3-39, it can be seen that the greater the current density, S, is (that is to say the *smaller* its number in mils per ampere), and the greater the flux density, B, is, the smaller the Wa product, or the smaller the core required is. For any given Wa product, however, a wide spread is possible in core weight and volume; a glance at a core catalog verifies this.

Putting the Equations to Work

Example 9—As an example of how these formulas are used and as a preliminary exercise in design, consider the following specification for a transformer: It is required to work from 110 volts, 60 hertz, and deliver 50 volts at 2.0 amperes (Fig. 3-17).

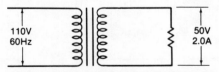

110V
60Hz

50V
2.0A

Fig. 3-17. Basic specification of worked example.

In small-transformer design, current densities on the order of 500 to 1200 CM/A (circular mils per ampere) are commonly used. For this exercise, a conservative 1000 circular mils per ampere is selected. The type of iron core most commonly found in small power transformers is silicon iron, and this material can be used at flux densities of around 13,000 gauss.

From Equation 3-39, the required Wa product can be determined to be

$$Wa = \frac{17.26 \times 1000 \times 100}{60 \times 13,000} = 2.2 \text{ in}^4$$

A core is on hand (salvaged from a junked TV set) that has the measurements shown in Fig. 3-18. The core cross-sectional area is $a = 2 \times 1.25 = 2.5$ inches^2. The window area is $W = 0.625 \times 1.875 = 1.17$ inches^2. The Wa product is then $Wa = 2.5 \times 1.17 = 2.925$ inches^4. This seems more than ample and can be reduced to 2.2 inches^4 by reducing the thickness of the stack of laminations. The required core area is given by the Wa product figure divided by the window area. It is

$$a = \frac{Wa}{W} = \frac{2.2}{1.17} = 1.88 \text{ sq in}$$

The stack thickness must then be the core area a divided by the dimension E, which is 1.25 inches:

$$\text{Stack} = \frac{a}{E} = \frac{1.88}{1.25} = 1.5 \text{ inches}$$

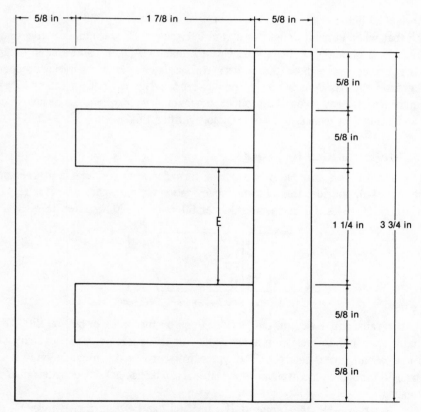

Fig. 3-18. Lamination size for worked example.

Turning now to the voltage equation in the form given by Equation 3-20:

$$\frac{N}{V} = \frac{10^8}{28.64 \times 60 \times 1.88 \times 13,000} = 2.38 \text{ turns per volt}$$

For the primary winding, the turns needed are then $V_1 \times N/V = 110 \times 2.38 = 262$ turns, and for the secondary winding, $V_2 \times N/V = 50 \times 2.38 = 119$ turns. The power output is $I_2 \times V_2 = 50 \times 2 = 100$ watts. Expect an efficiency of perhaps 90 percent; that is to say, the input power must be $P_{in} = 100/0.9 = 111$ watts, and the input current I_1 will be the input watts divided by the input volts V_1 or $I_1 = 111/110 = 1.0$ ampere.

At a current density of 1000 CM/A, the primary winding wire must have an area of $1000 \times 1.0 = 1000$ CM and the secondary winding wire must have an area $1000 \times 2 = 2000$ CM. A glance at the wire table in Chapter 6 (Table 6-1) reveals that the wire gauges with areas closest to these numbers are No. 20 AWG and No. 17 AWG, respectively.

From this point, the fit of the wires in the core window space is checked as discussed in Chapter 6. It is usually permissible to try other sizes of wire if necessary to achieve a good fit so long as other design criteria are met.

FOUR

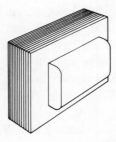

Losses

As the design progresses, it is usually necessary to make adjustments for losses in the transformer—power and voltage losses in the windings and power dissipated in heating the core. Reactances, both capacitive and inductive, also contribute to losses in various ways. In order to compensate for these factors, it is necessary to know how they modify the performance of the transformer.

This subject of losses is in fact central to good transformer design, and if taken to textbook extremes, it is perhaps the most complex part. A simple treatment sufficiently accurate for most purposes, however, is given in the following.

THE LOSSES DEFINED

The metal of the core, like the copper wires, is linked by the varying magnetic flux; thus, circulating currents are induced in it. These *eddy currents*, as they are called, together with an effect called *hysteresis* create a power loss in the form of heat in the iron core (Chapter 5); this is often called *iron loss*. Also, the current in the primary winding encounters the resistance of the copper wire, that results in a further I^2R loss and voltage drop. Because these losses are independent of load, they are part of the no-load loss.

In addition, the load currents through the resistances of the primary and secondary windings create I^2R losses that heat up the copper wires and cause voltage drops. These losses are of course load losses. The iron-core power loss and primary winding load currents are in phase and are therefore directly additive. They usually account for most of the power loss and indeed are often the only factors taken into account in the design.

Other sources of loss exist, however. An occasionally important one is due to the magnetizing current. This current depends on the reactance of the primary winding and is independent of the load. It is in fact the primary winding current as given by $I_M = V_1/X_{L1}$, where V_1 is the primary winding voltage and X_{L1} the primary winding reactance. Because I_M is a purely reactive current, it is not in phase with the iron-core power loss and load currents, and it cannot be added directly to them. When this current is taken into account (which is relatively seldom, in power transformers at least), it has to be added vectorially, as explained shortly.

The volt-amperes required to establish the reactive magnetizing current is sometimes referred to as the *apparent loss* because in itself it does not represent a real loss in power-consuming watts. Nevertheless, the flow of the magnetizing current through the resistance of the winding does create a real I^2R loss and voltage drop, although both are generally quite small.

STRAY CAPACITANCE AND LEAKAGE INDUCTANCE

Two factors that contribute to losses (and other undesirable phenomena) are stray capacitance and leakage inductance. Unplanned and therefore *stray* capacitance inevitably exists between turns, between one winding and another, and between windings and the core. These capacitances form shunt paths that modify the performance of the transformer. Generally, they need be considered only at relatively high frequencies, for then the capacitive reactance is low.

Leakage inductance arises because not all the flux links the windings via the core. Figure 4-1 shows an example in which some flux has leaked from the iron core and

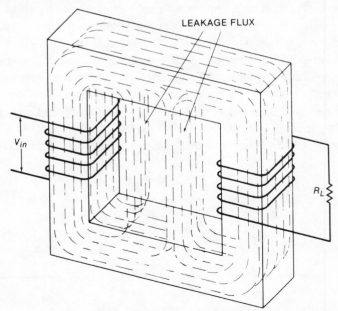

Fig. 4-1. Illustrating leakage flux from the core.

completed the magnetic circuit through the air. Such leakage is associated with both the primary and secondary windings. For convenience of illustration a core-type transformer with windings on separate limbs is shown. The principle, however, applies to any transformer (or inductor for that matter). If the windings are placed one on top of the other, as is more usual, there will still be leakage inductance, but probably to a lesser degree.

The *effect* of leakage inductance is as though a small part of the total inductance had been detached and placed in series with the winding, as shown schematically in Fig. 4-2, where L_P' and L_S' are the primary winding and secondary winding leakage inductances, respectively. Again, the effect is generally not important except at relatively high frequencies, for then the reactances are high, and being in series, have a marked effect on performance.

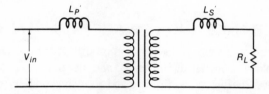

Fig. 4-2. Schematic diagram showing leakage inductances of the circuit in Fig. 4-1.

THE EQUIVALENT CIRCUIT

The classical approach in discussing the effects of losses on performance, and the one used here, is first to draw an equivalent circuit consisting of a *perfect* transformer (a transformer having no losses) with imaginary discrete components connected to it to represent the loss factors. One form of such a circuit is shown in Fig. 4-3. Here the perfect transformer (T) has connected to it resistors R_1 and R_2, representing the resistances of the primary and secondary windings—in other words, resistors connected in such a way as to have the same electrical effect as the actual copper resistances.

Similarly, the small series inductors L_P' and L_P' stand for the effects of the primary and secondary leakage inductances. The shunt inductor L_P represents the primary winding inductance, and the shunt resistor R_0 suggests the effects of the iron-core losses.

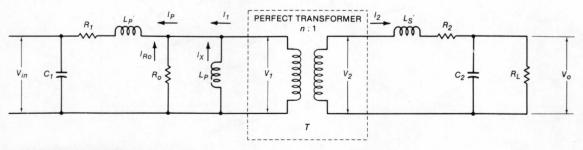

Fig. 4-3. Equivalent transformer circuit.

Stray capacitances are represented by the "lumped" capacitors C_1 for the primary winding and C_2 for the secondary. The load resistance is the resistor R_L.

Like the smile on the face of the Cheshire Cat, all that is now left of transformer T is the abstract functions of voltage and current transformation symbolized by the windings inside the dotted-line box. This transformer has a turns (or voltage) ratio of n:1 but no other electrical characteristics; these have all been assigned to the components outside the box.

Reactances

If Fig. 4-3 looks like a network of ghastly complexity, believe it. As a picture of losses, however, it isn't strictly true. One way or another, losses are a function of ohmic values in the circuit, and therefore the inductive and capacitive elements are best viewed as reactances X_L and X_C rather than L and C. Recall that reactance is a function of frequency as given by the equations $X_L = \omega L$ and $X_C = 1/\omega C$, where $\omega = 2\pi f$, L is the inductance in henrys, and C is the capacitance in farads. Thus, by relabelling the reactances as in Fig. 4-4, frequency is brought into the picture and all the elements are stated in ohms.

The values of the resistive elements (except R_0) can be considered as independent of frequency.

Because there are both shunt and series reactances all varying in value with frequency, it is evident that the circuit must perform differently at high and low frequencies. In fact, if separate circuits are drawn to represent low- and high-frequency conditions as will be done shortly, the circuits will be markedly dissimilar. (Note that although R_0 is frequency dependent, it is not reactive. This phenomenon is explained in Chapter 5.)

A Circuit Diagram in Motion

Suppose the frequency of the equivalent circuit input is gradually increased from 0 hertz (dc, in other words) to many thousands of hertz. Imagine, also, that through the magic of motion picture animation, the reactances can be made to fade in and out of the diagram in accord with their effects on the performance at various frequencies.

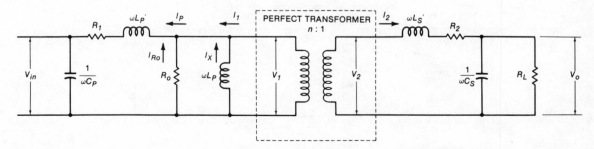

Fig. 4-4. Equivalent transformer circuit with reactances instead of inductance and capacitance.

For example, at zero frequency, the inductive reactances are 0 ohms ($X_L = 2\pi fL$ ohms; if $f = 0$, then $X_L = 0$). In effect, then, the leakage reactance $\omega L_P'$ is shorted out and acts simply as a series connection. The primary winding reactance ωL_P acts as a dead short across the circuit, effectively eliminating the primary and secondary circuits and R_o from consideration. The capacitive reactances are infinitely high, and being in shunt, have no effect on the circuit. Figure 4-5 then represents the circuit at zero frequency.

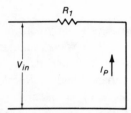

Fig. 4-5. Transformer when input is at zero frequency (dc).

The entire input voltage V_{in} appears across the usually small winding resistance R_1 of the primary winding. The primary winding current I_P is limited only by this small resistance and reaches a very high value. Power is dissipated in R_1 in accordance with $P = V^2/R$ or I^2R watts. (Smoke is now pouring from the transformer while water jets from the local fire brigade gracefully arch through the window, adding a charming touch to the scene.)

As the frequency increases to a few hertz, the picture changes to that in Fig. 4-6. The primary winding reactance ωL_P increases, but it is still not very high. The input voltage V_{in} is now distributed between V_{d1} and V_1, across R_1 and ωL_P, respectively. Voltage V_1 is the effective primary voltage, while V_{d1} is the primary winding voltage drop. Transformer action takes place, resulting in voltage V_2 across the secondary winding. A secondary load current I_2 flows. The output voltage V_o is equal to V_2 minus the voltage drop V_{d2} across the secondary winding resistance R_2. A "reflected" load current

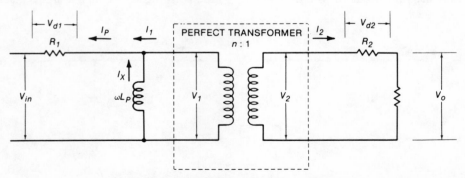

Fig. 4-6. Transformer at low frequency.

I_1 flows in the primary winding circuit, and together with I_X, makes up the total primary winding current through R_1. Power is dissipated (lost) in both R_1 and R_2. At this low frequency, the iron-core loss is small (therefore R_0 of Fig. 4-4 is a high resistance) so its shunt effect still does not show.

Because leakage inductance is usually quite small and the frequency is low, the inductive leakage reactance still does not show. Similarly, the shunt capacitive reactances are very high and ineffective.

Approaching, say, 60 hertz, the primary reactance ωL_P is relatively high. Its shunting effect is considerably less than before, so it begins to fade from the picture, but R_0, the iron-core loss resistance, begins to emerge because its value is relatively low compared with ωL_P. At this order of frequency, the leakage reactances are still very low and capacitive reactances are high, so they remain invisible.

The circuit now looks like that in Fig. 4-7. This diagram is in fact rather typical of a well-designed power transformer at 60 hertz and somewhat higher frequencies. As is explained later, excellent design results are achieved at the usual power transformer frequency ranges by considering only this relatively simple equivalent circuit.

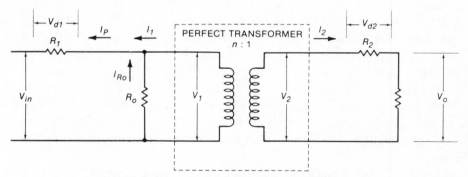

Fig. 4-7. Transformer at lower middle frequency.

The picture as it now stands is valid for a fairly wide range of frequencies, but as the upper audio levels are penetrated, the leakage reactances $\omega L_P'$ and $\omega L_S'$, and the capacitive reactances $1/\omega C_P$ and $1/\omega C_S$ begin to emerge as potent factors in the performance. The picture of Fig. 4-8 is then obtained. Here voltage is dropped across

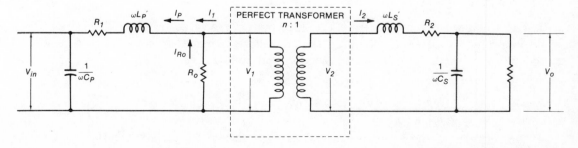

Fig. 4-8. Transformer at high frequency.

the relatively high leakage reactances, while the shunting effects of the decreasing capacitive reactances contribute to voltage loss. The primary reactance ωL_P, however, is high enough to have negligible effect and therefore vanishes from the picture.

In the design of wide-range components such as audio amplifier output transformers, all three diagrams (Figs. 4-6, 4-7, and 4-8), representing the effective circuits at low, mid, and high frequency, serve to underline the design parameters.

For transformers intended to run at one frequency or a narrow range of frequencies, one of the diagrams will be appropriate. In later chapters, some of the practical design points relating to transformers for low and high frequencies is covered. For the moment we will discuss losses with respect to transformers working in the lower middle range (about 60 to 400 hertz), as depicted in Fig. 4-7.

COMBINING CURRENTS AND VOLTAGES

Even though in power transformer design the shunt reactance ωL_P is usually expected to be high enough to be ignored, it is good to keep the effect of low reactance in mind. For example, air gaps in a poorly assembled core or running the core into flux saturation will reduce the primary winding inductance and therefore the reactance. This in turn leads to a high magnetizing current and consequent excessive voltage drop and copper loss in the primary winding. In any case, it is wise in critical applications to run a check on whether or not the primary reactance can be ignored. This can be easily done, as explained shortly.

Note again that although the equivalent iron-core loss resistance R_0 is represented as being frequency dependent, it is not reactive; it is purely resistive. The difference between reactive and resistive elements is very significant in combining the currents and voltages involved and should be clearly understood. Simple cases are discussed in the following.

Consider the transformer once more at a lower middle frequency (Fig. 4-7). Here, there are no reactive elements (keep in mind that the windings shown inside the dotted box represent only the function of transformation). In this case, the two currents I_1 and I_{R_0} are simply added together to obtain I_P. Thus

$$I_p = I_1 + I_{R_0} \tag{4-1}$$

with I_P and I_{R_0} as shown in Fig. 4-7.

In Fig. 4-9, however, consider the effect of the primary reactance, as shown, but not the iron-core loss current I_{R_0}. In this case, the reactive current I_X cannot be simply added to I_1 to make I_P. It has to be added vectorially to I_1. Thus

$$I_P = \sqrt{I_X{}^2 + I_1{}^2} \tag{4-2}$$

is the vector sum of I_X and I_1.

If the currents for the iron-core power loss (I_{R_0}), due to the primary winding reactance (I_X) and the load current (I_1), are all to be considered as in Fig. 4-10, they

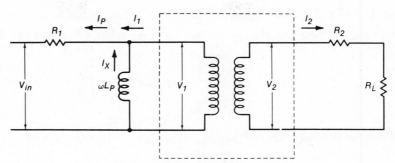

Fig. 4-9. Adding currents I_X and I_1.

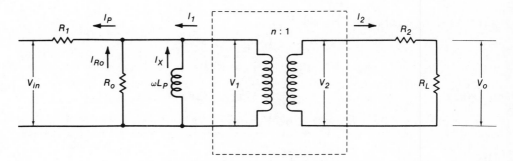

Fig. 4-10. Adding currents I_1, I_X, and I_{Ro}.

are combined as follows. The resistive currents I_1 and I_{Ro} are combined by straight addition and then are added vectorially to the reactive current. Thus

$$I_P = \sqrt{ I_X^2 + (I_1 + I_{Ro})^2 } \tag{4-3}$$

where the currents are as shown in Fig. 4-10.

Voltage relationships are expressed as follows. In the case of Fig. 4-7 again,

$$V_1 = V_{\text{in}} - V_{d1} \tag{4-4}$$

because $V_{d1} = I_P R_1$

$$V_1 = V_{\text{in}} - I_P R_1 \tag{4-5}$$

If necessary, this can be expressed in terms of I_{Ro} and I_2 substituting $I_1 + I_{Ro}$ for I_P and again substituting I_2/n for I_1. But this is not usually required in practice.

In the case of Fig. 4-9,

$$V_1 = \sqrt{ V_{\text{in}}^2 - (I_X R_1)^2 } - I_1 R_1 \tag{4-6}$$

and for Fig. 4-10,

$$V_1 = \sqrt{ V_{\text{in}}^2 - (I_X R_1)^2 } - R_1(I_1 + I_{Ro} \tag{4-7}$$

are the primary winding voltages.

VOLTAGE AND TURNS RATIOS_____

Referring again to the typical power transformer case in Fig. 4-7, Equation 4-4 states

$$V_1 = V_{in} - V_{d1}$$

From inspection of the circuit, it can also be said that

$$V_2 = V_{d2} + V_o \tag{4-8}$$

Equation 3-1 (Chapter 3) states

$$\frac{V_1}{V_2} = \frac{N_1}{N_2}$$

Therefore, substituting for V_1 and V_2,

$$\frac{N_1}{N_2} = \frac{V_{in} - V_{d1}}{V_o + V_{d2}} \tag{4-9}$$

Rearranging the equation gives the full-load output voltage

$$V_o = \frac{N_2(V_{in} - V_{d1})}{N_1} - V_{d2} \tag{4-10}$$

Similarly, the no-load output voltage can be shown to be

$$V \text{ (no load)} = \frac{N_2(V_{in} - V_{d1}(\text{no load}))}{N_1} \tag{4-11}$$

These useful results have a variety of applications in the design procedure.

Compensation for Voltage Drop

An example of a common use for Equation 4-9 is to adjust the turns ratio to compensate for voltage drop. Given the number of turns N_1 on the primary winding and the voltage drop in each winding, the number of turns N_2 required on the secondary winding is given by the following rearrangement of Equation 4-10:

$$N_2 = \frac{N_1(V_o + V_{d2})}{V_{in} - V_{d1}} \tag{4-12}$$

is the number of turns on the secondary.

The reverse situation (given N_2, what is N_1?) is just as easily arranged:

$$N_1 = \frac{N_2(V_{in} - V_{d1})}{V_o + V_{d2}} \tag{4-13}$$

is the number of primary turns.

Note that in practice, V_o and V_{in} are known by specification. Voltages V_{d1} and V_{d2} are easily calculated in the course of the design (Chapter 6), and preliminary numbers will have been calculated for N_1 and N_2 early in the design.

Example 1—Referring to the last worked example in Chapter 3, $V_{in} = 110$ volts, $V_o = 50$ volts, $N_1 = 262$ turns, and $N_2 = 119$ turns. When this design is continued, the voltage drops for the primary and secondary windings work out to approximately 2.0 volts and 1.0 volt, respectively. Using Equation 4-12, the revised number for the number of turns on the secondary in order to compensate for the voltage drop is

$$N_2 = \frac{262(50 + 1)}{110 - 2} = 123.7 \text{ turns}$$

Call it 124 turns.

AUTOTRANSFORMER LOSSES AND RATIOS

The losses in the autotransformer have the same causes as in the standard version, that is to say, copper loss due to current in the resistance of the windings and a loss due to heating in the iron core. For equal ratings, however, both the copper and iron losses tend to be substantially smaller in the autotransformer for reasons already discussed. The simple step-up equivalent circuit of Fig. 4-11 includes winding resistances R_1 and R_2. As in the two-winding case, the iron-core power loss could be represented as a resistance, but this value is rarely calculated in practical design procedures. Instead, the watts loss per pound of core material is read directly from specially prepared curves (Chapter 5).

For this case (Fig. 4-11), the turns-volts ratio equation is

$$\frac{N_1}{N_2} = \frac{V_{in} - V_{d1}}{V_o + V_{d1} - V_{d2}} \tag{4-14}$$

and for the step-down case it is

$$\frac{N_1}{N_2} = \frac{V_{in} - V_{d2} - V_{d1}}{V_o + V_{d1}} \tag{4-15}$$

where the voltages are as shown in Fig. 4-12.

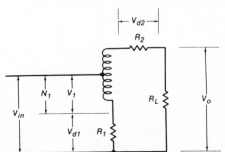

Fig. 4-11. Autotransformer equivalent circuit: step-up case.

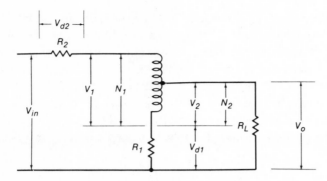

Fig. 4-12. Autotransformer: step-down case.

REGULATION

This parameter was defined earlier by Equation 3-13, Chapter 3, but its precise determination is rather complicated mathematically. Regulation depends on load, power factor, frequency, and temperature and is best considered to essentially be a measured quantity.

Nevertheless, it is good practice in the design to *aim* for good regulation, assuming that it is an important factor for the required application. This is done by taking account of things that strongly affect it. For instance, winding resistance has an important effect on regulation, and therefore those factors that make for low winding resistance must be considered if regulation is to be good. Other factors that will come under scrutiny are the shape of the core, magnetizing current, and iron-core loss current. A square core area rather than rectangular means less wire and therefore less resistance, and small loss currents mean small voltage drops and better regulation.

An approximate method of assessing regulation as a function of transformer efficiency is discussed a little later in this chapter in the section "Losses and Guesstimates."

TRANSFORMER RESISTANCE

It is often useful to know the transformer resistance as it appears to the load, or, as it is sometimes put, as it appears "looking into the secondary" from the standpoint of the load. The approximate value is given by

$$R_T = R_2 + R_1 \left(\frac{N_2}{N_1}\right)^2 \tag{4-16}$$

where

R_1 is the primary winding resistance,
R_2 is the secondary-winding resistance,
N_2/N_1 is the secondary-to-primary turns ratio.

The derivation of this equation is easily performed. In Chapter 3 we saw that a resistance connected across one winding appears as an equivalent resistance in the other winding equal to the resistance times the square of the turns ratio. This applies not only to resistance connected across the winding but also to the resistance of the windings themselves. Thus $R_1(N_2/N_1)^2$ is in fact the resistance of the primary winding referred to (or as seen "looking into") the secondary winding. This resistance added to the secondary winding resistance R_2 gives the total transformer resistance. The latter is sometimes expressed as

$$R_T = R_2 + R_1 \left(\frac{V_2 \text{ (unloaded)}}{V_{\text{in}}} \right)^2 \qquad (4\text{-}17)$$

Example 2—Using Equation 4-16 and the numbers of the last example with $R_1 = 2.0$ ohms, $R_2 = 1.0$ ohm, and the turns ratio $N_2/N_1 = 124/262$, then

$$R_T = 1.0 + 2.0 \left(\frac{124}{262} \right)^2 = 1.45 \text{ ohms}$$

is the total transformer resistance.

LOSSES AND GUESSTIMATES

Transformer design is largely a matter of cut and try. The starting points of the procedure are established from assumptions and estimates or even downright guesses (sometimes called "guesstimates" if any thought has gone into them).

In truth, of course, a thread of logic runs through the design procedure. Usually, the assumptions, estimates, and approximations that are used to get the design started are based on experience (our own or someone else's) rather than guesses snatched out of the blue. It *is* possible to get by with wild guesses, but the more logic and/or experience that goes into them, the quicker you get into the ball park. But the beauty of it all is that even if you are way off with the initial guess, the right answers must eventually come out. So if you are ever stuck for a number to assign to, say, the core cross-sectional area, or the transformer efficiency, or the core window area, choose a number—any number within reason—and plug it into the appropriate equation. The following comments illustrate how to do this.

Efficiency

One of the easiest and most useful numbers to assume is that for the efficiency. Efficiency is usually denoted by the Greek letter η. Transformers are very-low-loss devices as a rule, and most have efficiencies of from 0.75 to 0.95. So any number in this area is likely to be a good guess. If you wish, the number assigned can be viewed as an assertion rather than a guess, because you can guarantee to meet it in practice by modifying the design. For example, if 0.95 efficiency is what you want, then assume this figure and work to meet it in the design. And if, for one reason or another, you don't meet it, you will get as close to it as possible under the circumstances.

What does this guess or assertion do for us? First, it allows a figure to be calculated for the input power in watts from knowledge of the output power in watts. Thus

$$P_{in} = \frac{P_o}{\eta} \qquad (4\text{-}18)$$

This leads logically to a figure for the primary current

$$I_1 = \frac{P_{in}}{V_{in}} \qquad (4\text{-}19)$$

where

I_1 is the primary winding current,
V_{in} is the primary winding voltage.

A figure for the core area a can then be derived as shown in the design examples given earlier, and as was seen, this allows the numbers of turns to be calculated.

There is more. By making further basic and entirely valid assumptions, it is possible to leapfrog ahead of the design process using the estimated efficiency figure and arrive at good, approximate figures for voltage drops in the windings, the number of turns needed to compensate for the voltage drops, and the actual secondary winding voltage that will be obtained, taking account of the voltage drop. And all this can be done without even knowing the winding resistances or the currents in the windings. The first step is to assume that the following ideal distribution of losses in the transformer will be achieved in practice. Like the efficiency figure, this can be viewed as an assertion that will be brought to pass as the design progresses.

Distribution of Losses

As an approximation and a target to aim for in the design, observe that power transformers are generally most economical and most efficient in operation when:

1. The primary winding copper loss equals the secondary winding copper loss.
2. The iron loss is equal to the total copper loss.

Or to put it another way, when half of the total loss is in the iron and half is in the copper, with the copper loss split evenly between the primary winding and the sum of the secondary winding. This can be put into simple mathematical form, where P_{in} is the input power in watts, and η the efficiency of the transformer. Note, first of all, that efficiency, denoted by the Greek letter eta (η), is the ratio of the output power to the input power expressed either as a simple ratio or as a percentage, whichever is most convenient. Thus

$$\eta = \frac{P_o}{P_{in}} \qquad (4\text{-}20)$$

or

$$\% \ \eta = \frac{P_o}{P_{in}} \ 100 \qquad (4\text{-}21)$$

Also, η is a function of the losses and may be expressed as

$$\eta = \frac{P_{in} - P_L}{P_{in}} \tag{4-22}$$

or

$$\eta = 1 - \frac{P_L}{P_{in}} \tag{4-23}$$

where
  η is the efficiency of the transformer,
  P_o is the power output of the transformer,
  P_{in} is the power input of the transformer,
  P_L is the power loss of the transformer.

For Equations 4-22 and 4-23, multiply each right-hand side by 100 if the efficiency is required as a percent. It is important to note that the efficiency is a function of the power in watts, not of the volt-amperes. The losses, however, depend on the latter. For this reason, the efficiency with a reactive load expressed in volt-amperes depends on the load power factor and is highest when this is 1.0. Then:

Power output: $\qquad\qquad P_o = P_{in} \times \eta \tag{4-24}$

Total power loss: $\qquad\quad P_L = P_{in} (1 - \eta) \tag{4-25}$

Primary copper loss: $\qquad P_{LP} = P_{in} \left(\dfrac{1 - \eta}{4} \right) \tag{4-26}$

Secondary copper loss: $\quad P_{LS} = P_{in} \left(\dfrac{1 - \eta}{4} \right) \tag{4-27}$

Iron loss: $\qquad\qquad\quad P_{LI} = P_{in} \left(\dfrac{1 - \eta}{2} \right) \tag{4-28}$

When these desirable conditions exist and the efficiency is greater than 0.5, which is nearly always the case, the ratio of full-load output volts V_o to input volts (supply volts) V_{in} is related to the secondary-to-primary turns ratio as follows to a close approximation.

$$\frac{V_o}{V_{in}} = \sqrt{\eta} \frac{N_2}{N_1} \tag{4-29}$$

Example 3—If V_{in} is 117 volts and V_o is to be 5.0 volts and N_1 is 702 turns, what must N_2 be if η is 0.80? Rearranging Equation 4-29 gives

$$N_2 = \frac{V_o N_1}{V_{in}\sqrt{\eta}} = \frac{5 \times 702}{117 \times 0.894} = 33.55 \text{ turns}$$

Call it 34 turns.

The secondary winding *no-load* voltage V_2 and therefore the total voltage drop in the transformer referred to the secondary can also be deduced from

$$V_2 = \frac{V_o}{\sqrt{\eta}} \qquad\qquad (4\text{-}30)$$

In the example, $\eta = 0.80$, and therefore the no-load voltage on the secondary is

$$V_2 = \frac{5}{0.894} = 5.59 \text{ volts}$$

and the total voltage drop in the transformer all referred to the secondary winding is $5.59 - 5.0 = 0.59$ volt. The voltage drop in the transformer referred to the secondary can also be calculated as follows. The turns per volt for the transformer is $N_1/V_{in} = 702/117 = 6$. Therefore, the voltage across the secondary, unloaded, is $N_2/6 = 33.55/6 = 5.59$ volts; and the voltage drop, as before, is $5.59 - 5.0 = 0.59$ volt.

Losses Relative to Outputs

In a well-designed transformer where the losses are distributed as stated, the loss for each winding is proportional to the percentage of the total power that it delivers.

Example 4—Suppose a transformer has four secondary windings. If the total copper loss is 20 watts, 10 watts are lost in the primary winding and 10 in the combined secondary windings. If secondary winding N_2 delivers 10 percent of the total output, then it will have 10 percent of the secondary winding loss, or 1.0 watt. If the remaining windings deliver 20, 20, and 50 percent, respectively, the power lost in each will be 2, 2, and 5 watts, respectively.

THE GEOMETRY OF LOSSES

A sometimes useful approximation that emerges from the preceding is that for maximum efficiency, the available winding space should be divided equally between the primary winding and the sum of the secondary windings. In other words, the primary winding copper should occupy half of the winding space and the copper in all the secondaries combined should occupy the other half.

In this context, "winding space" refers to the cubic content, or volume, of the winding, not to the winding cross-sectional area. In the common power transformer configuration in which the secondary is wound directly on top of the primary, the winding areas for equal volume are not equal. This is shown in Fig. 4-13. Here the line l, representing the division between the two windings, runs exactly in the center of the distance between the outside and the inside edges of the coil. The rectangle in broken line is the total cross-sectional area a of the coil. The cross-sectional area of winding A is obviously the same as winding B, and just as obviously, the volumes are *not* equal.

In the less common configuration shown in Fig. 4-14, the windings are side by side, and equal winding area means equal winding volume.

WINDING CROSS SECTION AREAS

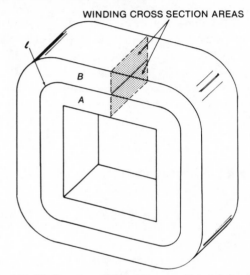

Fig. 4-13. Winding areas of **A** and **B** are equal, but their volumes are not.

WINDING CROSS SECTION AREAS

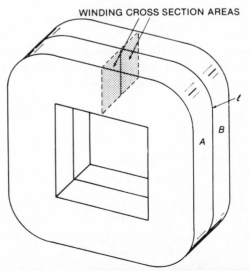

Fig. 4-14. Winding areas of **A** and **B** are equal and so are the volumes.

MORE ON REGULATION

It can be shown from the preceding that as an approximation,

$$\% \text{ Regulation} = 100 \left(\frac{1}{\sqrt{\eta}}\right) - 1 \qquad (4\text{-}31)$$

where
  η is the efficiency of the transformer expressed as a decimal fraction,
  regulation is expressed in percent.

This equation is useful in the design process, especially in power transformer design where conscious efforts are usually made to achieve a balance of losses. An example of an application for Equation 4-31 is to determine an approximate efficiency figure needed to achieve a given regulation, assuming that the losses will be balanced or nearly so in the course of the design. The equation may be used to show that

$$\% \ \eta = 100 \left(\frac{1}{\text{Regulation} + 1}\right)^2 \qquad (4\text{-}32)$$

where
  regulation is expressed as decimal fraction,
  $\% \ \eta$ is expressed as a percent.

Example 5—Suppose the regulation is to be about 0.05 (5 percent), then η should be

$$\% \ \eta = 100 \left(\frac{1}{0.05 + 1}\right)^2 = 90.7 \text{ percent}$$

from Equation 4-32.

LOSSES AND TEMPERATURE RISE

Temperature rise is the result of heat generated inside the transformer due to losses in the windings and the core. The load current passing through the resistance of the windings generates power loss in accordance with

$$P_L = I^2 R$$

where
  P_L is the power loss in watts,
  I is the current in the winding in amperes,
  R is the resistance of the winding in ohms.

Similarly, there are losses in the iron of the core due to eddy currents and hysteresis (Chapter 5). All of these losses create heat. The insulation in the windings and in the core tend to blanket these sources of heat and prevent the heat from escaping, like the insulation around the hot-water tank in the home. The temperature to which the windings rise is therefore a function of the amount of power dissipated in the transformer, the amount of insulation through which the heat must pass to the outside surface, and the amount of insulation on the outside surface.

Consider Fig. 4-15, which shows the cross section of a transformer with three windings. The heat generated in winding A must pass through windings B and C on its way to the outside surface, where it is dissipated. Similarly, the heat generated in winding B must pass through winding C, and the heat generated in winding C has the shortest distance to the outside.

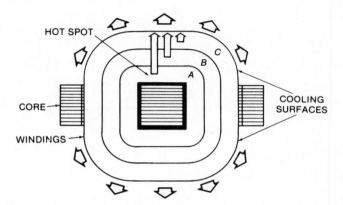

Fig. 4-15. Heat flow in windings.

Winding A, deep in the heart of the transformer, is the hottest. The so-called "hot spot" of the transformer is somewhere in this winding. The temperature rise is approximately the sum of the temperature drops across windings A, B, and C plus a temperature gradient from the cover to the outside air.

The hot spot in winding A can easily be dangerously hot although the outside cover feels cool to the touch, just as the insulation on a hot-water tank may feel cool despite scalding water temperatures inside.

Temperature rise is the main limiting factor to the power-handling capability of a transformer. By the same token, the reduction that can be achieved in the size of the transformer is limited by temperature rise. For a given power, a small transformer will usually run hotter than a large one, all other things being equal.

It should be understood that in design procedures in which temperature rise is not specifically part of the calculations, large safety margins are (or should have been) built into the method of choosing wire gauges and core sizes; in such cases, deviation from the rules of selection can result in disaster.

In more sophisticated procedures, provision is made for calculating the expected temperature rise. If, for example, a wire gauge is chosen by rule of thumb, the choice in terms of its effect on temperature rise can be checked at some later time in the design and suitable adjustments made if necessary. The choice of core size is intimately bound up with the temperature rise as well.

In the power capability equation (Equation 3-15) that relates current density, core size, and flux density to transformer power, very conservative choices must be made for the values of the factors in the equation if a calculation for temperature rise is not to be made.

Procedures for calculating temperature rise with a fair degree of accuracy for most conditions are given in Chapter 6.

FIVE

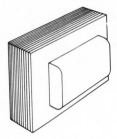

The Core

Up to now, the core material has been frequently referred to as "the iron." The truth is that it is usually not iron at all, although iron can be used. In general, core materials are alloys, and one class is so far removed from iron as to contain more than 85 percent of nickel plus only a few percent of iron and other things. Another material is not even metal, it is a true ceramic. The most common type of core is a steel composed mainly of iron with a few percent of other materials, notably silicon.

Nevertheless, regardless of the core composition, the term "iron" will continue to be used generically in the following pages. It is one of those technical colloquialisms that the user hopes confers on him an air of careless competence born of familiarity— but, more to the point, it is convenient.

PROPERTIES OF CORE MATERIALS

Generally speaking, there are five properties to be considered in core materials. These are: (1) permeability, (2) saturation, (3) electrical resistivity, (4) remanence, and (5) coercivity. The important property of hysteresis, which will be explained shortly, is taken care of in the list because it is dependent on the relative values of (4) and (5).

Permeability (μ)

In Chapter 3, permeability is described as the ability to conduct flux. Mathematically, it is the ratio of the flux density (B) to the magnetizing force (H) that causes B. In symbols

$$\mu = \frac{B}{H} \qquad (5\text{-}1)$$

When B is plotted against H, a curve is obtained variously called the *magnetization curve, saturation curve,* or simply (and truly) the *B-H curve,* as in Fig. 5-1. This is a *B-H* curve for a sample of material *that had been previously totally demagnetized* and then subjected to a gradually increasing magnetizing force while the flux density was measured. The slope of this curve at any given point gives the permeability at that point. When the permeability is calculated and plotted against B (or H) it is seen to be not constant. It varies, and therefore its value can be stated only at a given value of B (or H). This relative variation is shown by the dotted line in Fig. 5-2.

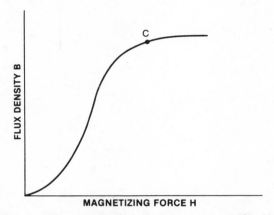

Fig. 5-1. Variation of *B* with *H* in a previously demagnetized specimen.

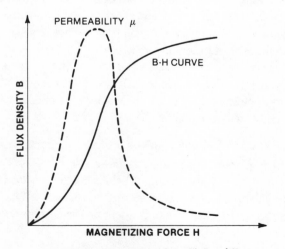

Fig. 5-2. Relative variation of μ with *B* and *H*.

The permeability at low values of H is called the *initial permeability.* The common grades of core materials, such as low-carbon steel and silicon steel, have low initial permeability. Many alloys, in particular the nickel-iron types, have been developed over the past several decades to provide high initial permeabilities, often incredibly high.

Another term met with frequently in transformer design is *incremental permeability*, sometimes called *apparent* or *ac permeability*. This is the permeability when an ac magnetizing force is superimposed on a dc magnetizing force—a common situation in some kinds of electronic circuitry. The effect of the dc is to bring the iron closer to its saturation point and so decrease (usually) the permeability for ac. Under these conditions, the permeability is improved by including an air gap of optimum size in the magnetic circuit. Figure 5-3 shows the effect of varying the gap of the core on the inductance of an iron-core coil. Three levels of dc current are shown for a fixed ac level.

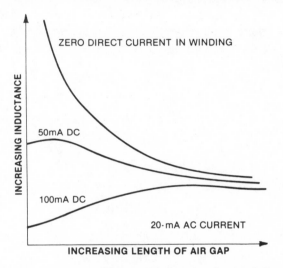

Fig. 5-3. Variation of inductance with three dc values in winding
and fixed ac.

Saturation

The *B-H* curve clearly shows the meaning of saturation. It can be seen that beyond a certain value of *H* (point *C* in Fig. 5-1), there is little increase in *B*; the iron is approaching saturation. Different materials saturate at different values of flux density. Note that at saturation, the permeability must be small or zero because there is little or no increase in *B* for an increase in *H*. This means that the inductance is very small when the iron is taken into saturation. Normally, care is taken not to run the iron beyond this point, although there are important exceptions as will be seen later.

Electrical Resistivity

The lines of flux that link the windings of the transformer also pass through the core itself and induce electrical currents in it. These *eddy currents* heat up the core, thus wasting power. If the electrical resistance of the core is high, the current will be low; therefore a feature of low-loss material is high electrical resistance. This type of loss is also reduced by building the core with thin laminations, each being insulated from the others.

Hysteresis (Remanence and Coercivity)

When an initially demagnetized sample of material is taken through a complete cycle of magnetization and B is plotted against H, a figure like that in Fig. 5-4 results. Note that this curve, like Fig. 5-1, is also a B-H curve. The difference is that in Fig. 5-1, the material was taken through a quarter cycle only, while Fig. 5-4 is a picture of the complete cycle that follows the first quarter cycle. The first quarter cycle is shown by the dotted line. Here the curve starts at 0. As H is increased, the flux density (B) increases along the dotted line to the saturation point, S. When H is now decreased and B plotted, the curve is found to follow a path such as SR; it does not return along the original path OS. Hence, when H has been reduced to zero, the core is still magnetized and a value R of flux density still remains in the sample. This amount R is appropriately called the *remanence*.

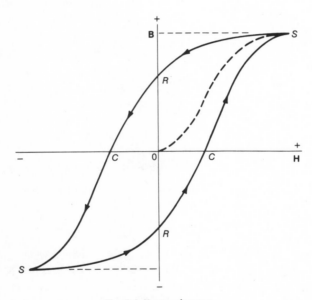

Fig. 5-4. Hysteresis curve.

The magnetizing force H is now reversed in polarity to give negative values. These are progressively increased, and the curve continues from R to C. At point C, the flux density is once more at zero, but H has a specific value. This value *(OC)* is called the *coercivity*; it is the amount of magnetizing force H_C needed to force (or coerce) the flux density back to zero. As H is taken through the remainder of the cycle, the curve continues to S, then to R, then to C, and back again to S.

This curve shows that after the initial magnetization OS, the flux density always lags behind the magnetizing force H. It also shows that the magnetized state of the material depends not only on the magnetizing force being currently applied but also on the previous magnetic state. This hysteresis loop represents energy lost in the core, a kind of magnetic friction, which is additional to eddy current losses. Manufacturers frequently lump *all* the core losses together in their data under the general heading of "iron loss." Typical loss curves are shown with the loss expressed in watts per pound of core material (Fig. 5-6).

An interesting point in all this is that in practice, the magnetization of a core in a working transformer *never* follows the *B-H* curve of Fig. 5-1. It can happen only once in the lifetime of a working core and that only for a fleeting instant if the core happens to be in a totally demagnetized state when the magnetizing force is first switched on (an unlikely event).

The area of the hysteresis loop is a measure of the loss.

KINDS OF ALLOYS

There are five main groups of magnetically soft alloys classified primarily by the chief constituents of the metal. These are: *low-carbon steel, silicon steel, nickel-iron* (permalloy), *cobalt-nickel-iron* (perminvar), and *cobalt-iron* (permendur). A number of variants exist within each group, each with its own unique characteristics created by subtle, and sometimes not so subtle, differences in composition and in heat, electrical, and physical treatments.

Each manufacturer has its own distinctive trade name for similar competing materials. Some examples are listed in Table 5-1 along with the generic descriptions of the alloys concerned. A glance at the list reveals some of the things that designers are searching for—high permeability, *B-H* loop squareness (of special interest to us later in the book), high flux density, and so on. Not mentioned in the table, but always important, are low losses and low cost.

Low-Carbon Steel

Sometimes described simply as *cold-rolled steel* or by such proprietary names as Hypertran (Magnetic Metals Corp.), this is the cheapest and least sophisticated of the alloys. Yet it is second only to the most sophisticated and one of the most expensive types in its high saturation point (vanadium-cobalt-iron), a fact sometimes overlooked by designers. Compared with most other materials, however, it has relatively high losses and low permeability, but for inexpensive general-purpose components it is often a good choice.

Silicon Steel

This was one of the first alloys to be used for cores and probably is still the most common, although it has been much improved since its introduction. Silicon steel consists mainly of iron with a small but significant addition of silicon in percentages of from about 1.0 to about 4.0 percent. This addition increases the electrical resistivity and thus reduces eddy current losses. It also improves material stability—the tendency for the characteristics not to change with age.

Table 5-1. Selected Types of Core Material, Compositions, and Characteristics

Material Code	Material Description**	Magnetic Metals Designations	Trade Names of Similar Materials*
46	Grain oriented 50% Nickel/Iron alloy processed for maximum B-H loop squareness	Square 50	Hy-Ra "40"[1], Orthonol[2], Delta-max[3], Hipernik V[6]
81	80% Nickel/Iron/Molybdenum alloy processed for B-H loop squareness	Square 80	Square Permalloy Hy-Ra "80"[1]
87	Special 80% Nickel/Iron/Molybdenum alloy processed for maximum B-H loop squareness	Supper Square 80	Exclusive to Magnetic Metals Company
107	High Flux-density vanadium/cobalt/iron alloy processed for maximum B-H Loop squareness	Supermendur	Supermendur[4]
33	Grain oriented silicon/ iron alloy	Microsil	Silectron Z[4] Oriented T-S[5]
86	Special 80% Nickel/Iron/Molybdenum alloy processed for highest initial permeability and lowest H_o	Supermalloy	Suppermalloy[4] Hymu "800"[1]
80	80% Nickel/Iron/Molybdenum alloy processed for high initial permeability	SuperPerm 80	4-79 Permalloy, HyMu "80"[1], Mu-metal[3], Hipernom[6]
49	Non-oriented 50% Nickel/ Iron alloy processed for high initial permeability	SuperPerm 49	High Permeability "49"[1], 48-Ni[5], 4750[3]

*Trade Names of: 1—The Carpenter Steel Co. 2—Magnetics Div., Spang Industries, Inc. 3—Alleghany Ludlum Steel Corp. 4—Arnold Engineering Co. 5—Armco Steel Co. 6—Westinghouse
**Compositions shown are approximated.

Silicon steel has a high saturation point, good permeability at high flux density, and moderate losses. An important member of this group is grain-oriented silicon steel. Grain-oriented material takes advantage of the fact that the easiest direction of magnetization is along the edges of the cube-shaped crystals of which the metal is composed. The material is treated so that the edges of the cubes all lie in the same direction, thus creating one direction in which the metal magnetizes easily. This leads to vastly improved permeability, lower losses, and a higher saturation point. The greatest advantage is gained in cores wound with strip steel so that the flux always flows in the direction of orientation. Other types of alloys are also produced in this way.

Silicon steel in all its forms is widely used in power transformers, audio output transformers, and many other applications.

Nickel-Iron (Permalloy)

This is an important metal alloy consisting primarily of nickel and iron, often with molybdenum and copper added. The nickel-iron content might range from about 45 percent to over 85 percent.

This alloy is characterized by exceedingly high initial permeability. That is to say, it has incredibly high permeability at low flux density. Saturation occurs, however, at a relatively low flux density. Losses are low.

Some nickel-iron alloys are processed to accentuate the squareness of the *B-H* loop and carry proprietary names like SuperSquare 80 (Magnetic Metals Corp.) and Square Permalloy Hy-Ra 80 (Carpenter Steel Co.).

Cobalt-Nickel-Iron (Perminvar)

In perminvar, cobalt has been added to nickel and iron to produce an alloy that exhibits a substantially constant permeability and low hysteresis loss at low flux density. In some cases, it might also have the rather odd property of negligible coercive force and remanence, although the hysteresis loop area is still substantially greater than zero.

Cobalt-Iron (Permendur)

The mating of cobalt and iron without nickel yields an alloy with high permeability at high flux densities and a very high saturation point. It also has a high incremental permeability, that is to say, high ac permeability in the presence of a large dc magnetizing force.

One of the newest alloys in this group is vanadium-cobalt-iron, which of all the alloys has the very highest saturation point together with low loss. But it is expensive.

THE SOFT FERRITES

In the past twenty-odd years, a rather unlikely substance has moved into the front rank of core materials. It isn't really a metal or even a bonded powdered-metal product like the tuning slugs in high-frequency coils. It is a ceramic that is manufactured using all the processes common to the production of ceramics, but different from most milk jugs in one remarkable respect—it is magnetic. This class of magnetic ceramic materials is called the *soft-ferrite ceramics*, or more commonly, simply *ferrites*.

Because the ferrite is an easily molded ceramic, it surpasses strip and sheet metals in its ability to take on intricate shapes, and it does not have the restriction on size common to the powdered-metal products that must be formed by high pressures. It appears in numerous shapes—bars, rods, hollow pots, toroids, cross format, Es, and Is—name it and you can have it along with the performance advantages that the choice of shape confers: shielding from extraneous fields, minimum stray field, low leakage inductance, and so on. This flexibility also provides advantages in mounting and installation methods. For instance, components are easily adapted to circuit-board mounting, printed and otherwise.

Most readers are familiar with at least one form of ferrite ceramic, because the antennas of battery-operated transistor radios are made from it. It has many telephone and computer applications. It is found in flyback transformers in TV sets. In short, it is best used as a component for the relatively high frequency region of the electromagnetic spectrum. In the case of power transformers, for example, ferrites are rarely used because power transformers traditionally operate in the low-frequency regions where the laminated metal alloys do the best job. But out there on the far edges of the audio spectrum and beyond, a battle is being fought. Standard laminations have become thinner and thinner, even to 0.001 inch (25.4 μm) and less in order to carry power transformation into the high audio frequencies—ferrite territory. And now the ferrites have entered power transformation.

This does not mean that ferrites have moved significantly down the frequency scale to challenge the traditional materials on their own ground. No, it means that ferrites are enticing the designers of certain kinds of power transforming circuitry to push their frequencies to beyond the 20-kHz border, where riches in the form of substantially reduced size and weight are found and where ferrites claim a vastly superior cost-related performance over alloys.

Power inverters and converters using laminated transformers (Chapter 10) former-ly operated in the 60- to 400-Hz area but then moved to beyond 2000 Hz and still higher into 10 kHz (and higher) as laminated cores improved magnetically at those frequencies. At the same time, these cores become more expensive and difficult to handle.

But just at those frequencies where the laminated core is becoming impractical, the ferrite, it is claimed, produces a lower price-to-performance ratio.

The performance comparisons in Tables 5-2 and 5-3 by Indiana General Corp. show both why ferrites have not moved down the frequency scale and why laminated alloys are having problems in moving higher; but alloy manufacturers claim in turn that operation as high as 40 kHz is feasible with thin-gauge tape-wound cores. In Table 5-2, which shows the dc properties, the alloys are ahead of the ferrites except for resistivity. It is the property of high resistivity that keeps the eddy current losses small in the ferrites and eliminates the need for laminated construction. The alloys, however, undoubtedly retain a heavy general advantage throughout the lower frequency range with useful performance at high frequency if the laminations are thin enough.

Table 5-2. DC Properties of Ferrites

Property	Alloys	Ferrites
μ_i	5000–300,000	5–16,000
B_s (kilogauss)	8–24	3–5
H_c (oersteds)	0.003–3	0.05–20
Resistivity (ohm-cm)	10^{-5}	$10-10^{11}$

Courtesy Indiana General, Div. of Electronic Memories & Magnetics Corp.

In Table 5-3, the picture at 25 kHz is quite different. Here the ferrites are shown to have a clear advantage in all departments, especially in those important characteristics of initial permeability μ_i, and maximum flux density B_M. Because the low-loss characteristic is well maintained, the data means excellent performance at high frequencies for ferrites.

Table 5-3. Ferrites at 25 kHz

	Alloys (0.004 in thick)		Ferrites	
μ_i	≤4000		≤16,000	
Temp. rise (°C)	40	65	40	65
Core loss density (watt/cm^3)	0.14	0.24	0.12	0.20
B_M (kilogauss)	0.9	1.2	2.3	3.0
B_S (kilogauss)	8.5	8.5	5.0	5.0
B_M/B_S	0.11	0.14	0.46	0.60

Courtesy Indiana General, Div. of Electronic Memories & Magnetics Corp.

Why High Frequency?

What are the specific advantages of high-frequency operation in terms of transformer design, and, specifically, how do they relate to, say, power transformers? First, at high frequencies, transformers are small and light; a glance at what is often referred to as the basic converter-inverter equation reveals why:

$$N = \frac{V \times 10^8}{4FfaB}$$

(Yes, it is identical with the basic transformer equation.) The symbols have the same meanings as before except that in converter-inverter use, the form factor F is assigned the value of 1.0 because these circuits are essentially square-wave types. Obviously, the number of turns N will be very much less at 20 kHz than at, say, 60 hertz. To put it another way, at high frequencies, the cross-sectional area a of the core can be very much smaller for a given number of turns.

Another good reason for using high frequencies, if possible, is in rectifier circuits, Here, the required smoothing components are smaller and lighter at the higher frequencies.

But nothing is perfect. There is no universal ferrite any more than there is a universal metal alloy. The bulk of ferrites are composed of manganese-zinc or nickel-zinc oxides. By varying the composition and adding percentages of other materials, a wide range of ferrites is obtained to cover the field of applications. These ferrites are temperature sensitive, a fact that must be taken into account when considering them.

RELATIVE PERFORMANCE OF CORES

The broad groups and the numerous grades of core material within them have exactly the same kinds of electromagnetic properties. The differences between them are ones of degree, and the problem of selection becomes one of establishing degrees of performance in various characteristics to match a stated need.

Performance, however, is governed not only by the composition of the alloy and its treatment in manufacture but also by the thickness of the sheet or strip from which the laminations are formed, the configuration of the core into which they are assembled, and the care and skill used in the assembly. The frequency at which the core is to perform, the flux density used, and even the temperature play roles in transformer operation.

All of this results in a wide overlap in degrees of performance over the different kinds of material. For instance, in a power transformer, a core of grain-oriented silicon iron might perform just as well as one of vanadium-cobalt-iron, simply because it is being used more effectively, despite the superior potential of the vanadium core. Again, a silicon-steel core might have a higher permeability for a given set of conditions than a nickel-iron core, because the nickel-iron core might be poorly assembled or because the flux density is too high.

Assuming, however, that due attention is paid to the operating conditions and quality of assembly, the relative performance of a core is governed chiefly by the type of material and the basic format of the core. This is well illustrated in the composite curves of Fig. 5-5, which combine the *B-H* and μ (permeability) characteristics. The curves

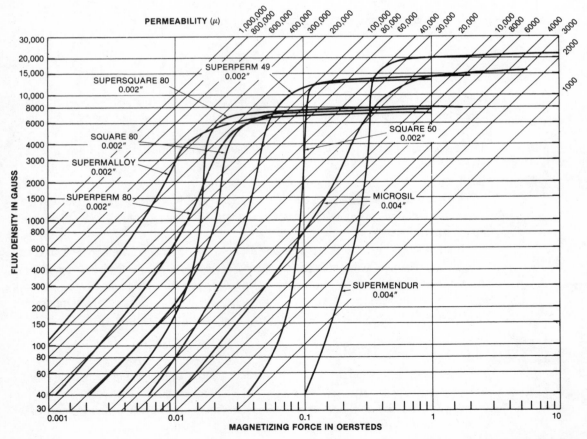

Courtesy Magnetic Metals Corp.

Fig. 5-5. *B-H* curves for various core alloys with permeability also shown along the top.

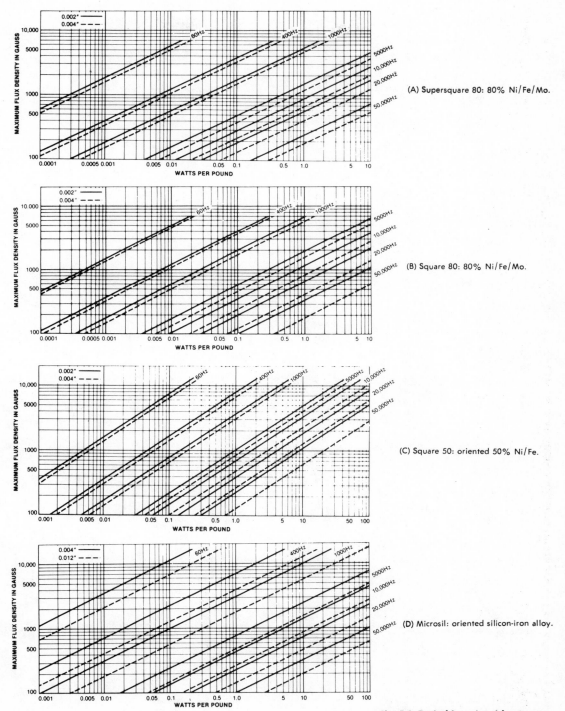

(A) Supersquare 80: 80% Ni/Fe/Mo.

(B) Square 80: 80% Ni/Fe/Mo.

(C) Square 50: oriented 50% Ni/Fe.

(D) Microsil: oriented silicon-iron alloy.

Fig. 5-6. Typical iron (core) loss curves.

Fig. 5-6 (continued).

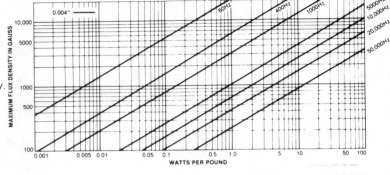

(E) Supermendur: special 49% Co/4% Fe/2% V.

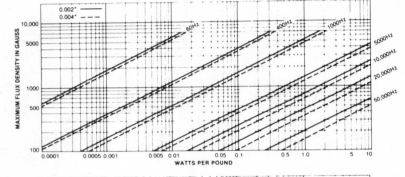

(F) Supermalloy: Special 80% Ni/Fe/Mo.

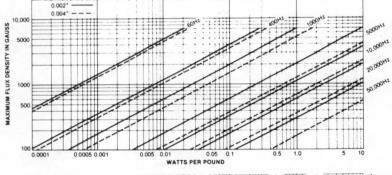

(G) Superperm 80: 80% Ni/Fe/Mo.

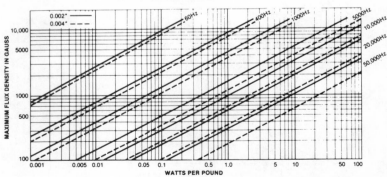

(H) Superperm 49: 50% Ni/Fe.

Fig. 5-6 (continued).

Courtesy Magnetic Metals Corp.

are plotted for tape-wound cores (toroids) constructed from the materials listed in Table 5-1. The tape-wound core results in a close-to-perfect magnetic circuit that eliminates the mechanical defects of other types and reduces the equivalent air gap to an absolute minimum.

IRON LOSSES

Power lost in the core due to eddy currents and hysteresis is dissipated in the core in the form of heat. Because this lost power has to be supplied from the power source, it manifests itself also as a loss current in the primary winding. The iron loss is sometimes referred to as the *no-load loss* because its magnitude does not depend on the load. There is, however, some no-load loss in the copper of the primary windings due to the loss current, but this is usually so small that we can consider it to be negligible.

Finding Loss From Curves

The useful curves shown in Fig. 5-6 illustrate the effects on core losses of flux density and frequency in various alloys and in lamination thicknesses of 0.002 and 0.004 inch (50 and 100 *v*m). The loss in terms of watts per pound of core weight is plotted against flux density. Thus, if the flux density at which the core is to be run and the weight of the core are known, the watts loss is easily determined.

Observe that losses increase with both increasing flux density and frequency. It is evident, then, that although increasing one or both these quantities results in a reduction in the number of turns required on a transformer, as can be seen from the basic equation, it is not all gain.

Nevertheless, in power transformer design, it is generally an advantage to run at as high as flux density as possible *below* the knee of the *B-H* curve. Up to this point, losses tend to be proportional to the square of the flux density and increase rather more rapidly above it. Working below the knee also avoids distortion of the output waveform, which can be important in some kinds of components.

Example 1—To illustrate the use of these curves, suppose a core made of 0.004-inch Microsil material is to be run at 10,000 gauss and 1000 hertz. To determine the core loss, find the curve in Fig. 5-6D representing 0.004-inch material (solid line) at 1000 hertz. From the 10,000-gauss point on the vertical axis of the graph, draw a horizontal line to intersect the 0.004-inch, 1000-hertz line. From the intersection, drop a perpendicular to the base line, and read the answer of 10 watts per pound.

If the core weighs 0.33 lb, then the watts lost will be 0.33 × 10 = 3.3 watts. This in turn can be expressed as a loss current in the primary winding. Thus

$$I_{Ro} = \frac{Wt \times W/lb}{V_{in}} \tag{5-2}$$

where

I_{Ro} is the loss current in amperes,
Wt is the core weight in pounds,
W/lb is the watts per pound,
V_{in} is the volts on primary winding.

If the input is 110 volts, then

$$I_{Ro} = \frac{0.33 \times 10}{110} = 0.03 \text{ ampere}$$

is the loss current in the primary winding.

Note that for specified flux density, frequency, and lamination thickness, the iron loss in any given class of alloy will vary widely, depending on the grade of material. For instance, at 15,000 gauss, a high-quality grade of silicon iron might have a loss of as low as 0.65 watt per pound and range up through lower grades to as high as 2.0 or 3.0 watts per pound at 60 hertz.

If high accuracy is required, it is therefore necessary to use manufacturers' figures. If the loss is not known, a figure of around 1.5 watts per pound is usually good enough for silicon iron at, say, 14,000 gauss. A "C" core (see the section "Configurations and Proportions" in this chapter) salvaged from a power transformer would probably be made from grain-oriented silicon iron for which a figure of around 1.0 watt per pound would be reasonable at 14,000 gauss.

Calculating Loss Without Curves

Iron loss is approximately proportional to the square of the flux density. Given the loss per pound for a material at a specific flux density and frequency, a constant k can be derived:

$$k = \frac{W/lb}{B^2} \tag{5-3}$$

where

W/lb is the watts loss per pound,
B is the flux density in gauss,
k is a constant.

With the derived constant k, the loss per pound can now be calculated for any other value of flux density at the same frequency, provided that the density is below the knee of the curve. Thus

$$W/lb = kB^2 \tag{5-4}$$

Example 2—If a material has a known loss of 3.0 watts per pound at 14,000 gauss, the constant is

$$k = \frac{3.0}{14,000^2} = 1.53 \times 10^{-8}$$

What will be the loss at 6000 gauss?

$$kB^2 = 1.53 \times 10^{-8} \times 6000^2 = 0.55 \text{ W/lb}$$

At 6000 gauss, the loss is 0.55 watt per pound.

It was pointed out earlier that loss increases with both flux density and frequency. In a wideband transformer, such as an audio component, however, the iron loss tends to remain constant with varying frequency up to about 2500 hertz. The reason can be seen by referring again to the basic transformer equation, $V = 4FfaBN \times 10^{-8}$. As the frequency f is increased, the flux density B decreases. Therefore, the tendency to increased loss due to increased frequency is offset by a reduction due to decreased flux density.

Note also that because flux density B is directly proportional to voltage V, the loss must be approximately proportional to the square of the applied voltage:

$$\text{W/lb} = cV^2 \tag{5-5}$$

using Equation 5-4.

Loss curves for a common grade of silicon steel (such as that used on radio and TV transformers) are given in Fig. 5-7A and for C core of grain-oriented silicon steel of two lamination thicknesses in Fig. 5-7B.

APPARENT LOSS

In order to establish the magnetic flux in the core, it is necessary to supply a magnetizing current to the primary winding. The magnitude of the magnetizing current is governed by the reactance of the primary winding. This reactance, of course, depends on the inductance and frequency. The inductance, in turn, depends on the permeability of the core at the given flux density and frequency.

As the flux density is increased into saturation, the permeability drops and the magnetizing current increases much more than proportionally to the increase in flux density. Because the flux density is proportional to voltage, a plot of magnetizing current against voltage has the same shape as the *B-H* curve, as shown in Fig. 5-8.

The relationship between magnetizing current and flux density can be stated in terms of volt-amperes per pound of core material in a manner similar to the watts per pound used for real core loss. Examples of curves are given in Fig. 5-9. These curves are all for the same kind of material, Microsil, but for different lamination thicknesses and frequencies. The effects of air gaps in the core are not included in the curves; this is the next topic of discussion.

Real Loss Versus Apparent Loss

Although the apparent loss curves seem to be in much the same terms as the real loss curves, it is important not to confuse one with the other. As we have seen, watts are not the same as volt-amperes.

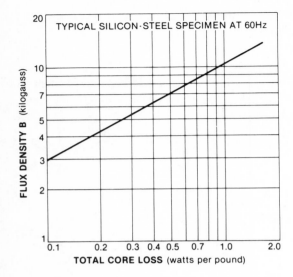

(A) Salvaged silicon-iron cores can be expected to have losses of this order.

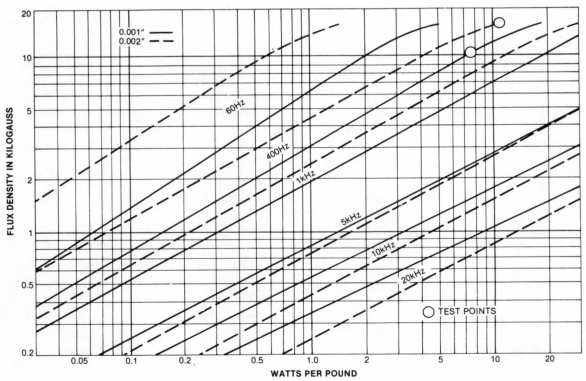

Courtesy Magnetic Metals Corp.

(B) For grain-oriented silicon steel with lamination thickness 0.001 inch (dotted line) and 0.002 inch (solid line).

Fig. 5-7. Core loss curves.

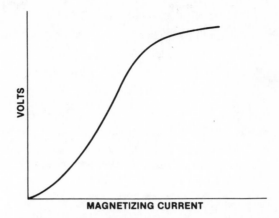

VOLTS

MAGNETIZING CURRENT

Fig. 5-8. Volts/magnetizing current curve has same shape as *B-H* curve.

Correction for Gaps

These curves, as they stand, are fine for toroidal cores, which are virtually gapless. Other kinds, however, have air gaps in the magnetic circuit due to imperfections in the assembly. In some cases, a gap might be deliberately introduced into the core. As discussed earlier, even a small gap can drastically reduce the effective permeability, and therefore the inductance, and increase the exciting (magnetizing) current.

The following equation is used in conjunction with the curves; it has a correction factor for the gap built into it:

$$I_M = \frac{\text{Wt} \times \text{VA/lb}}{V_{\text{in}}} + \frac{1.43\ Bgs}{N} \tag{5-6}$$

where

Wt is the weight of core in pounds,
VA/lb is the volt-amperes per pound, taken from curves,
V_{in} is the voltage across primary in volts,
B is the flux density in gauss,
s is the stacking factor,
N is the number of turns on primary winding,
I_M is the magnetizing current in amperes,
g is the gap distance in inches.

Example 3—Consider first a core without a gap, such as a toroid, with the same weight, frequency, lamination thickness, and flux density used in the core loss example. What is the magnetizing current?

The graph of Fig. 5-9B contains the information for the 0.004-inch (100-μm) lamination thickness. Find the 10,000-gauss point on the vertical axis and draw a line to intersect the 1000-hertz curve. From the intersection, drop a line to the base to find the figure of approximately 18.0 VA/lb. Therefore

$$I_M = \frac{\text{Wt} \times \text{VA/lb}}{V_{\text{in}}} = \frac{0.33 \times 18}{110} = 0.054 \text{ ampere}$$

is the magnetizing current.

Suppose a core assembled from laminations, or a C type configuration (discussed in the next section), has a gap of 0.003 inch. For comparison, take the same frequency, weight, lamination thickness, flux density, and voltage as in the preceding example.

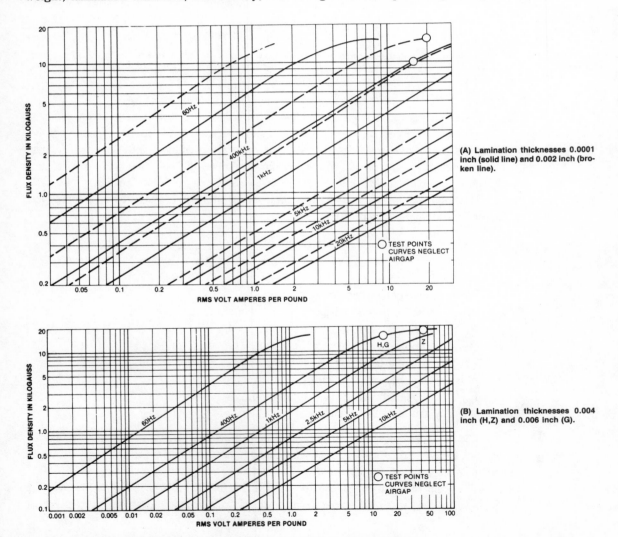

(A) Lamination thicknesses 0.0001 inch (solid line) and 0.002 inch (broken line).

(B) Lamination thicknesses 0.004 inch (H,Z) and 0.006 inch (G).

Fig. 5-9. Apparent loss in volt-amperes per pound for grain-oriented silicon steel (Microsil).

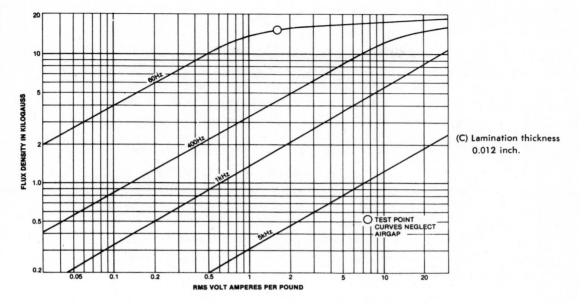

Fig. 5-9. (continued).   Courtesy Magnetic Metals Corp.

Suppose the stacking factor is 0.90 and the number of turns on the winding is 250. The figure derived from the curve will be the same as before, namely, $I_M = 0.054$ ampere. Modified by the gap, it becomes

$$I_M = 0.054 + \frac{1.43 \times 10{,}000 \times 0.003 \times 0.90}{250}$$

$$= 0.054 + 0.154$$

$$= 0.208 \text{ ampere}$$

Total No-Load Current

The loss current and magnetizing current together make up the total no-load current. The loss current is a power component of the total current, while the magnetizing current is not. In other words, the two components are not in phase. It is necessary, therefore, to add them vectorially rather than arithmetically. Thus

$$I_o = \sqrt{I_M{}^2 + I_L{}^2}$$

where

I_o is the total no-load current,

I_L is the loss current,

I_M is the magnetizing current.

The total no-load current was also discussed in Chapter 4.

CONFIGURATIONS AND PROPORTIONS_____
Stampings
By far the most common method of assembling cores is from stampings—thin, flat shapes punched out of sheet or coil metal in presses. Two stampings are used to form one complete lamination having the general shape shown in Fig. 5-10. Figure 5-11 shows some of the combinations of shapes used. For example, configuration C is the popular E and I type. The reason for these arrangements is to permit the coils to be wound separately from the core and then assembled to it later. Each lamination has three "breaks" in it, at points *A, B,* and *C.* It is important to note two facts about this. First, these "breaks" are in fact gaps in the magnetic circuit that can never be eliminated. In most constructions, the edges of the metal are supposed to butt tightly together at these points, but due to imperfections and burrs at the butting edges, an effective air gap always remains. With careful assembly, this might be kept to an equivalent of perhaps 0.003 inch (76.2 μm) with standard laminations and less for C-type cores, which have specially finished butting surfaces. This might not seem like very much, but the effect of gaps of this order has already been demonstrated.

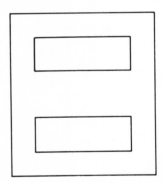

Fig. 5-10. A "shell" type lamination.

The second point is that the total equivalent air gap in the magnetic circuit is *twice* the break distance, as illustrated in Fig. 5-12. The length of the magnetic circuit is the distance *l* around *one* window only. In this distance, there are two breaks in series: one at point *A* and one at *B,* or *B* plus *C.* The total gap is the sum of the two breaks. As we have seen, a gap can be used to good effect when the windings are carrying dc current as well as ac; otherwise, it should be viewed with loathing, as something to be reduced to the absolute minimum.

One method of reducing the effect of the gap is to assemble the laminations in alternate order as in Fig. 5-13. Each break is then shunted magnetically by the preceding and following laminations.

Note the essential point that each lamination is electrically insulated from its neighbor by a coating of some kind. Often, this is simply the oxide film formed on the metal during heat treatment. Its purpose is to reduce eddy current loss in the core. Because this

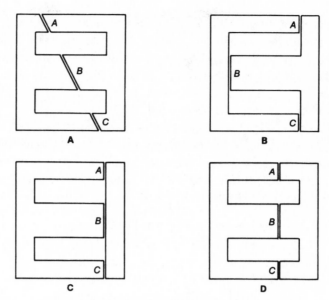

Fig. 5-11. A few of the many configurations of laminations.

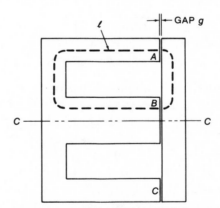

Fig. 5-12. The effective gap is 2g.

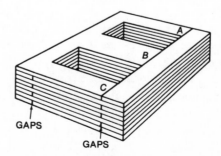

Fig. 5-13. E and I laminations assembled in alternate order.

type of loss increases with frequency, the laminations are made very thin at high frequencies to as little as 0.0005 inch (12.7 μm). As can be imagined, metal this thin needs careful handling; at standard power frequencies, the laminations are usually of the order of 0.014 inch (0.355 mm) or thicker.

Shapes and Proportions

The shape just described is the now-familiar "shell" type core. The "core" type shape illustrated in Fig. 5-14 has the disadvantage of increased leakage inductance, but for some applications it has advantages. An important advantage is that the primary and secondary windings can be totally separated by putting each on a different leg. Other advantages are discussed in detail in Chapter 6.

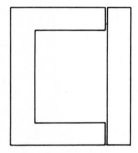

Fig. 5-14. Core type shape.

Observe from Fig. 5-10 that in the shell-type core, the center leg is wider than the outside legs. In fact, it is usually about twice as wide. The reason is that all the flux is carried by the center leg but only half of the total in each of the side legs. In other words, the center leg is the main flux artery, as it were, and the flow divides to right and left around the magnetic circuit. If you wish to prove this, wind a small coil around a side leg and measure the voltage across the coil; the turns per volt will be found to be half of that for the center coil, because it is threaded by only half of the flux. In the core shape, the width is the same all the way around because the flux is the same value around the core. (NOTE: The legs of three-phase transformers of the shell type are usually all the same width. These, however, are not dealt with in this book.)

Not all iron-core transformers are power transformers, and the proportions most suited to power transformers are not necessarily the best for other purposes. In the design of inductors, such as smoothing chokes, or in the design of some types of audio transformers, many of the available lamination shapes lead to bulky and wasteful designs. But this is usually of more concern to the professional designer than the individual experimenter who often uses only what he or she can get. From the experimenter's point of view, most shapes can be used for most purposes. Power transformer laminations are convertible to inductors and vice versa. Even if the design is not the most efficient, it will usually be adequate.

Quite literally, there are many hundreds of sizes and shapes available, and choosing one by formula alone to suit a given task is not possible. Design has to be done by trial and error, using formulas to get into the ball park initially.

It is also useless to use hypothetical shapes and sizes in your paper designs unless you can afford to have the laminations made. Nor is it sensible to use a standard catalog model as the basis for a design unless you are sure that it can be obtained. Your design must be based on a core that is actually in your hand or one that you are certain you can get. Obviously, then, it is not always possible to achieve optimum results, whether you are professional or amateur. Compromise is necessary to an extent, but there is usually a wide choice of shapes available even to the amateur whose main source might be salvaged material.

If there is a choice, then remember that some shapes are considered to be more desirable than others. Table 5-4 lists a number of useful approximate proportions. The dimensions are given as multiples of dimension *B* in Fig. 5-15. While Fig. 5-15 is intended to serve as a general reference for Table 5-4, it is in fact proportioned according to line 1 of the table. This shape has a special significance in that it is not only useful but sometimes cheaper, because it is less wasteful to produce in manufacturing than other shapes.

Table 5-4. Useful Core Proportions (See Fig. 5-15)

A	B	C	D	E	F
1	1	2	3	5	6
0.5	1	1	2.5	3	4
0.73	1	1.36	1.36	2.72	4.8
0.33	1	0.58	1.75	2.33	3.25

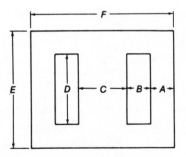

Fig. 5-15. Reference diagram for Table 5-4.

This shape is the so-called waste-free design, and the reason for the name is shown in Fig. 5-16. It is seen here that the pieces stamped out of the window spaces are exactly the right size to form the "I" pieces. Because two Is are stamped out for every two Es, waste is dramatically reduced. The window is rather narrow, so this core is best utilized in larger sizes where the thickness of the coil bobbin does not use so much of

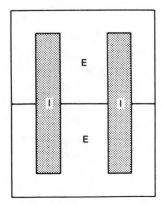

Fig. 5-16. Stamping out two Es and two Is with no waste.

the winding space. This design is usually a good choice for power transformers and large audio components. The proportions of line 4 of Table 5-4 are usually good in the smaller sizes for impedance matching, such as input and coupling transformers. All of these shapes are generally good for inductors.

Cut Cores

The cut (sometimes called the C type) core is shown in Fig. 5-17. This type of core is wound from continuous metal tape and then bonded into a solid uniform mass and cut into two parts. The mating surfaces are specially treated to achieve a highly accurate butt joint with the very minimum gap when they are brought together. The gap dimension is claimed to be of the order of 0.001 inch (25.4 μm) for cross sections up to 2.25 square inch (14.5 sq cm) or when dimension E (Fig. 5-17) is less than 1.0 inch (2.54 cm) and 0.002 inch (50.8 μm) for cross sections greater than 2.25 square inches, or E greater than 1.0 inch.

As with conventional stampings, the coils are wound separately and then assembled to the core. The core halves are then strapped together. One assembly forms a C type configuration; two assemblies put together as in Fig. 5-18A form a shell type configuration. Toroid shapes are also produced as in Fig. 5-18B.

This type of core was originally devised to take advantage of the superior characteristics of grain-oriented silicon steel, but other alloys are now supplied in this style.

From the design point of view, the procedures and equations are the same as for any other kind of core. Table 5-5 lists typical applications for different materials, lamination thicknesses, and core shapes.

Although these cores are factory built, note from Table 5-6 that the stacking factor is still significant and must be taken into account. In Table 5-6, KG means kilogauss, or thousands of gausses, or thousands of lines (maxwells) per square centimeter.

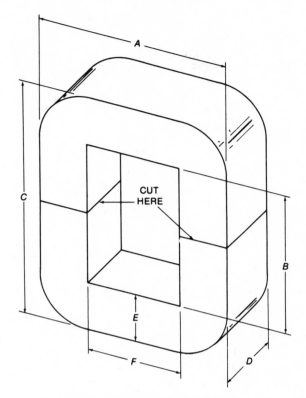

Fig. 5-17. Tape-wound cut core (C core).

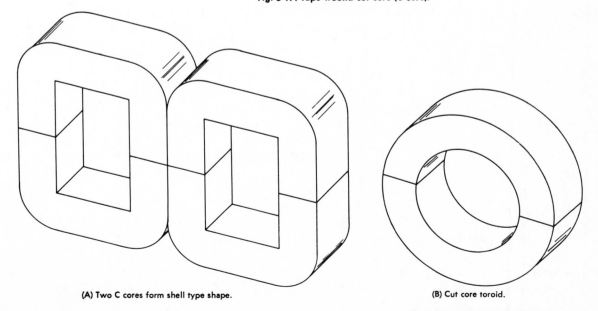

(A) Two C cores form shell type shape.

(B) Cut core toroid.

Fig. 5-18. Cores made with C-type cores.

Table 5-5. Materials and Applications

Material	Gage	Shapes	Sizes	Application
Microsil	1, 2 mil M, L	C, G, R	All standard sizes up to A × B = 34″ × 28″	At high frequencies where eddy current loss is significant, pulse transformers, chokes.
Microsil	4 mil H	C, E, G, R	All standard sizes up to A × B = 34″ × 28″	Up to 16 KG, at 400 Hz; transformers, filter chokes, reactors, amplifiers, pulse transformers.
Microsil	4 mil Z	C, E, G, R	All standard sizes up to A × B = 34″ × 28″	At inductions from 16 KG to 18 KG in similar applications as 4 mil H.
Microsil	6 mil G	C, E, G, R	All standard sizes up to A × B = 34″ × 28″	At 400 Hz to 800 Hz as a lower cost substitute for 4 mil H.
Microsil	12 mil A, S	C, E, G, R	All standard sizes up to A × B = 22″ × 22″ Max. wgt. 35 lb.	Up to 18 KG at 60 Hz for transformers, filter chokes, reactors, magnetic amplifiers (exciting current increases rapidly above 15 KG).
Superflux Supermendur (2V-Co-Fe)	4 mil V	C, E, G, R	All standard sizes up to A × B = 12″ × 12″	Up to 21 KG at 400 Hz or higher where weight reduction is important. For improved performance up to 400 Hz stacks of 6 mil or 12 mil Superflux laminations are recommended.*
Square 50 50% NiFe	1, 2, 4 mil BM, BL, BH	C, E, G, R	All standard sizes up to A × B = 12″ × 12″	Pulse transformers requiring higher pulse permeability and lower core loss than Microsil.
Superperm 80 80% NiFe	1, 2, 4 mil CM, CL, CH	C, E, G, R	All standard sizes up to A × B = 12″ × 12″	Low noise inverters and pulse transformers requiring low remanence, highest pulse permeability and lowest loss.

* See Magnetic Metals Transformer Lamination Catalog A1

Courtesy Magnetic Metals Corp.

UNCUT CORES (TAPE-WOUND CORES OR TOROIDS)

A doughnut is a torus or toroid-shaped object with a permeability about the same as that of air. It probably has low electrical resistivity (depending on how moist it is), no hysteresis loss, a completely linear *B-H* curve, and an infinitely high saturation point. (If only the permeability were better!)

The word "toroid" describes a shape such as is shown in Fig. 5-18B; there are, however, toroids and toroids. The toroids about to be discussed are toroids with magnetic *oomph*. The oomph, let it be understood, does not stem only from the materials used to construct the toroids; the toroids may be made from exactly the same materials used for stampings and cut cores. What then?

Like the cut core, the toroid is wound from metal tape. This permits the easiest direction of magnetization to be utilized, giving toroids the same advantage over stampings possessed by cut cores. Then there's the shape—the magnetic field of the toroid is contained almost entirely within the toroid. Leakage inductance and stray capacitance are very small. Note also that when the toroid is placed in an interfering field, the interfering forces tend to act equally all around the toroid, resulting in the cancellation of interfering voltages in the coils. These points are illustrated in Fig. 5-19.

Table 5-6. Stacking Factors, Typical Working Flux Densities, Loss and Apparent Loss for Various Materials of Cut Cores

Material	Gage	Stacking-Factor	Material Code	Bmax Kilogauss	Test Points		Typical Data	
					B (KG)	f (Hz)	W/lb*	VA/lb
Microsil†	1 mil	.83	M	16	10	400	8.0	16.5
Microsil	2 mil	.89	L	17	15	400	12.0	20.0
Microsil	4 mil	.90	H	18	15	400	10.0	13.1
Microsil	4 mil	.90	Z	18.5	17.6	400	16.9	48.0
Microsil	6 mil	.95	G	18	15	400	11.0	14.0
Microsil	12 mil	.95	A (≤ 10 lbs.)	18	15	60	.9	1.7
Microsil	12 mil	.95	S (> 10 lbs.)	18	15	60	.9	1.7
Superflux†	4 mil	.90	V	22	21	400	15.0	36.0
Square 50†	1 mil	.83	BM	15	12	400	3.0	3.5
Square 50	2 mil	.89	BL	15	12	400	3.8	4.0
Square 50	4 mil	.90	BH	15	12	400	4.2	4.5
Superperm 80†	1 mil	.83	CM	7.5	4	400	.18	.70
Superperm 80	2 mil	.89	CL	7.5	4	400	.20	.70
Superperm 80	4 mil	.90	CH	7.5	4	400	.25	.75

* To find W/core see equation (3)

† Specific Weights γ (lb/in³). Microsil $\gamma = .276$, Superflux $\gamma = .296$
Square 50 $\gamma = .298$, Superperm 80 $\gamma = .316$

Courtesy Magnetic Metals Corp.

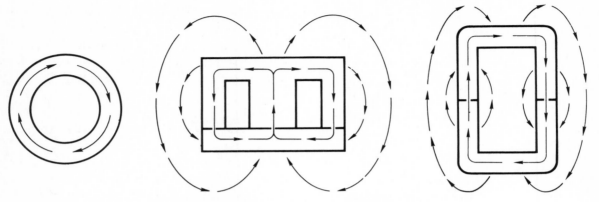

(A) Leakage flux is small compared with that of other types of cores.

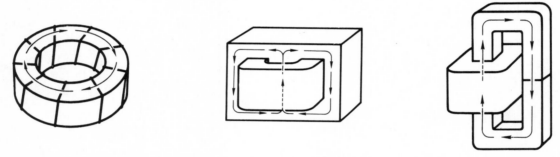

(B) External field is small compared with that of other types of cores.

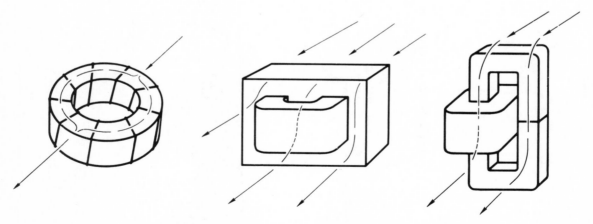

(C) Interfering fields tend to cancel in a toroidal core compared with other types of cores.

Courtesy Magnetic Metals Corp.

Fig. 5-19. Advantages of toroid cores.

Cut cores also have most of these advantages in the toroidal form, but the big plus for the toroid (using the word now to describe a class of component) is that it is not cut; there is no gap in the magnetic circuit and therefore there is very little depreciation of the fine magnetic qualities of the alloys used. This is illustrated by the curves of Fig. 5-5 for the Magnetic Metals range of tape-wound cores (a better term, perhaps, than "toroids").

These cores are constructed with insulated metal tape wound on to a mandrel under closely controlled tension, then annealed and treated to develop fully the required magnetic characteristics. Unlike other types of core, the characteristics of toroids are never dependent on the way the user assembles the cores; they are assembled and treated at the factory for optimum results. The finished core is then permanently encased in plastic or aluminum to protect it from the strains of winding and other external forces. A damping medium fills the space between the core and the case to further protect it from shock and vibration; your cherished offspring doesn't receive greater care in its cradle. This is also a clue as to the great sensitivity of high-performance materials to mechanical abuse. Twisting, banging, and other tortures of cores result in deterioration of performance. A cutaway view is shown in Fig. 5-20.

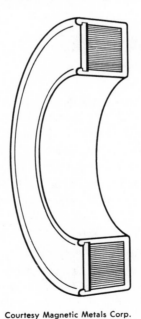

Courtesy Magnetic Metals Corp.

Fig. 5-20. Cutaway view of an aluminum-cased tape-wound core.

There is, or course, a snag to all this. Because the core is an unbroken ring, how does one put the wire on it? The answer, as the comedians say, is, "with difficulty." It is not possible as with other core types to wind the coils separately and then assemble them to the core. The wire must be wound directly into the core. For commercial purposes, this requires specially designed machines. But this should not deter the experimenter from considering them; in many applications, the number of turns needed are so small that hand-winding does not present great problems.

Tape-wound cores are used in all the applications common to other core types—power transformers, current transformers, output transformers, saturating-core transformers, switching devices, and so on. But due to higher cost, they are generally confined to situations that demand high performance. For this reason, the alloys used in them tend to be high-performance types.

Basically, the design criteria in using tape-wound cores are no different than for other types. It is worth noting that manufacturers' literature, especially that of Magnetic Metals Corp., contains so much information both of a general nature and specific to these products as to constitute a minicourse in design. Struggling through data of this kind can be highly educational. An excellent example is the chart with appended design data shown in Chart 5-1. (I refer to it later in the book.)

Ferrites (Molded Ceramic Cores)

Ferrites come in all the shapes associated with the traditional metal stampings and tapes, plus many more besides. Some of these shapes are shown in Fig. 5-21. If you must have a shape that is not a manufacturer's standard, and if you have enough money, you can have the shape you wish. This is one of the beauties of using ferrites.

However, the scope of this book permits us to touch only the fringe of the possible applications for ferrites. Our interest is mainly in power applications at lower frequencies—the so-called high-flux applications.

Common shapes in ferrites are the traditional Us, Es, Is, and toroids, plus two unique shapes: the pot and cross cores. The manufacturer's data sheets for two specific cores of the latter types are given in Figs. 5-22A and 5-22B. These illustrate the core shapes and the very complete design data supplied by Indiana General Corp. Note that the shape pictured on the right of the sheet is actually half of the core. Two such pieces put together as shown in the small picture on the left of the sheet form a complete core.

As with the metal cores, the ferrite toroid gives maximum performance because there is no break in the magnetic circuit. And as with other homogenous cores (such as tape-wound metal cores as distinct from stampings assembled by the user) the manufacturer can and does guarantee the magnetic characteristics of the core. The nomographs supplied for each size and shape permit a total design to be derived using little more than a straightedge.

Note the absence of terms like flux density, oersteds, gauss, etc. This is not because ferrites obey unique magnetic laws; they don't. As discussed in Chapter 1, there are several systems of units and terminology in general use. In the present case, the manufacturer has elected to present the data in terms of volts-time-turns rather than gauss (or webers and teslas). The manufacturer's stated reason is that volts-time-turns are the daily fare of the circuit designer and therefore easier to use than less familiar magnetic terms.

THE *Wa* PRODUCT

Whatever type of core is considered, an essential step in the design process is to find a figure for its required cross-sectional area. This number together with frequency *f*, flux density *B*, and voltage *V*, plugged into the basic transformer equation enables the number of turns *N* to be calculated for each winding. With *N* fixed and suitable wire gauges selected on the basis of the currents to be carried, it is now necessary to establish a value for the area of the core window needed to accommodate the windings. This in turn leads to the selection of a core having both a cross-sectional area *a* and window area *W* to suit the design.

It is easy to see that the equation (Equation 3-14) can be satisfied for any value for *a* so long as *N* is adjusted. In other words, an *increase* in the core area results in a *decrease* in the number of turns required and vice versa. Thus if all other factors are fixed including the sizes of wire, the required window size must also vary inversely with variation in core area; a reduction in core area means an increase in the required window size and vice versa.

It was seen earlier in the power capability equation (Equation 3-39) that the product *Wa* is directly proportional to the power capability of a transformer. For convenience, it is repeated again below:

$$Wa = \frac{17.26SP}{fB}$$

Although *Wa* is fixed for given values of *P, S, f,* and *B*, an infinite number of values can be assigned to *W* and *a* individually so long as a proper relationship between them is maintained. If, for instance, *Wa* = 4, then theoretically any positive number can be assigned to *a* and a corresponding number found for *W* that will result in a product of 4.

It can be seen, then, that the *Wa* product is useful only if there is some way of establishing a number for either *a* or *W* (given the product and one number, the other number can be found). But how is this conundrum to be solved?

The perhaps surprising answer is by stabbing a pin into the pages of a vendor's catalog of laminations or core sizes and trying out whatever one hits for size. True, some educated guesswork and/or experience helps to select the page to stab, but basically the procedure is one of cut and try.

If the designer is contemplating the use of a factory-built core such as a tape-wound toroid or cut core, the *Wa* product for each core is usually given in the catalog, which expedites matters somewhat. Or if a core on hand is being considered, the *Wa* product can be measured directly and decisions made immediately as to suitability.

In the case for laminations, which the designer must stack together for himself or herself, the situation is a little more fluid. The trick here is to select a size that, when stacked together to the thickness needed for the *Wa* product, results in a nicely proportioned component. This is not nearly as difficult as it sounds. Generally speaking, if the ratio of the stack thickness to the width of the lamination center limb (in other words the measurements *D* and *E*, which give the core area when multiplied) is of the

Chart 5-1. Vendor's Design Chart

CORE AND CASE SIZES AND CHARACTERISTICS

Core No.	Gross Core Area CM²[1]	Mean Path Length CM	Core I.D. In.	Core O.D. In.	Core Height In.	Gross Core Wt. Lbs ×10⁻²[2]	Plastic I.D.	Plastic O.D.	Plastic Ht.	Plastic Wire Min[3]	Plastic Wire Max	Metal I.D.	Metal O.D.	Metal Ht.	Metal Wire Min[3]	Metal Wire Max	T/V 60Hz[4]	T/V 400Hz	μh/Turn[5]	Ni/Oersted[6]	WA CA[7]
47	.0504	4.49	.500	.625	.125	411	440	685	195	640	1.31	437	688	197	650	1.33	746	112	1.41	.357	.0012
2	.101	4.99	.500	.750	.125	.914	440	820	195	775	1.69	435	825	197	790	1.71	371	56	2.54	.397	.0023
5	.101	6.18	.650	.900	.125	1.134	585	975	195	785	1.71	585	975	197	790	1.73	371	56	2.06	.491	.0042
9	.101	8.98	1.000	1.250	.125	1.646	915	1.340	195	820	2.44	915	1.340	197	825	2.45	371	56	1.41	.715	.0103
79	.201	6.98	.750	1.000	.250	2.56	665	1.085	340	1.105	2.22	665	1.085	327	1.080	2.21	185	28	3.62	.556	.0108
30	.201	8.98	1.000	1.250	.250	3.29	915	1.340	330	1.070	2.70	915	1.340	327	1.085	2.72	185	28	2.81	.715	.0205
37	.227	6.48	.625	1.000	.188	2.68	570	1.085	272	1.065	1.88	570	1.085	265	1.050	1.89	165	25	4.41	.515	.0090
7	.227	7.48	.750	1.125	.188	3.09	665	1.215	262	1.070	2.54	665	1.215	265	1.075	2.55	165	25	3.81	.593	.0122
3	.302	6.48	.625	1.000	.250	3.57	570	1.085	340	1.200	2.04	570	1.085	327	1.175	2.03	124	18.6	5.87	.515	.0120
10	.302	9.48	1.000	1.375	.250	5.21	925	1.455	320	1.175	2.93	925	1.455	327	1.190	2.95	124	18.6	4.01	.755	.0315
16	.302	14.46	1.625	2.000	.250	7.95	1.525	2.110	330	1.255	4.21	1.525	2.110	327	1.250	4.21	124	18.6	2.63	1.152	.0856
39	.403	9.97	1.000	1.500	.250	7.31	925	1.570	330	1.310	2.86	925	1.570	327	1.305	2.86	93	14.0	5.08	.793	.0420
13	.403	11.97	1.250	1.750	.250	8.78	1.170	1.822	340	1.240	3.60	1.170	1.820	327	1.215	3.60	93	14.0	4.24	.952	.0672
11	.604	9.97	1.000	1.500	.375	10.97	925	1.570	455	1.570	3.09	925	1.570	452	1.555	3.11	62	9.3	7.62	.793	.0630
96	.604	14.46	1.625	2.000	.500	15.91	1.525	2.100	590	1.775	4.73	1.525	2.100	607	1.810	4.72	62	9.3	5.24	1.152	.1712
62	.807	9.97	1.000	1.500	.500	14.63	915	1.595	621	1.927	3.17	925	1.570	607	1.865	3.17	46	6.9	11.80	.793	.0840
29	.807	11.97	1.250	1.750	.500	17.55	1.170	1.822	620	1.900	4.16	1.170	1.820	607	1.870	4.16	46	6.9	8.46	1.107	.1344
14	.807	15.96	1.500	2.000	.500	20.5	1.400	2.110	620	1.960	4.41	1.400	2.110	607	1.940	4.41	46	6.9	7.23	1.107	.1924
17	.807	19.96	2.000	2.500	.500	26.3	1.860	2.652	610	2.025	5.36	1.860	2.652	607	2.025	5.36	46	6.9	5.63	1.428	.340
18	.807	21.94	2.500	3.000	.500	32.2	2.360	3.152	620	2.145	6.47	2.360	3.152	607	2.125	6.47	46	6.9	4.61	1.742	.547
75	.908	12.97	1.250	2.000	.375	21.4	1.170	2.110	445	1.840	3.91	1.170	2.110	452	1.855	3.93	413	6.2	14.78	.952	.1512
60	1.210	12.97	1.250	2.000	.500	28.5	1.170	2.110	620	2.190	4.23	1.170	2.110	607	2.170	4.23	31	4.6	11.52	1.032	.202
15	1.613	15.96	1.500	2.500	.500	46.8	1.400	2.600	610	2.430	4.69	1.400	2.600	607	2.425	4.71	23	3.5	12.72	1.269	.385
76	1.613	19.95	2.000	3.000	.500	58.5	1.890	3.120	621	2.485	5.73	1.910	3.120	607	2.440	5.71	23	3.5	10.16	1.586	.716
19	1.613	23.94	2.500	3.500	.500	70.2	2.313	3.688	698	2.670	6.74	2.313	3.688	607	2.605	6.71	23	3.5	8.46	1.909	1.050
168	2.420	14.96	1.500	2.250	1.000	65.8	1.400	2.360	1.110	3.190	5.77	1.420	2.340	1.135	3.200	5.79	15.5	2.3	20.25	1.188	.594
58	3.226	19.95	2.000	3.000	1.000	117.0	1.860	3.152	1.198	3.705	6.80	1.860	3.152	1.135	3.580	6.80	11.6	1.74	20.35	1.586	1.359
20	3.226	23.94	2.500	3.500	1.000	140.4	2.313	3.688	1.188	3.765	7.74	2.313	3.688	1.135	3.665	7.69	11.6	1.74	16.96	1.909	2.101
21	5.040	24.94	2.500	3.750	1.250	228.6	2.313	3.938	1.448	4.535	8.38	2.313	3.938	1.385	4.410	8.33	7.5	1.12	25.51	1.983	3.283
22	6.050	24.94	2.500	3.750	1.500	274.3	2.313	3.938	1.688	5.015	8.88	2.313	3.938	1.685	5.015	8.83	6.2	0.93	30.60	1.993	3.939
23	6.050	30.92	3.250	4.500	1.500	340.1	3.062	4.688	1.688	5.020	10.54	3.062	4.688	1.685	5.055	10.55	6.2	0.93	24.55	2.465	6.904
25	8.064	36.91	4.000	5.250	2.000	541.2	3.813	5.438	2.188	6.020	13.61	3.813	5.438	2.195	6.040	13.63	4.65	0.70	27.35	2.935	14.27
77	8.468	32.92	3.250	5.000	1.500	506.9	3.062	5.188	1.688	5.525	10.83	3.062	5.188	1.685	5.530	10.84	4.43	0.67	32.35	2.620	9.665
78	12.90	39.90	4.000	6.000	2.000	936.2	3.813	6.188	2.188	6.775	13.95	3.813	6.188	2.195	6.795	14.05	2.92	0.44	40.55	3.175	22.84

NOTE 1. For net core area of each of the following strip thicknesses use the indicated stacking factor:

.0005"—.50	.006"—.90
.001"—.75	.012"—.95
.002"—.85	.014"—.95
.004"—.90	

NOTE 2: To obtain net weight, multiply by the stacking factor of Note 1 and the following material factors:

Super Square 80	1.059
Square 80	1.059
Supermalloy	1.059
Superperm 80	1.059
Square 49	1.000
Superperm 49	1.000
Supermendur	0.988
Microsil	0.927

NOTE 3: Wire length per turn in inches

Min.—First layer, nominal length of wire per turn.

Max.—Outside layer, approximate length for fully wound core based on the minimum ID increasing for larger size cores.

NOTE 4: Turns/volt for 10 kilogauss Bm for gross core area. Divide by factors of Note 1 to obtain turns per volt for actual core thickness. Can be used for any core having indicated cross-sectional area

NOTE 5: Microhenries/turn for 10,000 permeability for gross core area. Multiply by factors of Note 1 for each metal thickness.

NOTE 6: Use this to find ampere-turns to saturate core. Multiply by oersteds required from core material curves. Can be used for any core having these path lengths.

NOTE 7: Product of window area x core area in (inches)[4]

These are IEEE Publication No. 104 recommended core sizes plus a few others. For more complete listing of available sizes, see core catalog. For other sizes, contact your Magnetic Metals sales representative.

WIRE TABLE

Wire Size AWG	Wire Diameter With Heavy Insulation (Inches)	Area Circular Mils Nominal	Resistance In Ohms Per 1000 Ft	Weight In Pounds Per 1000 Ft	Layer Winding In Turns Per Inch	Random Winding In Turns Per Inch*	Machine Winding Minimum Wound I.D. (Inches)	Maximum Turns For Case I.D. 0.5 In	Maximum Turns For Case I.D. 1.0 In
8	.132	16510	.628	50.4	6	42	2.000	—	—
9	.118	13090	.793	40.0	7	52	1.750	—	—
10	.106	10380	.999	31.7	8	75	1.500	—	—
11	.094	8230	1.26	25.2	9	95	1.250	—	—
12	.084	6530	1.59	20.1	11	130	1.000	—	—
13	.075	5190	2.00	15.9	12	159	.875	—	—
14	.067	4110	2.52	12.6	13	193	.875	—	—
15	.060	3260	3.18	10.0	15	248	.875	—	—
16	.054	2580	4.02	7.95	17	316	.750	—	120
17	.048	2050	5.05	6.32	19	394	.750	—	180
18	.043	1620	6.39	5.02	21	487	.750	—	260
19	.039	1290	8.05	3.99	23	596	.500	60	360
20	.035	1020	10.13	3.16	26	792	.500	80	450
21	.031	812	12.77	2.51	29	982	.500	90	560
22	.028	640	16.20	1.99	32	1210	.438	150	680
23	.025	510	20.30	1.59	36	1260	.438	180	850
24	.022	404	25.67	1.26	40	1550	.313	180	1040
25	.020	320	32.37	1.01	45	1940	.313	250	1310
26	.018	253	41.02	.799	50	2700	.300	310	1560
27	.016	202	51.44	.634	55	3550	.300	370	1870
28	.014	159	65.31	.504	62	4180	.300	620	2600
29	.013	128	81.21	.401	68	5160	.300	620	3250
30	.012	100.0	103.7	.318	77	6560	.250	750	4000
31	.011	79.2	130.9	.254	85	8090	.250	920	5050
32	.010	64.0	162.0	.202	94	10000	.250	1250	6870
33	.009	50.4	205.7	.161	105	12500	.250	1510	8740
34	.008	39.7	261.3	.127	119	16250	.218	1920	10620
35	.007	31.4	330.0	.101	133	20600	.218	2440	13120
36	.0060	25.0	414.8	.0803	145	25000	.218	2930	16250
37	.0055	20.2	512.1	.0641	161	30900	.218	3500	19370
38	.0049	16.0	648.2	.0509	181	39300	.218	4300	23750
39	.0043	12.2	846.6	.0403	205	51500	.218	5300	30000
40	.0038	9.61	1079	.0319	226	72000	.218	7450	42500
41	.0034	7.84	1323	.0252	250	89800	.218	9850	52500
42	.0030	6.25	1659	.0200	283	113500	.218	12600	72500
43	.0027	4.84	2143	.0159	315	143000	.187	14900	85000
44	.0025	4.00	2593	.0127	340	168500	.187	17400	100000

Courtesy Magnetic Metals Corp.

1. To determine the number of turns required by a transformer winding operating at a specific voltage and frequency:

   a. Use:
   $$N = \frac{E \times 10^8}{4.44 f A_c B_m}$$

   Where:
   N = turns required for winding
   E = voltage applied to winding in volts rms
   f = frequency of applied voltage.
   A_c = net cross-sectional area of core in cm²
   B_m = maximum flux density for material selected in gausses

   OR

   b. Use turns/volt for 10 kilogausses in table 1. Make certain to adjust turns/volt for actual flux level and applicable stocking factor.

2. To determine the inductance of a winding on a core:

   a. Use:
   $$L = \frac{4\pi\mu N^2 A}{10^8 l}$$

   Where:
   L = inductance (henries) of winding
   N = number of turns on winding
   A = net core area in cm²
   l = core length in cm

   OR

   b. Use table 1 for inductance in microhenries per turn for 10,000 permeability and adjust for net core area and appropriate core permeability. Inductance is proportional to square of turns.

3. To determine degree of saturation for current through a winding:

   a. Calculate
   $$H = \frac{4\pi NI}{l}$$

   Where:
   H = magnetizing force in oersteds
   N = turns on winding
   I = RMS current
   l = core length in cm

   THEN determine flux level (B_m) from B–H loop or magnetization curve.

   OR

   b. From desired flux level (B_m), determine magnetizing force (H) in oersteds from B–H loop or other material performance curves. Then use "NI/Oersteds" from table 1 to determine the ampere-turns required to attain desired flux level.

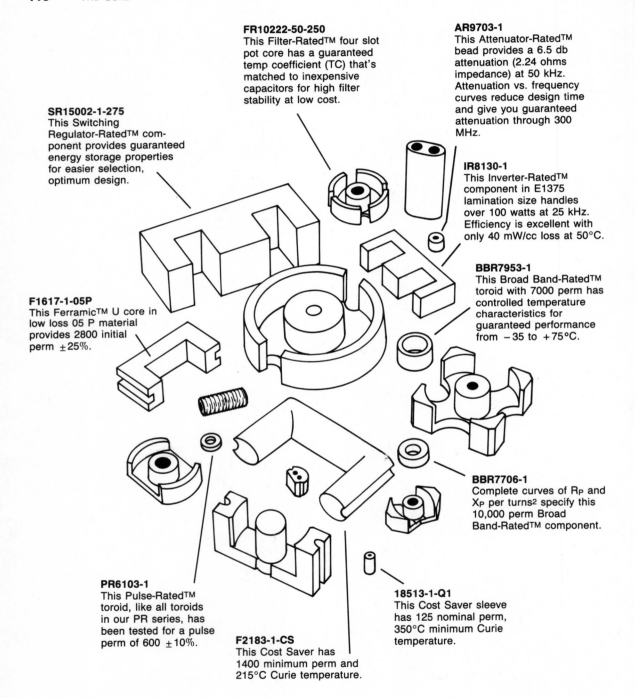

FR10222-50-250
This Filter-Rated™ four slot pot core has a guaranteed temp coefficient (TC) that's matched to inexpensive capacitors for high filter stability at low cost.

AR9703-1
This Attenuator-Rated™ bead provides a 6.5 db attenuation (2.24 ohms impedance) at 50 kHz. Attenuation vs. frequency curves reduce design time and give you guaranteed attenuation through 300 MHz.

SR15002-1-275
This Switching Regulator-Rated™ component provides guaranteed energy storage properties for easier selection, optimum design.

IR8130-1
This Inverter-Rated™ component in E1375 lamination size handles over 100 watts at 25 kHz. Efficiency is excellent with only 40 mW/cc loss at 50°C.

F1617-1-05P
This Ferramic™ U core in low loss 05 P material provides 2800 initial perm ±25%.

BBR7953-1
This Broad Band-Rated™ toroid with 7000 perm has controlled temperature characteristics for guaranteed performance from −35 to +75°C.

BBR7706-1
Complete curves of R_P and X_P per turns^2 specify this 10,000 perm Broad Band-Rated™ component.

PR6103-1
This Pulse-Rated™ toroid, like all toroids in our PR series, has been tested for a pulse perm of 600 ±10%.

F2183-1-CS
This Cost Saver has 1400 minimum perm and 215°C Curie temperature.

18513-1-Q1
This Cost Saver sleeve has 125 nominal perm, 350°C minimum Curie temperature.

Courtesy Indiana General, Div. of Electronic Memories & Magnetics Corp.

Fig. 5-21. A selection of ferrite shapes.

order of 1:1 to 3:1 or even a bit more, the proportion is reasonably good. If there is a choice, it should lean towards a squarish cross-sectional area to obtain minimum winding resistance and therefore the best regulation.

Using this and a "feel" for proportions as a guide, it is a simple matter to judge whether a lamination can be used to meet a given Wa figure. For a specific lamination, the required cross-sectional area is obtained by dividing the Wa product by the window area, which is usually given in the catalog or can be measured.

Alternatively, a cross-sectional area can be postulated and a lamination with an appropriate window area selected to obtain the required Wa number.

It should be understood, however, that although it is desirable, it is not at all necessary to find the Wa product. It is quite feasible to find a suitable core size by simply applying trial designs of likely looking laminations or cores. It should also be appreciated that factors such as flux density and current density are not in fact fixed. They can be adjusted as the design progresses to make the design fit the core. For example, an initially too small window could become acceptable if the wires are made smaller or flux density increased to reduce the number of turns needed.

Also, cores having the same Wa product can differ considerably in weight as well as dimensions. This gives an opportunity to adjust iron losses relative to copper loss for a given Wa product by careful selection of core.

Window Space Utilization

While the power capability equation is a useful tool in core selection, it is much less accurate for quite small transformers than for large ones. This is because window space utilization is generally much less efficient in small transformers. That is to say, the bobbin and insulation and layer margins tend to take up a large percentage of the available winding space, thus leaving less percentage of space for the copper. This results in rather optimistic numbers for the Wa product, which then has to be increased as the design proceeds on small transformers. The equation puts you in the ball park but not into your seat. Despite this, it is still a useful tool.

In Chapter 6, this question of winding area utilization is discussed—frequently referred to as the K factor—and another equation for the Wa product is presented that is often more practical and only slightly more complex than the one already discussed.

ARITHMETIC OF CORE DIMENSIONS

When a core is on hand but the manufacturer's dimensioned data is not, it is a simple matter to work out measurements from the material itself. Start with a rough sketch and enter the basic dimensions using whatever system of units is comfortable for you. Then proceed as outlined in the following.

Stampings and Cut Cores

Mean Magnetic Path Length—The magnetic path length is needed as a step in finding the volume and weight of a core in order to determine losses, and also for calculating inductance.

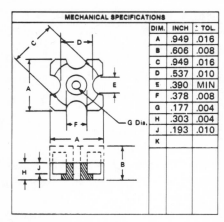

MECHANICAL SPECIFICATIONS

DIM.	INCH	± TOL.
A	.949	.016
B	.606	.008
C	.949	.016
D	.537	.010
E	.390	MIN
F	.378	.008
G	.177	.004
H	.303	.004
J	.193	.010
K		

ELECTRICAL SPECIFICATIONS

V_i	160	VOLTS	
t_1	20	μSEC	
t_d	25	μSEC MAX.	
N_1	117	TURNS	NOTE 1
$V_i t_1/N_1$	27.4	VOLTS · μS/TURNS	
N_2	78	TURNS	
$V_i t_1/N_2$	41.0	VOLTS · μS/TURNS	
i_{m1}	46	mA MAX.	
i_{m2}	121	mA MAX.	NOTES 2,3
A_{L1}	5130	nH/TURNS² MIN.	
A_{L2}	4341	nH/TURNS² MIN.	
P (typ.)	78	WATTS	NOTES 3,4

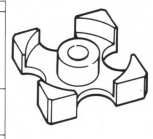

NOTES

1. TEST CONDITIONS 25°C
2. RATINGS AT 25°C
3. RATINGS AND P VALUES REFER TO PAIRS OF CORES
4. P CORRESPONDS TO 40°C HOT SPOT TEMPERATURE RISE. IN STILL AIR WITH NO HEAT SINK

IR 8202

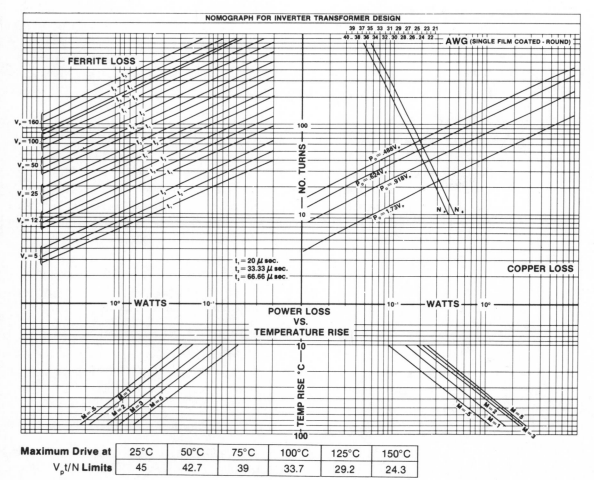

Maximum Drive at	25°C	50°C	75°C	100°C	125°C	150°C
$V_p t/N$ **Limits**	45	42.7	39	33.7	29.2	24.3

Fig. 5-22. Vendor's data sheets with nomographs.   (A) Cross type core.

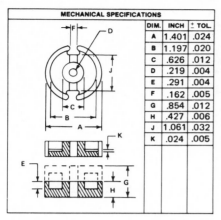

MECHANICAL SPECIFICATIONS

DIM.	INCH	± TOL.
A	1.401	.024
B	1.197	.020
C	.626	.012
D	.219	.004
E	.291	.004
F	.162	.005
G	.854	.012
H	.427	.006
J	1.061	.032
K	.024	.005

ELECTRICAL SPECIFICATIONS

V_i	160	VOLTS	
t_1	20	μSEC	
t_d	25	μSEC MAX.	
N_1	43	TURNS	NOTE 1
$V_1 t_1 / N_1$	74.4	VOLTS · μS/TURNS	
N_2	29	TURNS	
$V_1 t_1 / N_2$	110	VOLTS · μS/TURNS	
i_{m1}	154	mA MAX.	
i_{m2}	403	mA MAX.	NOTES 2,3
A_{L1}	11238	nH/TURNS² MIN.	
A_{L2}	9407	nH/TURNS² MIN.	
P (typ.)	173	WATTS	NOTES 3,4

NOTES

1. TEST CONDITIONS 25°C
2. RATINGS AT 25°C
3. RATINGS AND P VALUES REFER TO PAIRS OF CORES
4. P CORRESPONDS TO 40°C HOT SPOT TEMPERATURE RISE, IN STILL AIR WITH NO HEAT SINK

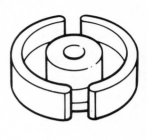

IR 8204

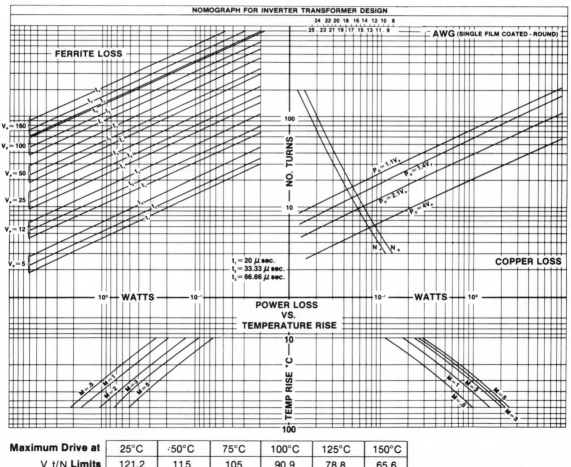

Maximum Drive at	25°C	·50°C	75°C	100°C	125°C	150°C
$V_p t/N$ **Limits**	121.2	115	105	90.9	78.8	65.6

Fig. 5-22 (continued).　　(B) Pot type core. Courtesy Indiana General, Div. of Electronic Memories & Magnetics Corp.

In the shell core shape of Fig. 5-23A, the mean magnetic path length is shown as the heavy dotted line. Note that it is the distance around *one* window only. In the usual case, the center limb is twice the width of the side limbs, and the top and bottom members are the same as the sides. From inspection, the length of the mean magnetic path length is then

$$l = 2G + 2F + 4E$$

This is true also for the core shape shown in Fig. 5-23B.

Sometimes the corners are assumed to be rounded as in Fig. 5-23C and reckoned to be quarter circles. The mean path length then becomes

$$l = 2G + 2F + \pi E$$

The lengths are different, but not so different that it matters. The choice of method is up to the individual.

In the toroid shape (Fig. 5-23D), the mean magnetic path length is a circle of diameter $(D + d)/2$ and length $\pi(D + d)/2$.

Area (Cross Section)—The cross-sectional area of a core is indicated by the shaded portion in Fig. 5-24. Ostensibly the area is given by $D \times E$. However, in laminated cores (which means all types except the molded kind), the laminations can never be perfectly close due to imperfections and insulation between the laminations. The actual area of iron contained within the area DE is usually somewhere between 0.85 and 0.95 of $D \times E$; in other words, $D \times E$ must be multiplied by a stacking factor S to determine the actual area of the iron. Thus, in equations such as the basic transformer voltage and power capability formulas, the factor a is really $S (D \times E)$. For example, if $D \times E = 2.0$ square inches, then if $S = 0.9$, $a = 0.9 \times 2.0 = 1.8$ square inches. But for calculating the physical fit of bobbins and so on in the core the dimension, $D \times E$ should be used.

The actual value of S depends on a number of factors and is supplied by vendors for their material. When an official figure is not available, a value of 0.9 is good enough for most purposes.

Area (Core Window)—This is an important dimension, which together with other parameters such as core cross-sectional area determines the power capability of the transformer. For the shapes in Figs. 5-23A and 5-23B, it is given by $W = G \times F$, and for the toroid in Fig. 5-23D it is $W = 0.785d^2$, where d is the inside diameter of the window.

The area dimensions are usually expressed in square centimeters or square inches, but for toroids, the circular mil is frequently used.

Volume—The core volume v is given by the cross-sectional area a multiplied by the mean magnetic path length l. Thus $v = a \times l$.

Weight—This is the last step before determining core losses from curves. The weight is given by the volume v multiplied by the specific weight γ of the material. Thus, Wt $= v \times \gamma$.

The specific weight, commonly in pounds per cubic inch but sometimes in other units, is supplied by the manufacturer. The specific weights of commonly used core materials are given in Table 5-7.

Table 5-7. Density or Weight per Unit Volume
(Specific Weight) of Various Alloys

Alloy	Grams/cu cm	Pounds/cu in
Super Perm 80, Super Q80	8.74	0.316
Super Perm 49	8.25	0.298
Microsil, silicon	7.65	0.276
Super Flux	8.20	0.296
Hypertran	7.85	0.284

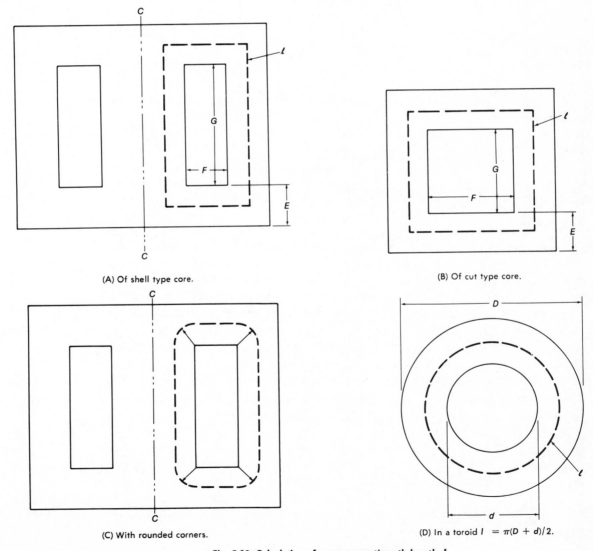

(A) Of shell type core.

(B) Of cut type core.

(C) With rounded corners.

(D) In a toroid $l = \pi(D + d)/2$.

Fig. 5-23. Calculation of mean magnetic path length l.

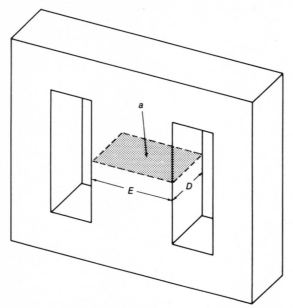

Fig. 5-24. The physical cross-sectional area is **D** $\times$ **E** but the area of iron is **D** $\times$ **E** multiplied by stacking factor **S**.

MANUFACTURER'S DATA SHEETS

Stampings

A manufacturer's data sheet for a waste-free type lamination is shown in Fig. 5-25. (One can see from the dimensions that this is a waste-free type.) Note that not only is the lamination dimensioned in considerable detail, but the data for a stack of the laminations having a square cross section is also given. All the kinds of alloys in which the lamination is produced are listed at the bottom of the page. Dimensions are conveniently presented in both English and metric systems.

By using a simple ratio factor, much of the data can be referenced to stacks of different thicknesses. For example, because the data is for a stack with a square C.S.A. (cross-section area), the thickness of the stack is the same as the center limb width, in this case $1\frac{1}{8}$ inches. Then a stack x inches thick will weigh $x/1\frac{1}{8}$ times the weight given. For silicon, the weight of the stack is given as 2.27 lbs. Then a stack $1\frac{11}{16}$ inches thick weighs $(1\frac{11}{16}/1\frac{1}{8}) \times 2.27 = 3.4$ lbs.

Similarly, the volume and cross-sectional area are easily found for various stack thicknesses. The weight figure is used to find the core loss and magnetizing current as we have discussed. Formulas are provided directly on the sheet (Fig. 5-25) under "Magnetic Design Formulas" for calculating flux density, magnetizing force, and inductance for this specific core. As a matter of interest, the B_{max} formula is yet another arrangement of the basic voltage equation of Chapter 3. Because it expresses flux

LAMINATION TYPE 112 MH

CHARACTERISTICS OF A CORE STACK HAVING A SQUARE CROSS SECTION

VOLUME AND WEIGHT

VOLUME	– 8.56 in.³	– 140 cm.³
WINDOW AREA	– .949 in.²	– 6.12 cm.²
WT. SUPER Q 80	– 2.71 lb.	– 1232 g.
WT. SUPERPERM "49"	– 2.52 lb.	– 1147 g.
WT. SUPERFLUX	– 2.49 lb.	– 1129 g.
WT. SILICON	– 2.27 lb.	– 1029 g.

MAGNETIC DESIGN FORMULAE

Properties of Core Stack with Winding of "N" Turns

$$B_{max} = \frac{46.0 \times 10^3}{K_1 N} \text{ Gausses Per Volt at 60 Hertz}$$

$$H_o = (.073 \times 10^{-3})N \text{ Oersteds}$$
(Gilberts per centimeter) per milliampere of direct current

$$L_a = (.5957 \times 10^{-6})K_1 N^2 \mu_{ac} \text{ Henries}$$

MAGNETIC PATH DIMENSIONS

l = 6.75 in.	17.15 cm.
A = 1.27 in.²	8.16 cm.²

K₁ (STACKING FACTOR)

Thickness	Butt Jointed	Interleaved one per layer
.004″	.90	.80
.006″	.90	.85
.014″	.95	.90
.0185″	.95	.90

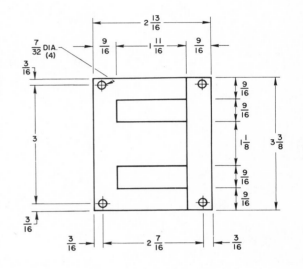

PERFORMANCE DESIGNATION	MATERIAL TYPE	THICKNESS (Inches)	CATALOG NUMBER	WEIGHT AND COUNT	
				LBS. /M PCS.	PCS./ LB .
SUPERPERM 80	HyMu 80	.004	112MH8404	9.27	108
SUPERPERM 80	HyMu 80	.006	112MH8406	13.72	72.9
SUPERPERM 80	HyMu 80	.014	112MH8414	32.44	30.8
SUPER Q 80	HyMu 80	.004	112MH8004	9.27	108
SUPER Q 80	HyMu 80	.006	112MH8006	13.72	72.9
SUPER Q 80	HyMu 80	.014	112MH8014	32.44	30.8
SUPERTHERM 80	HyMu 80	.006	112MH7406	13.72	72.9
SUPERTHERM 80	HyMu 80	.014	112MH7414	32.44	30.8
SUPERPERM 49	49	.004	112MH4904	8.62	116.
SUPERPERM 49	49	.006	112MH4906	13.06	76.5
SUPERPERM 49	49	.014	112MH4914	30.49	32.8
SUPERFLUX	PERMENDUR	.006	112MHVP06	12.90	77.5
SUPERFLUX	PERMENDUR	.010	112MHVP10	21.56	46.3
MICROSIL	Gr. Or. Silicon	.004	112MH3304	8.06	124
MICROSIL	Gr. Or. Silicon	.006	112MH3306	12.09	82.7
MICROSIL	Gr. Or. Silicon	.014	112MH3314	28.22	35.4
SILICON	Non Or. Silicon*	.014	112MH**14	28.22	35.4
SILICON	Non Or. Silicon*	.018	112MH**18	36.29	27.6
SILICON	Non Or. Silicon*	.025	112MH**25	50.40	19.8
HYPERTRAN	Low Carbon	.025	112MH2125	51.71	19.3

* Customer to designate AISI grade of material desired.
** See "How To Order Section" for Code Number.

Fig. 5-25. Example of manufacturer's data sheet.   Courtesy Magnetic Metals Corp.

density in terms of gauss per volt at 60 hertz for this specific core, the formula is derived by simply plugging the known numbers into the basic equation and assigning a value of 1.0 for volts. Factor K_1 is identical in meaning with what we have called the S factor in this book, and numbers are given on the sheet for different thicknesses of laminations and the two assembly methods. Thus, the basic equation becomes (using Equation 3-28 and adding the stacking factor K_1)

$$B_{max} = \frac{V \times 10^8}{25.8FfaNK_1}$$

$$= \frac{1.0 \times 10^8}{25.8 \times 1.11 \times 60 \times 1.27 \times K_1 \times N}$$

$$= \frac{46.0 \times 10^3}{K_2 N}$$

where

V is the voltage across winding in volts rms,
F is the form factor,
f is the frequency in hertz,
a is the core cross-sectional area,
N is the turns on winding.

Cut Cores

Each cut core consists of two matched halves as shown in Figs. 5-17 and 5-18A. The dimensional data is provided in the form of tables such as Table 5-8, which relates the dimensions to the letters in the core diagram. The first column of Table 5-8 is the manufacturer's catalog number. Columns D, E, F, and G are in fact the dimensions of the window and core cross-sectional areas, respectively (Equation 3-40, Chapter 3) in which F and G define the window, and E and D the cross-sectional area. The fifth column is what we have referred to as the Wa product.

These columns are followed successively by Weight (pounds), Leg Area (square inches), which is the cross-sectional area, Watts Loss, Ampere Turns, and, finally our old friend, Turns Per Volt, already worked out. Note carefully that the figures in the last three columns are valid only for the conditions stated at the top of the sheet. The numbers to be used for other values of flux density, frequency, and air gap are obtained from additional design data sheets.

Tape-Wound Cores (Toroids)

Tape-wound cores differ from other types in that they are usually encased for protection and therefore the dimensions of the core cannot be measured directly (refer to Fig 5-20). Only the case dimensions can be checked by the user. An example of manufacturer's data is given in Table 5-9. This table is a selection from a much larger range, and because I intend to return to it later, some explanations are again in order.

Table 5-8. Page From Vendor's Cut Core Catalog

SINGLE-PHASE CUT CORES

MATERIAL- MICROSIL 12 MIL, STACKING FACTOR .95

- NOTE, THIS LISTING SEQUENCED ON CATALOG NUMBER -

CORE LIMITS BASED ON 1.70 VA/LB & .90 WATTS/LB @ 60 HZ, 15.0 KILOGAUSS
AMPERE TURNS ASSUMES .001 IN. AIRGAP, EXCEPT .002 IN. WHERE STARRED*

MAGNETIC METALS CATALOG NUMBER	D	E	F	G	DEFG PRO-DUCT	WEI-GHT LBS.	LEG AREA SQ.IN.	WATTS LOSS	AMPERE TURNS	TURNS PER VOLT	MAGNETIC METALS CATALOG NUMBER
MS-26A	4.125	1.250	.875	5.000	22.6	20.9	4.90	18.8	68.9*	.791	MS-26A
MS-43D	3.750	1.000	1.375	4.500	23.2	14.4	3.56	13.0	67.5*	1.09	MS-43D
MS-44B	1.375	1.375	2.750	7.000	36.4	11.7	1.80	10.5	83.7*	2.16	MS-44B
MS-70	3.375	1.375	2.000	5.437	50.5	23.1	4.41	20.8	75.3*	.879	MS-70
MS-77	2.500	1.500	3.000	5.000	56.3	20.2	3.56	18.1	78.1*	1.09	MS-77
MS-77B	3.000	2.000	3.000	5.000	90.0	34.7	5.70	31.2	80.9*	.680	MS-77B
MS-78D	2.000	1.156	3.000	3.875	26.9	10.4	2.20	9.37	72.0*	1.77	MS-78D
MS-78F	3.000	1.500	3.000	3.250	43.9	20.1	4.28	18.0	71.7*	.907	MS-78F
MS-79G	2.875	1.750	2.000	5.500	55.3	26.8	4.78	24.1	77.7*	.811	MS-79G
MS-81N	1.500	1.500	5.500	4.000	49.5	13.9	2.14	12.5	83.5*	1.81	MS-81N
MS-82C	1.500	1.500	4.000	5.000	45.0	13.3	2.14	11.9	81.7*	1.81	MS-82C
MS-87B	4.000	2.625	2.125	7.750	173.	76.5	9.98	68.8	91.3*	.389	MS-87B
MS-122	2.500	1.000	1.375	4.875	16.8	10.1	2.38	9.10	68.8*	1.63	MS-122
MS-122A	2.500	1.000	2.250	4.625	26.0	10.9	2.38	9.84	71.1*	1.63	MS-122A
MS-122D	2.500	1.750	5.000	2.500	54.7	23.3	4.16	20.9	77.7*	.933	MS-122D
MS-122G	2.500	1.625	2.500	5.000	50.8	21.2	3.86	19.1	77.0*	1.00	MS-122G
MS-122J	.625	1.500	2.250	4.750	10.0	4.55	.891	4.09	74.4*	4.35	MS-122J
MS-123A	3.125	1.156	2.500	5.812	52.5	19.0	3.43	17.1	77.2*	1.13	MS-123A
MS-148G	1.750	1.750	3.000	10.000	91.9	25.1	2.91	22.6	97.7*	1.33	MS-148G
MS-149H	2.000	2.000	4.000	7.000	112.	29.4	3.80	26.5	91.8*	1.02	MS-149H
MS-154B	1.500	1.500	6.000	8.000	108.	19.2	2.14	17.3	99.9*	1.81	MS-154B
MS-154C	2.750	1.093	1.625	4.187	20.5	11.7	2.86	10.5	67.8*	1.36	MS-154C
MS-154K	2.000	1.000	6.000	6.000	72.0	14.1	1.90	12.7	89.8*	2.04	MS-154K
MS-154P	2.375	1.250	.875	6.500	16.9	14.4	2.82	12.9	74.4*	1.37	MS-154P
MS-154T	1.500	1.000	7.000	7.750	81.4	12.8	1.43	11.5	99.8*	2.72	MS-154T
MS-156	2.812	1.093	2.000	5.000	30.7	13.9	2.92	12.5	72.1*	1.33	MS-156
MS-156L	1.500	1.500	2.000	5.000	22.5	10.9	2.14	9.82	74.4*	1.81	MS-156L
MS-156M	1.625	1.625	2.000	5.000	26.4	13.1	2.51	11.8	75.1*	1.55	MS-156M
MS-156N	2.250	1.125	2.000	4.875	24.7	11.3	2.40	10.2	71.8*	1.61	MS-156N
MS-157	4.000	1.562	2.125	6.125	81.3	34.7	5.94	31.2	79.3*	.653	MS-157
MS-158	2.812	1.032	3.062	9.375	83.3	21.2	2.76	19.1	91.5*	1.41	MS-158
MS-159A	1.500	1.500	6.000	6.000	81.0	16.8	2.14	15.1	92.6*	1.81	MS-159A
MS-161	3.375	1.343	2.125	6.062	58.4	24.2	4.31	21.8	77.8*	.900	MS-161
MS-161B	3.125	1.343	2.125	6.062	54.1	22.4	3.99	20.2	77.8*	.972	MS-161B
MS-162C	2.750	1.375	.750	4.375	12.4	14.2	3.59	12.8	66.9*	1.08	MS-162C
MS-162D	2.750	1.375	.750	4.875	13.8	15.2	3.59	13.7	68.7*	1.08	MS-162D
MS-162F	3.250	1.312	.750	4.250	13.6	15.5	4.05	14.0	66.1*	.957	MS-162F
MS-163A	1.875	1.875	.812	2.250	6.42	10.9	3.34	9.79	62.2*	1.16	MS-163A
MS-165	3.125	1.250	1.812	4.875	34.5	17.5	3.71	15.7	71.9*	1.05	MS-165
MS-166	2.812	1.093	1.750	4.687	25.2	13.0	2.92	11.7	70.0*	1.33	MS-166
MS-171A	4.125	1.250	1.000	4.625	23.8	20.2	4.90	18.2	68.0*	.791	MS-171A
MS-171C	4.000	1.250	1.000	2.000	10.0	12.9	4.75	11.6	58.6*	.816	MS-171C
MS-174A	4.000	1.000	.750	3.750	11.3	12.5	3.80	11.3	62.5*	1.02	MS-174A
MS-177A	3.125	1.218	1.875	5.812	41.5	18.9	3.62	17.1	75.3*	1.07	MS-177A
MS-178	4.000	1.656	2.375	7.312	115.	42.3	6.29	38.1	85.1*	.616	MS-178
MS-178E	4.000	2.000	2.500	6.000	120.	48.4	7.60	43.5	82.7*	.510	MS-178E
MS-178F	2.500	3.500	3.500	13.500	413.	103.	8.31	92.5	122.*	.466	MS-178F
MS-178Q	3.000	1.000	2.500	5.750	43.1	15.3	2.85	13.8	76.1*	1.36	MS-178Q
MS-179C	3.000	1.500	2.500	4.750	53.4	22.4	4.28	20.2	75.3*	.907	MS-179C
MS-183	3.750	1.406	2.250	6.562	77.8	30.2	5.01	27.2	80.5*	.774	MS-183

MATERIAL- MICROSIL 12 MIL

Table 5-9. Example of Vendor's Data for Tape-Wound Cores (Toroids)

CORE SIZE	CORE DIMENSIONS IN			CASE DIMENSIONS IN						NET CORE AREA CM²			MEAN PATH LENGTH		CASE WINDOW AREA IN²		RATIO ID/OD	GR. CORE WEIGHT SQUARE 50*		PRODUCT Wa X Ca IN⁴
	ID	OD	HT	ID Metal	ID Plastic	OD Metal	OD Plastic	HT Metal	HT Plastic	.001 SF=.75	.002 SF=.85	.004 .006 SF=.90	IN	CM	Metal	Plastic		Pounds	Grams	
59	2.000	2.500	1.000	1.860	1.890	2.652	2.620	1.135	1.152	1.210	1.371	1.452	7.069	17.95	2.717	2.806	.800	.527	238.87	.679
225	2.500	3.000	1.000	2.410	2.390	3.100	3.120	1.135	1.120	1.210	1.371	1.452	8.639	21.94	4.562	4.486	.833	.644	291.95	1.140
164	1.250	2.000	1.000	1.170	1.150	2.090	2.110	1.135	1.110	1.815	2.056	2.177	5.105	12.97	1.075	1.039	.625	.570	258.77	.403
168	1.500	2.250	1.000	1.420	1.400	2.340	2.360	1.135	1.110	1.815	2.056	2.177	5.890	14.96	1.584	1.5	.667	.658	298.58	.594
491	1.625	2.375	1.000	1.545	1.515	2.465	2.495	1.135	1.162	1.815	2.056	2.177	6.283	15.96	1.875	1.803	.684	.702	318.49	.703
494	2.000	2.750	1.000	1.910	1.890	2.850	2.870	1.135	1.162	1.815	2.056	2.177	7.461	18.95	2.865	2.806	.727	.834	378.20	1.074
340	1.250	2.250	1.000	1.170	1.150	2.340	2.280	1.135	1.162	2.419	2.742	2.903	5.498	13.96	1.075	1.039	.556	.819	371.57	.538
169	1.500	2.500	1.000	1.410	1.390	2.600	2.620	1.135	1.120	2.419	2.742	2.903	6.283	15.96	1.561	1.517	.600	.936	424.65	.781
97	1.750	2.750	1.000	1.650	1.650	2.850	2.850	1.135	1.120	2.419	2.742	2.903	7.069	17.95	2.138	2.138	.636	1.053	477.73	1.069
58	2.000	3.000	1.000	1.860	1.860	3.152	3.152	1.135	1.198	2.419	2.742	2.903	7.854	19.95	2.717	2.717	.667	1.170	530.81	1.359
20	2.500	3.500	1.000	2.313	2.313	3.688	3.688	1.135	1.188	2.419	2.742	2.903	9.425	23.94	4.202	4.202	.714	1.404	636.98	2.101
199	3.000	4.000	1.000	2.900	2.880	4.110	4.130	1.135	1.130	2.419	2.742	2.903	10.996	27.93	6.605	6.514	.750	1.638	743.14	3.303
492	1.750	3.000	1.000	1.660	1.640	3.100	3.120	1.135	1.162	3.024	3.427	3.629	7.461	18.95	2.164	2.112	.583	1.390	630.34	1.353
495	2.000	3.250	1.000	1.900	1.880	3.360	3.380	1.135	1.162	3.024	3.427	3.629	8.247	20.95	2.835	2.776	.615	1.536	696.69	1.772
496	2.250	3.500	1.000	2.150	2.130	3.610	3.630	1.135	1.162	3.024	3.427	3.629	9.032	22.94	3.630	3.563	.643	1.682	763.04	2.269
133	1.750	2.750	1.500	1.850	1.650	2.850	2.850	1.685	1.688	3.629	4.113	4.355	7.069	17.95	2.138	2.138	.636	1.580	716.60	1.604
441	2.000	3.000	1.500	1.910	1.890	3.100	3.120	1.635	1.662	3.629	4.113	4.355	7.854	19.95	2.865	2.821	.667	1.755	796.22	2.149
200	3.000	4.000	1.500	2.900	2.880	4.110	4.130	1.685	1.640	3.629	4.113	4.355	10.996	27.93	6.605	6.514	.750	2.458	1114.71	4.954
22	2.500	3.750	1.500	2.313	2.313	3.938	3.938	1.685	1.688	4.536	5.141	5.444	9.817	24.94	4.202	4.202	.667	2.743	1244.09	3.939
23	3.250	4.500	1.500	3.062	3.062	4.688	4.688	1.685	1.688	4.536	5.141	5.444	12.174	30.92	7.364	7.364	.722	3.401	1542.68	6.904
185	2.000	3.500	1.500	1.900	1.880	3.610	3.630	1.685	1.640	5.444	6.169	6.532	8.639	21.94	2.835	2.776	.571	2.896	1313.76	3.190
201	3.000	4.500	1.500	2.870	2.880	4.640	4.630	1.685	1.640	5.444	6.169	6.532	11.781	29.92	6.469	6.514	.667	3.950	1791.49	7.278
203	3.250	4.750	1.500	3.120	3.100	4.890	4.910	1.685	1.660	5.444	6.169	6.532	12.566	31.92	7.645	7.548	.684	4.213	1910.93	8.601
442	3.500	5.000	1.500	3.370	3.350	5.140	5.160	1.635	1.662	5.444	6.169	6.532	13.352	33.91	8.920	8.814	.700	4.476	2030.36	10.035
211	4.500	6.000	1.500	4.370	4.350	6.140	6.160	1.685	1.660	5.444	6.169	6.532	16.493	41.89	14.999	14.862	.750	5.529	2508.09	16.873
497	2.250	3.500	2.000	2.150	2.130	3.610	3.630	2.195	2.182	6.048	6.855	7.258	9.032	22.94	3.630	3.563	.643	3.643	1526.09	4.538
499	3.000	4.250	2.000	2.870	2.850	4.390	4.410	2.195	2.182	6.048	6.855	7.258	11.388	28.93	6.469	6.379	.706	4.242	1924.20	8.087
24	3.250	4.500	2.000	3.062	3.062	4.688	4.688	2.195	2.188	6.048	6.855	7.258	12.174	30.92	7.364	7.364	.722	4.535	2056.90	9.205
25	4.000	5.250	2.000	3.813	3.813	5.438	5.438	2.195	2.188	6.048	6.855	7.258	14.530	36.91	11.419	11.419	.762	5.412	2455.01	14.274
190	2.250	4.000	1.500	2.150	2.130	4.110	4.130	1.685	1.640	6.351	7.198	7.621	9.817	24.94	3.630	3.563	.563	3.840	1741.73	4.765
500	3.000	4.750	1.500	2.870	2.850	4.890	4.910	1.635	1.662	6.351	7.198	7.621	12.174	30.92	6.469	6.379	.632	4.761	2159.75	8.491
77	3.250	5.000	1.500	3.062	3.062	5.188	5.188	1.685	1.688	6.351	7.198	7.621	12.959	32.92	7.364	7.364	.650	5.069	2299.08	9.665
501	3.250	5.250	1.500	3.120	3.100	5.390	5.410	1.635	1.662	7.258	8.226	8.710	13.352	33.91	7.645	7.598	.619	5.968	2707.15	11.468
502	4.000	6.000	1.500	3.870	3.850	6.140	6.160	1.635	1.662	7.258	8.226	8.710	15.708	39.90	11.763	11.642	.667	7.021	3184.88	17.644
26	4.500	6.500	1.500	4.313	4.312	6.688	6.688	1.685	1.688	7.258	8.226	8.710	17.279	43.89	14.610	14.603	.692	7.724	3503.37	21.915
202	3.000	5.000	2.000	2.870	2.850	5.140	5.160	2.195	2.170	9.677	10.968	11.613	12.566	31.92	6.469	6.379	.600	7.490	3397.20	12.938
78	4.000	6.000	2.000	3.813	3.813	6.188	6.188	2.195	2.188	9.677	10.968	11.613	15.708	39.90	11.419	11.419	.667	9.362	4246.50	22.838
503	4.000	6.500	2.000	3.840	3.820	6.670	6.690	2.195	2.182	12.097	13.710	14.516	16.493	41.89	11.581	11.461	.615	12.288	5573.54	28.953
504	4.500	7.000	2.000	4.340	4.320	7.170	7.190	2.195	2.182	12.097	13.710	14.516	18.064	45.88	14.793	14.657	.643	13.458	6104.35	36.984

* For gross core weights of other materials, see Table 5-10.

Courtesy Magnetic Metals Corp.

Core and Case Dimensions—In the columns of Table 5-9 under "Core Dimensions" and "Case Dimensions" the terms ID, OD, and HT mean inside diameter, outside diameter, and height. Because the case may be in either metal or plastic with somewhat different dimensions, both are listed. Dimensions are in inches.

Net Core Area—Three columns are necessary under this heading for the same core size number, because the different thicknesses of tape in which the core is produced have different stacking factors; this results in different net areas, that is, the area of the actual metal in the core. Although the stacking factors (called SF in this case) are given for the different thicknesses of tape, there is no need to use them for calculations, because they are already incorporated in the net area. The area is given in square centimeters; if square inches are preferred, divide by 6.452.

Mean Path Length—This has the same meaning as in other types of cores. It is the mean magnetic path length around the core. If a core is not encased, this term can be calculated from direct measurements as explained previously.

Case Window Area—This important dimension is given for both metal and plastic cases in square inches. Unlike standard cores built up from stampings or in the squarish shape of many cut cores, the toroidal form presents an unusual limitation. In this shape, it is not practical to fill the entire window with copper and insulation as is usually done with other types, because winding is done directly onto the core (buffered by the casing) and the core is commonly wound over the entire 360 degrees. As the winding is built up, the space available for threading wire through the window diminishes. If winding is taken too far, it becomes impossible to continue because the shuttle (if the winding is machine wound) cannot get through. Therefore, for a given method of winding, it is essential to establish at the design stage a maximum area that can be assigned to the area while leaving sufficient space to maneuver.

Ratio ID/OD—This is the ratio of the *core* dimensions, not the case dimensions.

Gr. Core Weight—Square 50—The gross weight of a core composed of the Magnetic Metals' material called Square 50. It is given in both pounds and grams. To find the gross weights of cores in other materials, conversion factors are supplied in Table 5-10.

Product Wa × Ca—This is the window area times cross-sectional area, which we have identified previously as the *Wa* product. In this listing, note that it is the product

**Table 5-10. Weight Figures in Magnetic Metals'
Core Listings Are Multiplied by the
Appropriate Factor**

Designation	Factor
SuperSquare 80	1.059
Square 80	1.059
Square 50	1.000
Microsil	0.927
Supermendur	0.988
Supermalloy	1.059
SuperPerm 49	1.000
SuperPerm 80	1.059

of the case window area and the gross core area—that is to say, the core area before applying the stacking factor. The gross area is not supplied in a separate column; if you want it, derive it by dividing the Net Core Area by the SF figure at the top of the Net Core Area column when converting it to square inches. Multiplying this by the case window area, which is already in square inches, gives the $Wa \times Ca$ figure in this column. As with other core configurations, consideration of this figure leads to the selection of a proper core size for a given design.

A glance down these columns confirms a fact mentioned earlier: that identical $Wa \times Ca$ product numbers do not imply identical weights or dimensions.

CORE SELECTION

All this discussion about alloys, laminations, toroids, cut cores, metal powders, ceramics, and the many obviously overlapping characteristics is very fine—interesting even—but sooner or later the moment of truth must arrive. A choice must be made from the welter of possibilities. It is necessary to cut through the tangle of data and capture the one core or lamination that exactly meets your requirements. The choice is not necessarily easy, nor for that matter necessarily difficult. It depends on specific cases: on the demands of the circuit in which it is to be used, and, surprisingly perhaps, on individual preference, possibly even prejudice.

To illustrate the kind of thinking used in making a choice, let us discuss a simple case. Suppose you want to design a power transformer. This statement immediately narrows the search, because power transformation is a high-flux application. That is, it is generally accepted that the core material for a power transformer should be run at as high a flux density as possible to keep the transformer small and less costly.

The high-flux materials are generally considered to be low-carbon steel, silicon steel, and cobalt-iron alloys, all of which have theoretical maximum flux density capabilities higher than about 18,000 gauss. This short list just happens to include the least expensive (low-carbon steel) and the most expensive (cobalt-vanadium-iron) of the metal alloys. But a list of high-flux materials cannot exclude some of the ferrites. At high frequencies, their flux capability is likely to be considerably greater than that of the high-flux alloys even though the maximum figure may be only, say, 3000 gauss. In other words, at high frequencies, ferrites have high flux density capability while the so-called high-flux alloys might be entirely unsuitable.

At low and intermediate frequencies, there is a spread in the overall performances of the high-flux alloys and the various core configurations. Not the least consideration is that of cost.

Suppose the transformer is to work at 60 hertz and drive an ordinary doorbell. In this application, the primary winding of the transformer is constantly energized and the bell push switches in a low-voltage secondary circuit that drives the bell. The transformer thus isolates the bell push from the relatively high voltage supply and reduces the probability of finding your visitors in a limp, electrocuted heap on the door mat; it also permits the use of inexpensive low-voltage wiring.

The transformer is under load for only a few seconds at a time, but a no-load current passes through the primary winding at all times. It is likely that the size and weight of such a transformer is relatively unimportant, regulation does not have to be very good, and core losses are unimportant—or are they? Core loss is not a function of the load, so a small but constant power drain due to core losses is inseparable from the so-called idling current. If the transformer is to be sold commercially, the least expensive material (low-carbon steel) will probably be chosen; the core will be in the form of cheap stampings, and it will probably be run at very high flux density.

On the other hand, you might want to save energy, and bearing in mind that the sum of all the energy wasted by the world's bell transformers must add up to a few kilowatts per year, you might want to set an example by installing an efficient transformer. So you could choose silicon-steel stampings, which have a lower loss factor. If you feel very strongly about it, you might even consider grain-oriented silicon steel in cut-core form. Of if you are a total freak, you might go to a grain-oriented silicon-steel toroid or even to a toroid of cobalt-vanadium-iron, the most expensive and efficient of them all.

The point of this tongue-in-cheek digression is that guidelines can be only of a general nature. The designer is in the driver's seat. In general, then, the energizing of bells, the lighting of lamps, the powering of toy roadracing games, and so on may be considered as noncritical applications for which stampings in low-carbon or silicon steel are adequate; but if higher-grade material is handy, salvaged perhaps from junked equipment, then by all means use it if you want. Transformers for powering radio gear and the like commonly use somewhat higher grades of silicon steel, and where weight and efficiency are more critical, cut cores in grain-oriented silicon steel are frequently used.

If the transformer is required for higher frequency sine-wave operation, the choice is still centered on high-flux material. The emphasis, however, will shift to consideration of lamination thickness as well as the kind of material, because thinner metal tends to reduce losses at the higher frequencies.

A somewhat different case is the power transformer that is to serve the additional purpose of a saturable-core switching device in a converter or inverter (Chapter 10). Here we will usually look for a high-flux material with a square loop characteristic. Certain materials are processed to emphasize the square loop specifically for such purposes. In converter/inverter applications, the efficiency of the circuit as a whole is determined almost entirely by the transformer, which therefore tends to be considered very carefully indeed from the point of view of core losses. This leads to the use of high-performance materials in expensive formats, such as cut cores and toroids.

Nevertheless, very nice converter/inverter circuits can be put together using ordinary silicon-steel stampings. In short, there is no need to be intimidated by the more glamorous alloys.

For some kinds of transformers, notably output transformers in audio and other types of low-frequency amplifiers, high primary winding inductance is often required. In such cases, the search is likely to be for a high-flux material that also possesses high permeability at high flux density. This permits high inductance to be achieved with the minimum number of turns. Good-grade silicon-steel stampings, carefully assembled, will produce good results. Grain-oriented silicon steel in cut cores is even better and, of course, toroids are at the top of the heap, with cobalt-vanadium-iron material as the prime choice—at a price.

Note, however, that the presence of dc current in the windings might under some conditions rule out the use of toroids entirely because it is not possible to gap the core.

Finally, there are many applications in which nickel-iron cores of various kinds and in various formats are more suitable than the high-flux varieties. These low-flux materials generally give very high performances at low flux densities in terms of high permeability—often enormously high—and incredibly low losses.

There are alloys, like the well-known Mumetal by Alleghany Ludlum and Magnetic Metals' Superperm 80 or Supermalloy, in which permeabilities of over 100,000 are claimed at very low magnetizing force. These alloys find important applications where the input voltage is low and inductance has to be relatively high, such as the input or between the stages of low-frequency amplifiers, or where accuracy is required, as in some current transformers used in measuring instruments.

SIX

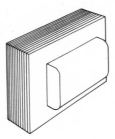

The Windings

The following discussion is slanted initially to the shell and core type shapes, i.e., the squarish core shape in which the coils are wound separately from the core and assembled to it later. The winding peculiarities of the uncut toroid are dealt with separately. But a great deal of what is written about the standard format applies also to toroids.

BASIC FORMAT

Iron-core transformers and inductors are commonly wound with solid copper wire insulated with enamel—the type of wire sometimes known as *magnet wire*. Generally, the wire is wound on a bobbin that insulates and protects the wire from the iron core and permits the coils to be wound separately from the core. The turns are usually laid side by side in neat rows or layers, and after the first layer, each layer is laid on top of the preceding layer, with insulating paper or other material between the layers (Fig. 6-1). The bobbin may or may not be fitted with cheeks to contain the wire.

Another type of winding sometimes used is the *random-wound* coil in which the wire is run on to the bobbin in a more-or-less random fashion (Fig. 6-2). In this case, cheeks are needed to contain the wire. This is much easier and quicker than layer winding, but it is most suitable for low-voltage coils or higher-voltage coils if double-insulated wire is used.

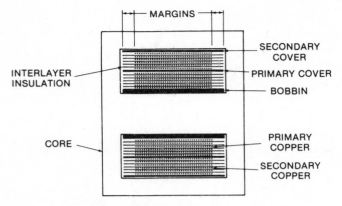

Fig. 6-1. Typical layer-wound construction.

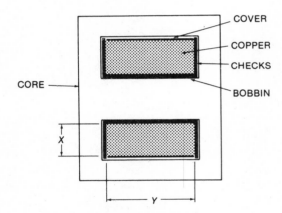

Fig. 6-2. Random-wound coil.

Once the number of turns of wire has been established, you then begin choosing wire gauges and fitting them (on paper) into the available core space. In doing this, attention is paid to achieving maximum working efficiency (that is, minimum power loss), adequate protection against voltage breakdown, and a temperature rise in the windings that is confined to safe limits.

The latter point is particularly important because the insulating materials can be easily destroyed by excessive heat. The power rating of a transformer is largely determined by the temperature rise. It is possible, though, by using special techniques and materials to design a transformer that will run so hot that solder laid on the core will melt. But if such a temperature is accidentally reached in your design, you are likely in deep trouble—and the smoke pouring from the windings will confirm this.

Because deliberately making a transformer to run this hot is an expensive way of melting solder, one might ask why we should do so. The answer is that the hotter it runs, the smaller and lighter it will be for a given power rating. In many applications, reduction in size and weight is a prime factor in the design. Furthermore, smaller and lighter often means cheaper, and this is *always* a factor to be considered. Often, therefore, the aim is to achieve the highest safe temperature, and the choice of wire is of prime importance in meeting this requirement.

COPPER WIRE TABLES

One of the most important design tools in accomplishing all this is the copper wire table. It looks simple enough, and in fact, so it is; but danger lurks here for the unwary. Many a strong man has been reduced to tears when his windings failed to fit the core because he used the table improperly.

Wire tables are published in many reference books and in manufacturers' literature. Some are general in nature and some are slanted to the specific purpose of transformer design; still others are even more specialized in being slanted to a specific shape of transformer (such as the uncut toroid). Table 6-1 is slanted to general-purpose transformer design.

The first thing to accept about wire tables, including this one, is that there are no absolute figures. Every figure is conditional. Usually, the conditions are not stated and there are strings attached. You obviously need to know what these are. Explanatory notes are given below for each column of Table 6-1 in turn. We will use this table frequently.

Wire Size AWG

AWG is the American Wire Gauge (B&S). There are other kinds of wire gauges, but AWG is standard for transformers in North America.

Circular Mils

This is the measure of wire cross-sectional area (discussed at some length in Chapter 1). The unstated condition here is that these figures are true only for 20°C (68°F). At other temperatures, they will be different. Nitpicking? Well, maybe just a little in this case, but keep the principle in mind. The effect of temperature can be important with reference to other quantities, as is soon explained.

The figures are for *bare* wires because they are used in relation to the current-carrying capacity. We will return to this point shortly.

Diameter (in inches)

The term (in) means inches. The comments about temperature with regard to area also apply here. The "Single Insulation" figures can be considered as the standard ones. The "Heavy Insulation" figures refer to double-thickness enamel that is safer under certain conditions. In some tables, the diameter is stated in mils, which is thousandths of an inch; for example, gauge No. 20, single insulation, which is given in this table as 0.0334 inch, would be stated as 33.4 mils. Other tables give the figure in millimeters—not to be confused with mils.

Table 6-1. General-Purpose Wire Table

Wire Size AWG	Circular Mils	Diameter (in) Single Insulation	Diameter (in) Heavy Insulation	Resistance ohms/1000 ft.	Weight lbs/1000 ft.	Random Winding Single Turns/in'	Random Winding Heavy Turns/in'	Layer Winding Single Turns/in	Layer Winding Heavy Turns/in	Layer Winding Layer Insulation (in)	Layer Winding Edge Distance (in)	Wire Size AWG
10	10384		.106	.999	31.7	86	75	9	8	.010	.250	10
11	8226		.094	1.26	25.2	108	95	10	9	.0100	.250	11
12	6529		.084	1.59	20.1	133	130	11	11	.0100	.250	12
13	5184		.075	2.00	15.9	162	159	12	12	.0100	.250	13
14	4109	.0658	.067	2.52	12.6	212	193	14	13	.0100	.188	14
15	3260	.0587	.060	3.18	10.0	255	248	15	15	.0100	.188	15
16	2581	.0524	.054	4.02	7.95	324	316	17	17	.0100	.188	16
17	2052	.0468	.048	5.05	6.32	405	394	19	19	.0070	.188	17
18	1624	.0418	.043	6.39	5.02	525	487	22	21	.0070	.125	18
19	1289	.0373	.039	8.05	3.99	641	596	24	23	.0070	.125	19
20	1024	.0334	.035	10.13	3.16	850	792	27	26	.0050	.125	20
21	812.3	.0298	.031	12.77	2.51	1055	982	30	29	.0050	.125	21
22	640.1	.0266	.028	16.20	1.99	1340	1210	34	32	.0050	.125	22
23	510.8	.0238	.025	20.30	1.59	1370	1260	38	36	.0050	.125	23
24	404.0	.0213	.022	25.67	1.26	1730	1550	42	40	.0020	.125	24
25	320.4	.0190	.020	32.37	1.01	2150	1940	47	45	.0020	.125	25
26	252.8	.0170	.018	41.02	.799	2990	2700	53	50	.0020	.125	26
27	201.6	.0152	.016	51.44	.634	3700	3550	59	55	.0020	.125	27
28	158.8	.0136	.014	65.31	.504	4680	4180	66	62	.0015	.125	28
29	127.7	.0122	.013	81.21	.401	5900	5160	73	68	.0015	.125	29
30	100.0	.0109	.012	103.7	.318	7500	6560	82	77	.0015	.093	30
31	79.21	.0097	.011	130.9	.254	9270	8090	91	85	.0015	.093	31
32	64.00	.0088	.010	162.0	.202	11400	10000	100	94	.0013	.093	32
33	50.41	.0078	.009	205.7	.161	14500	12500	113	105	.0013	.093	33
34	39.69	.0070	.008	261.3	.127	18800	16250	128	119	.0010	.093	34
35	31.36	.0062	.007	330.7	.101	24000	20600	144	133	.0010	.093	35
36	25.00	.0056	.0060	414.8	.0803	29650	25000	158	145	.0010	.093	36
37	20.25	.0050	.0055	512.1	.0641	37400	30900	177	161	.0010	.093	37
38	16.00	.0045	.0049	648.2	.0509	46700	39300	198	181	.0010	.062	38
39	12.25	.0039	.0043	846.6	.0403	62700	51500	226	205	.0007	.062	39
40	9.61	.0035	.0038	1079	.0319	89600	72000	262	226	.0007	.062	40
41	7.84	.0031	.0034	1323	.0252	107800	89800	274	250	.0007	.062	41
42	6.25	.0028	.0030	1659	.0199	133500	116500	304	283	.0005	.062	42
43	4.84	.0025	.0027	2143	.0159	167000	143000	340	315	.0005	.062	43
44	4.00	.0022	.0025	2593	.0127	217000	168500	386	340	.0005	.062	44

Resistance (Ohms/1000 ft)

Here again the unstated condition is "at 20°C." This time, however, the effect of temperature change cannot be ignored. In many transformers, it is likely that the wire will be operating at temperatures considerably higher than 20°C because it is buried in the heart of a heat-producing coil; resistance increases with temperature and in practice can be more than 20 percent higher at operating temperature than the figure given here. This leads to significantly increased voltage drops and power loss and must usually be taken into account. These situations are discussed further in specific design procedures later.

Weight (lbs/1000 ft)

Weight is a necessary characteristic for calculating the amount of wire needed.

Random Winding (Turns/in²)

This refers to the number of turns that can be accommodated to the square inch of winding area, the area being a cross section of the winding $x \times y$ in Fig. 6-2. Of necessity, this is an approximate figure, because some people (or machines) are more adept at winding than others; some windings might be tight and others slack, and so on. Random winding figures tend toward the low side of average, and with reasonable care, should be easily met in practice.

Layer Winding (Turns/in)

This figure includes a winding factor to account for the fact that wires cannot be laid with absolute precision side by side. Again, what is actually achieved depends to an extent on the machine or person performing the operation. Also, variations in the wire diameter and thickness of the insulating coating play a part. The winding factors allowed by the compilers of such tables depend on the wire gauge. The best winding efficiency in terms of turns per inch is obtained in gauges from about No. 10 to No. 22, where it is about 90 percent. In the middle gauges, from about 23 to 29, it drops to about 75 percent, then rises again to about 85 percent in the finer gauges of No. 40 and above. However, with reasonable care, these figures in the table should work out quite well.

Layer Insulation (in)

The thickness of the paper used as insulation between layers is chosen more for mechanical reasons than electrical. It is therefore possible to relate it directly to the wire gauge as has been done in this table. The figures given here are used in standard commercial practice.

Edge Distance (in)

In coils without cheeks, it is necessary to maintain a certain distance between the end turns in each layer and the core, as shown in Fig. 6-1. Here the edge distance is defined as *margin*. As with the interlayer insulation, however, the actual distance is related to mechanical considerations rather than electrical. Again, this makes it possible to relate the distance to the wire gauge as has been done.

Current Ratings

In the wire chart of Table 6-1, a column for current rating is conspicuously absent, and not without reason. It is a surprisingly common misconception that every wire gauge has associated with it a fixed current rating in the same way that it has "turns per inch" and "ohms per 1000 feet" figures. This view is reinforced somewhat by the fact that some wire tables include such a column. But the current that a wire can safely carry depends on the temperature to which it will rise, and this in turn depends on the way in which the wire is used. For example, a coil of No. 20 gauge wire buried in many layers of insulation might get too hot for safety with 0.25 ampere passing through it and yet handle a cool 1.0 ampere or more under conditions where the heat generated in the wire can be easily dissipated.

The question of temperature rise in coils and the calculations required to determine it are discussed at some length in this chapter. However, this is the last step in the design procedure, if it is done at all. In the meantime, it is essential to select wire gauges for the windings and to do this usually right at the start of the design. How else can we figure the fit of the windings to the core? But if the gauges cannot be selected on the basis of the current to be carried, how else can it be done?

Of course, wire selection has to be done after all on the basis of the current, but only in a very broad sort of way. The reasoning is as follows. It is known from experience that in most small transformers designed in the conventional way, using the standard design rules for insulation and having reasonable efficiency and safe temperature rise, the wire is commonly run at current densities in the approximate range of 500 to 1000 circular mils per ampere, give or take a hundred or so circular mils at either end of this range. As a rule, the smaller current densities (1000 circular mils per ampere) will be found in conservatively rated cool-running transformers or those designed in such a way that the heat cannot be easily dissipated. The higher current densities (500 circular mils per ampere) are found in transformers that run at fairly high temperatures or in windings where heat dissipation is easy.

Because this is the range of current density for most transformers, it makes sense for us to work in the same range. Of course, 500 to 1000 circular mils per ampere is quite a broad range. From the wire table, it is the difference between, say, No. 20 gauge and No. 23, or No. 23 and No. 26, or No. 40 and No. 43, in other words, a spread of four gauge numbers.

Generally speaking, 1000 circular mils per ampere can beconsidered a safe density and 500 circular mils per ampere a figure to be used with caution. The amateur is usually advised to work at no less than about 800 circular mils per ampere or higher, especially if he or she is not too sure of the temperature situation.

Suppose a winding is to carry 3.0 amperes. At 1000 circular mils per ampere, the wire gauge must have a cross-sectional area of $3 \times 1000 = 3000$ circular mils. From the wire table, this is about No. 15 gauge. If the wire is to be say, 700 circular mils per ampere, the required area is $3 \times 700 = 2100$ circular mils, which is about No. 17 gauge.

It is, of course, necessary to also consider the wire resistance in relation to voltage drops and power loss. But these factors, together with temperature rise, fall into place as the design proceeds. The figures given are used in standard commercial practice.

INSULATION

Although insulated wire is used in the winding, it is necessary to provide additional insulation for both mechanical and electrical reasons. The covering of the wire must be protected from the abrasive surfaces of the core. For this reason, a bobbin, or former, must be used to protect the inside of the windings and preserve the shape of the coil under stress so that it can be eventually assembled to the core. The outside of the winding must be covered to protect it mechanically as well as electrically. The interlayer paper provides a flat bed on which to lay the windings and prevents turns from sinking into the layer below as well as protecting the winding against electrical stresses.

Figure 6-1 shows a typical insulation plan on a two-coil transformer. Although the thickness of insulation is governed primarily by mechanical considerations, the designer must always be aware of the voltage conditions, which can sometimes be the determining factor.

The methods of constructing bobbins is dealt with elsewhere in this book. Materials used to construct them can be almost any common insulator—heavy cardboard, fiber sheet, plastic, and so on. Between layers, the professional generally uses an electrical-grade paper known as "kraft paper," but for the part-time experimenter, any good-quality paper is fine, including waxed kitchen paper. But be sure you know how thick it is; this is of top importance in winding calculations. Use a micrometer on it or have someone measure it for you. As we have seen, the insulation between layers is commonly between 0.0005 and 0.010 inch (12.7 μm and 0.254 mm) thick. To give you an idea of thickness, the page of this book is approximately 0.003 to 0.005 inch thick. A sample of waxed paper designated heavy-duty measured 0.0015 inch! Hence, a little 1-inch or ½-inch micrometer is handy to have around for this kind of work.

Between windings, as distinct from between layers, somewhat more insulation is normally used, if not for electrical reasons then to provide a good base for the next winding. At these points a good, tight layer of adhesive insulating tape holds everything together, and certainly a winding of tape around the finished coil is a good idea. However, it should not be overdone. Remember that the more insulation there is in the winding, the hotter it is likely to run; and not only that, the transformer will be less efficient because insulation is occupying space that could be better used for copper. In other words, insulation should be viewed as a necessary evil.

Voltage Distribution

When a transformer is in a steady operational condition, the voltage is distributed evenly throughout the windings. At the end of each layer where the wire rises from one layer to the next (Fig. 6-3), there is little voltage difference between the layers, but it increases along the layers laterally until at the other end it is equal to twice the voltage across one complete layer. That is, the maximum voltage between layers is

$$V_L = 2N_L \frac{V_W}{N_W} \qquad (6\text{-}1)$$

where

V_L is the maximum voltage between layers,
N_L is the number of turns per layer,
N_W is the total number of turns in the considered winding,
V_W is the voltage across the winding.

At other points in the transformer, the voltage differentials depend on the winding layout and the kind of circuit into which it is working, among other things.

On first switching the power on, the initial voltage distribution is not even due to the inductive nature of the transformer and the inherent distributed capacitance of the windings. At this time, large voltages appear briefly at normally low voltage points and stress the insulation. In transformers designed to work at fairly high voltages, allowance must be made for this.

Good working rules for choosing insulation are as follows. Design for voltages between the windings and between the windings and the core of twice the working voltage plus 1.0 kV (1000 volts), and allow 0.002 inch (0.025 mm) of paper for each 100 volts. For example, if the normal working voltage between two windings, or the winding and core, is 117 volts, insulate for 1234 volts. This requires no less than 0.025 inch of paper. Keep in mind that electrically, several layers of thin paper are much better than one layer of thick paper.

Other insulating materials—and there are many—can be used, but care must be taken to ensure they can withstand the temperatures inside the transformer. The word "withstand" here means "withstand" in every respect. For example, the melting point

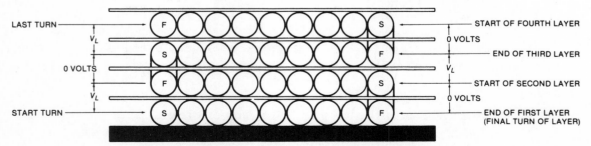

Fig. 6-3. Voltage between layers when V_L is given by Equation 6-1.

of a plastic might be in the range 300°F to 400°F, but it might soften at less than 212°F (100°C, the boiling point of water).

Insulation can break down in one of three ways: (1) by puncturing through the paper; (2) by tracking across the surface; and (3) by flashing through the air. The first possibility has just been discussed. The second possibility can occur if unsuitable paper is used, if moisture is present in the windings, or if the margins between the end turns and the core are not sufficient. Insufficient margins could also account for the third possibility if high voltages are present.

The margin (edge) distance given in the wire tables should be considered to be the minimum allowable distance for the wire gauge in question. Most of the time, this is adequate for the voltage, but not always. To check whether it is okay or not, a good rule is to double the working voltage (that will appear across the margin), add 1000 volts to it, then allow 0.062 inch (1.575 mm) of margin for each kilovolt; choose the larger of the wire table figure and the calculation. Expressed as an equation, this gives a formula of approximation:

$$\text{Margin} = \frac{V_W + 500}{8000} \text{ inches} \qquad (6\text{-}2)$$

where V_W is the working voltage.

Example 1—For example, if a coil is to be wound with No. 38 gauge wire, Table 6-1 advises a minimum margin of 0.062 inch (1/16 inch). If a voltage of 250 volts is expected to appear across the margin under working conditions, the calculated margin is

$$\text{Margin} = \frac{V_W + 500}{8000} = \frac{250 + 500}{8000} = 0.094 \text{ inch}$$

The table figure in this case should be considered insufficient and the margin increased to at least 0.1 inch.

If the finished windings are to be dipped in varnish to exclude dampness and to help hold the windings together (not essential but some people like to do this), use only electrical-grade products. That great old standby, shellac, for instance, must be avoided because it could contain acid that can attack the insulation on the wire. Even the acid from the fingers of a winding operator has been known to corrode fine-gauge wire, resulting in open circuits.

BUILD

The "build" is an important dimension of the windings. It is the thickness of the winding indicated by b in Fig. 6-4. Note that dimension e will usually be greater than b because the leads are arranged to come out on these faces; the anchoring technique for the leads (Chapter 14) tends to create a bulge here. Also, the overlaps of insulating paper are arranged on these faces of the coil. In other words, b is reserved for copper and the essential insulation only; it is always the thickness of winding *inside* the window.

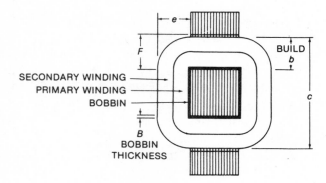

Fig. 6-4. Transformer cross section.

Equally important is that dimension c is not too great. If b is right but c is too large, the winding still will not fit the core. The available space for the winding is, of course, dimension F, the width of the lamination window. An important part of the design process, then, is to ensure that the windings will fit into the core.

Transformer design is largely a cut-and-try affair. Selection of lamination sizes and wire gauges is made as first approximations based on experience or rule of thumb. Then the mechanical fits and electrical characteristics are checked on paper. Adjustments and readjustments are made as required until the desired results are achieved. As an example of the procedure, suppose a designer has selected a lamination and core area, has worked out the turns needed for the primary and secondary windings, and made a preliminary selection of wire gauges. He or she must now check the fit of the windings into the available window space.

Fitting the Windings

Example 2—The designer first decides what space to assign to the bobbin. Refer now to Fig. 6-5. Because the space taken by the bobbin is a necessary evil—its presence, like the rest of the insulation, reduces the valuable copper space—the designer uses material that is just thick enough to support the mechanical and electrical stresses and can be easily fabricated. The thickness is usually on the order of $\frac{1}{32}$ inch to $\frac{1}{16}$ inch— say in this case $\frac{1}{16}$ inch. Because it is much easier to add up decimals than fractions of an inch, the thickness is designated as 0.063 inch. Some looseness must be allowed for in order to fit the core into the bobbin, say 0.032 inch, making a total bobbin space B of 0.095 inch, as shown in Fig. 6-5.

For the length of the bobbin, the designer allows, say, 0.032 inch less than the length G of the lamination window; if the window is 2.5 inches long, the bobbin will be 2.468 inches long.

A winding margin appropriate to the voltage and wire gauge is selected, say, 0.125 inch on each side. Subtracting the two margins from the window length gives 2.5 − 0.25 = 2.25 inches. This is the winding length. Note that the margins are subtracted from the window length, not the bobbin length.

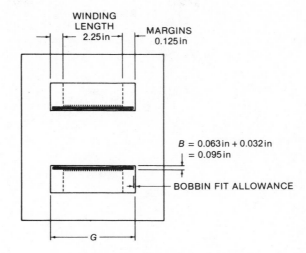

WINDING
LENGTH
2.25 in

MARGINS
0.125 in

$B = 0.063\,\text{in} + 0.032\,\text{in}$
$= 0.095\,\text{in}$

BOBBIN FIT ALLOWANCE

G

Fig. 6-5. Fit factors.

The number of turns per layer is established by multiplying the turns per inch figure (from the wire table) by the winding length. Suppose the wire is No. 32, heavy insulation; the wire table gives the turns per inch as 94. The number of turns per layer is therefore $94 \times 2.25 = 211$ turns. The number of layers needed is then the total turns for the winding divided by the turns per layer. Suppose there are 1200 turns in the winding. Then the number of layers is $1200/211 = 5.68$. For the purpose of this calculation, a part of a layer counts as a whole one; therefore the number of layers is six.

Six layers with 0.0013 inch of paper (from the wire table) between each layer means five layers of paper for a total paper thickness of $5 \times 0.0013 = 0.0065$ inch. From the table, No. 32 gauge wire with heavy insulation has a diameter of 0.010 inch. Therefore, six layers of wire have a total thickness of $6 \times 0.010 = 0.060$ inch, making a winding thickness so far of $0.060 + 0.0065 = 0.0665$ inch. A cover of, say, 0.015 inch of paper on top of the winding brings the total winding thickness to $0.0665 + 0.015 = 0.0815$ inch. In a like manner, the secondary winding might work out to be 0.075 inch, including the outside cover, making the thickness for both windings $0.0815 + 0.075 = 0.156$ inch.

There is an inevitable bulge in the winding that must be allowed for. A figure of 15 percent is commonly used, which brings it up 0.023 inch, for a total of 0.179 inch. Now adding the bobbin thickness to this gives the overall space needed inside the window, which is $0.179 + 0.095 = 0.274$ inch. If the window space in the lamination is, say, 0.500 inch, then there is surplus space of 0.226 inch. Generally, this would be considered excessive; either the wire gauges chosen are too small or the lamination is too large. A review of the design will indicate which, and suitable changes will be made. The "hows" and "whys" of this alteration are dealt with in Chapter 8.

In this phase of the design process, it is necessary to calculate the resistance of the windings, the power loss in the copper, and the voltage dropped. The first step of this phase is to determine the length of wire on the winding.

Calculation of Wire Length

Figure 6-6 is a cross-sectional view of a three-winding transformer coil in which W_1, W_2, and W_3 are the thicknesses, respectively, of the primary and two secondaries, and M_1, M_2, and M_3 are the mean lengths of the turns on each winding. The total length of wire on a winding is then $M \times N$, where N is the number of turns on the considered winding.

From inspection of Fig. 6-6, it can be seen that M_1 is equal to the sum of the sides of the bobbin plus the circumference of a circle of radius $W_1/2$ (the circle being made up of the quarter circles at each corner).

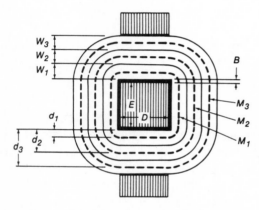

Fig. 6-6. Calculation of mean length of turn, M.

In terms of the core dimensions and the bobbin thickness (including the fit allowance), two sides of the bobbin are $E + 2B$ (each) and two sides are $D + 2B$ (each) for a total around the bobbin of $2(E + 2B) + 2(D + 2B) = 2(E + D + 4B)$. Add to this the circumference of the circle which is $2\pi W_1/2$. Then

$$M_1 = 2(E + D + 4B) + 6.28 W_1/2 \qquad (6\text{-}3)$$

is the mean length of the primary turns.

Similarly, M_2 consists of the distance around the bobbin plus the circumference of a circle of radius $W_1 + W_2/2$, which is $2\pi(W_1 + W_2/2)$. Therefore

$$M_2 = 2(E + D + 4B) + 6.28(W + W_2/2) \qquad (6\text{-}4)$$

is the mean length of the first secondary winding turns.

Also, M_3 consists of the distance around the bobbin plus the circumference of a circle of radius $W_1 + W_2 + W_3/2 = 2\pi(W_1 + W_3/2)$. Therefore, M_3 is

$$M_3 = 2(E + D + 4B) + 6.28(W_1 + W_2 + W_3/2) \qquad (6\text{-}5)$$

for the mean length of the second secondary turns.

The term $2(E + D + 4B)$ is a constant; call it K. Call the distances from the bobbin to each mean turn d_1, d_2, and d_3, as in Fig. 6-6, where $d_1 = W_1/2$, $d_2 = W_1 + W_2/2$, and $d_3 = W_1 + W_3/2$, Thus

$$M_i = K + 6.28d_i \tag{6-6}$$

where i = 1, 2, or 3. The equation is easier to work with in this format.

The lengths of wire on each winding, where l_1, l_2, and l_3 are the lengths and N_1, N_2, and N_3 are the numbers of turns on each winding, respectively, are

$$\begin{aligned} l_1 &= M_1 \times N_1 \\ l_2 &= M_2 \times N_2 \\ l_3 &= M_3 \times N_3 \end{aligned} \tag{6-7}$$

Winding Resistance

Example 3—The length of mean turn is usually determined in inches, or divide by 12 for feet. The wire table gives resistance in ohms per 1000 ft for each wire gauge. Therefore, dividing the table figure by 1000 and then multiplying by M in feet gives the resistance. For example, suppose a winding has 4800 inches of No. 32 wire. The table gives the resistance of this gauge as 162 ohms per 1000 ft. The resistance of the winding is therefore $(4800/12) \times (162/1000) = 64.8$ ohms.

But this gives the resistance at 20°C because the table figure for the resistance is at this temperature. The transformer temperature will increase in operation and so will the resistance. In a conservatively designed power transformer, for example, a resistance increase of the order of 20 percent can be expected. This represents a temperature rise of about 50°C, and although many transformers will go higher than this, it is a good figure as a first approximation.

This approximate ''hot'' temperature figure is easily obtained right away by dividing the winding length in inches by 10 instead of 12 when converting to feet—thus obtaining, as it were, ''hot feet.'' The hot resistance then is 77.76 ohms.

Voltage Drops

The currents in the windings will usually have been determined in the early stages of the design so that voltage drop in the windings can now be calculated from Ohm's law: $V_d = I \times R$. This voltage drop is compensated for by adjusting the number of turns using Equation 4-9 of Chapter 4.

Copper Loss

The copper loss—that is, the power lost in heating up the windings—is directly related to the resistance of the windings and the load current passing through them. In the primary winding, the watts lost are given by

$$V_{Pd} \times I_P \quad \text{or} \quad I_P^2 R_P$$

where V_{Pd} is the voltage dropped in the primary. In the secondary winding the watts loss is

$$V_{Sd} \times I_S \quad \text{or} \quad I_S{}^2 R_S$$

where V_{Sd} is the voltage dropped in the secondary winding. For instance, if R_2 in Fig. 6-7 is 0.1 ohm and the current passing through it is 2.0 amperes, the watts loss in the secondary winding is $I_2{}^2 R_2 = 2.0^2 \times 0.1 = 0.4$ watt.

The total copper loss in the transformer is the sum of the losses in each winding. Thus

$$W_{\text{total}} = I_1{}^2 R_1 + I_2{}^2 R_2 + \ldots + I_n{}^2 R_n$$

for n windings. It is this wattage dissipated in the resistance that is largely responsible for the temperature rise of the windings.

Example 4—Suppose a transformer is to be designed to work from 117 volts and to deliver 20 volts from the secondary (Fig. 6-7). If the turns per volt figure is, say, 6.0, then the turns of the primary and secondary are tentatively $6.0 \times 117 = 702$ and $6.0 \times 20 = 120$, respectively. If our calculations show that the voltage dropped in the primary and secondary windings will be 3.88 volts and 0.2 volt, then compensation can be made by reducing the turns of the primary winding to

$$
\begin{aligned}
N_1 &= \frac{N_2(V_{\text{in}} - V_{d1})}{V_o + V_{d2}} \\
&= \frac{120\,(117 - 3.88)}{20 + 0.2} \\
&= 672 \text{ turns}
\end{aligned}
$$

Or the secondary turns of the winding can be increased to

$$
\begin{aligned}
N_2 &= \frac{N_1(V_o + V_{d2})}{V_{\text{in}} - V_{d1}} \\
&= \frac{702(20 + 0.2)}{117 - 3.88} \\
&= 125.36 \text{ turns}
\end{aligned}
$$

say 125 turns.

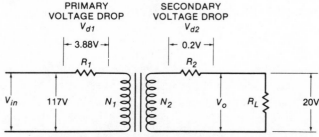

Fig. 6-7. Circuit for copper loss calculation and Example 4.

Another simple approach is to multiply the voltage drop in the primary winding by the turns-per-volt figure. Thus $V_{d1} \times 6.0 = 3.88 \times 6.0 = 23.28$ turns, and deduct this from N_1 to give $702 - 23.28 = 678.72$, say, 679. Then multiply the secondary voltage drop by 6.0. Thus $0.2 \times 6.0 = 1.2$, and *add* this to 120 to give 121.2, say, 121. This approach divides the compensation between both windings instead of applying it all to one or the other.

In any case, the turns ratio for the transformer will work out to be the same after compensation is applied. In case No. 1, $N_1/N_2 = 672/120 = 5.6$; in case No. 2, $N_1/N_2 = 702/126 = 5.6$; and in case No. 3, $N_1/N_2 = 679/121 = 5.6$. The original ratio without compensation is $702/120 = 5.85$.

THE *K* FACTOR

Maximum efficiency and minimum size are achieved in a transformer when the maximum amount of core window space is filled with the copper wire. However, it is obviously not possible to fill the window entirely with copper, because space must be left for the bobbin, insulation, and margins. In addition, there are the inevitable space wastage that exists between adjacent round wires and the wastage due to fit factors and clearances.

The ratio of the total cross-sectional area of copper in the winding to the area of the core window is known as the *K factor*. In equation form, it is stated as

$$K = \frac{w_T}{W} \tag{6-8}$$

where
   w_T is the total cross-sectional area of copper in the windings,
   W is the area of the core window.

The cross-sectional area of the copper in any one winding is given by the cross-sectional area of a single conductor in the winding multiplied by the number of turns in the winding. Thus, for a primary winding,

$$w_P = A_1 N_1 \tag{6-9}$$

where
   w_P is the area of copper in primary winding,
   A_1 is the area of a single strand of primary winding wire,
   N_1 is the number of turns in primary winding.

The combined area of copper on all the windings is

$$w_T = A_1 N_1 + A_2 N_2 + A_3 N_3 + \ldots + A_n N_n$$

where A_1, A_2, A_3, etc., are the areas of the single strands of wires on the windings, and N_1, N_2, N_3, etc., are the numbers of turns on each winding, respectively. The wire area may be expressed in any appropriate unit.

If the window area W is expressed in the same unit as the wire, then

$$K = \frac{A_1N_1 + A_2N_2 + A_3N_3 + \ldots + A_nN_n}{W} \qquad (6\text{-}10)$$

gives the K factor. The K factor would be 1.0 if the entire window area could be filled with copper. Because this is manifestly impossible, the K factor is always less than 1.0.

A convenient form for Equation 6-10 includes conversion factors so that the wire areas may be entered directly in circular mils from the wire table, while the window area is in familiar square inches. Thus

$$K = \frac{(A_1N_1 + A_2N_2 + A_3N_3 + \ldots + A_nN_n)\, 7854 \times 10^{-10}}{W} \qquad (6\text{-}11)$$

where

$A_1, A_2, \ldots, A_n$ are in circular mils,
W is in square inches.

When the core is built with standard type laminations or is a cut core, the value of K tends to follow naturally from the "build" parameters that are always slanted toward obtaining the highest possible value of K consistent with insulation and mechanical requirements. In other words, K is not usually a "selected" figure and might not enter specifically into the build calculations. The K figures for small transformers range up from about 0.25 and improve as the transformers become larger.

That is to say, only 0.25 of the total window space is occupied by copper. The balance, 0.75 of the window space, is taken up with insulation, margins, bobbin thickness, just plain, old, empty space between wires, and so on.

Example 5—This shameful but unavoidable waste of good winding space makes transformers larger and more costly in relation to their ratings. For example, consider Fig. 6-8, which shows the lamination window of a small transformer. The window

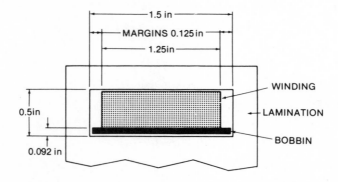

Fig. 6-8. Most of the window space is occupied by insulation and air, with only about 20 to 30 percent available for copper.

measured $1.5 \times 0.5 = 0.75$ square inch in area. A glance at the wire table of Table 6-1 shows that a typical margin would be ⅛ inch; this means that the available window length is reduced by $2 \times 0.125 = 0.25$ inch, making it 1.25 inch. A bobbin made from $\frac{1}{16}$-inch material can occupy a space about 0.092 inch thick, including the fit allowance. This must be deducted from the depth of the window, leaving 0.408 inch. The area available for the windings has now shrunk to 0.51 square inch. Then there is the interlayer insulation and the outside cover. This can easily amount to another 0.05 inch, thus reducing the window depth to $0.408 - 0.050 = 0.358$ inch. The available winding area is now down to $1.25 \times 0.358 = 0.447$ square inch from 0.75 square inch, or $K = 0.6$. Then there is the inevitable spaces between the turns of the wire; the area of a round wire fills only 0.78 of the square of space that it occupies in a layer-wound coil (Fig. 1-2, Chapter 1), thus reducing the effective winding area from 0.447 to $0.447 \times 0.78 = 0.348$ square inch, or $K = 0.46$. Finally, the bulge in the windings and imperfect alignment of the wire can easily reduce the K factor to 0.25 or 0.3.

The moral, then, if you would design small components, is to keep space wastage to a minimum.

In the case of toroids, it is usually necessary to limit K to a preselected maximum value to permit easier winding of the core; this is often on the order of only 0.15 to 0.35. This peculiarity of toroids is dealt with later in this chapter.

I explain shortly, however, that an estimate of the K factor can be valuable as a step in calculating the Wa product. When used for this purpose, it is convenient to express the total copper area in terms of the primary winding turns N_1 and the turns ratios of the primary to secondary windings, rather than in terms of numbers of turns in each winding as was done in Equation 6-10. The conversion is done as follows. If r_1 is the turns ratio of the primary winding to the first secondary, then $N_1/N_2 = r_1$. Similarly, $N_1/N_3 = r_2$, and so on. From these expressions $N_2 = N_1/r_1$ and $N_3 = N_1/r_2$. Substituting these expressions for N_2 and N_3, etc., in Equation 6-10 gives

$$K = \frac{N_1 A_1 + N_1 \dfrac{A_2}{r_1} + N_1 \dfrac{A_3}{r_2} + \ldots + N_1 \dfrac{A_n}{r_{n-1}}}{W}$$

or

$$K = \frac{N_1 \left(A_1 + \dfrac{A_2}{r_1} + \dfrac{A_3}{r_2} + \ldots + \dfrac{A_n}{r_{n-1}} \right)}{W} \tag{6-12}$$

where

$A_1, A_2, \ldots, A_n$ are the cross-sectional areas of the strands of $N_1, N_2, \ldots, N_n$,

W is the window area in the same units as $A_1, \ldots, A_n$.

A rearrangement of the equation puts it into the useful form below. From this, given the number of turns in the primary, the wire sizes in the windings, the turns ratios of the primary to secondaries, and the K factor, the window area required to accommodate the windings may be calculated:

$$W = \frac{N_1 \left(A_1 + \dfrac{A_2}{r_1} + \dfrac{A_3}{r_2} + \ldots + \dfrac{A_n}{r_{n-1}} \right)}{K} \qquad (6\text{-}13)$$

Again, a further variant permits circular mils to be used for wire area and square inches for the window area. So

$$W = \frac{N_1 \left(A_1 + \dfrac{A_2}{r_1} + \dfrac{A_3}{r_2} + \ldots + \dfrac{A_n}{r_{n-1}} \right) 7854 \times 10^{-10}}{K} \qquad (6\text{-}14)$$

where

W is in square inches,

$A_1, \ldots, A_n$ are in circular mils.

MORE ON THE *Wa* PRODUCT

In Chapter 5, the Wa product was shown to be a useful parameter in choosing a core to handle a given power requirement. The Wa product can also be defined in terms of the K factor just explained. This is especially useful in dealing with toroids where the winding limitations require that a K factor be selected at the start of the design. The relationship is achieved by simply combining the basic transformer equation and Equation 6-13.

First, the basic transformer equation suitably arranged gives an expression for a. Thus

$$a = \frac{V \times 10^8}{4FfNB}$$

Second, Equation 6-13 has an expression for W. Obviously by combining these two equations we have an expression for the Wa product:

$$Wa = \frac{V_1 \times 10^8}{4FfN_1B} \times \frac{N_1(A_1 + A_2/r_1 + A_3/r_2)}{K}$$

if only two secondaries are involved. Here N_1 top and bottom cancels out, leaving

$$Wa = \frac{V_1(A_1 + A_2/r_1 + A_3/r_2)10^8}{4FfBK} \qquad (6\text{-}15)$$

for the Wa product.

Right at the start of the design process, a number can be assigned to each symbol on the right-hand side of Equation 6-15. The input voltage and frequency, V_1 and f, are known by specification. The flux density (B) is chosen to suit the type of core material and application. The cross-sectional areas of conductors (A_1, A_2, and A_3) are derived from the wire gauges. The primary-to-secondary turns ratios r_1 and r_2 are defined by the specified voltages, and a tentative number suitable to the type of transformer is easily assigned to K.

In using this equation, as with other equations, care must be taken to apply appropriate conversion factors if necessary. As it stands, the window area W is in whatever units are used to define the wire areas A_1, A_2, and A_3. This is clear by referring to Equation 6-13. The core area a, on the other hand, is defined by the unit used for the flux density B. For example, if the wire areas are stated in circular mils, then W must be in circular mils. If at the same time B is in gauss, then a must be in square centimeters. If this Wa is to be compared to a vendor's Wa product figures for the purpose of selecting a core, then the vendor's figures must be in the same terms in order for the comparison to be valid.

Some vendors, however, give the Wa figure in inch^4, that is to say, W in inch^2 times a in inch^2 gives Wa in inch^4. In this case, the required Wa figure can still be calculated using the convenient units of circular mils for the wire areas and gauss for the flux density, and then converting the equation to inch^4 by multiplying by the conversion factor 1217×10^{-10}. (Multiplying by 7854×10^{-10} converts W in circular mils to square inches, while multiplying by 0.155 converts a from square centimeters to square inches.) The two multiplied together, 7854×10^{-10} times 0.155, gives the conversion factor 1217×10^{-10}. (Refer to conversions in Chapter 1.)

The equation with the conversion factors included and 1.11 entered for the form factor F, for a sine wave input, is

$$Wa = \frac{V(A_1 + A_2/r_1 + A_3/r_2)\,2.741}{KfB} \text{ inch}^4 \qquad (6\text{-}16)$$

where

A_1, A_2, A_3 are the conductor areas in circular mils,
r_1, r_2 are the primary-to-secondary turns ratios,
K is the ratio of copper to window area,
f is the frequency in hertz,
B is the flux density in gauss,
W is the window area in square inches,
a is the core area in square inches.

THE TEMPERATURE RISE

The temperature of transformer windings is a gradient ranging from relatively cool on the outside to a point deep in the heart of the windings known as the "hot spot." Discussion of temperature refers only to this hot spot, for this is where damage to the insulation first occurs if the temperature is too high. The temperature of the outside of a transformer offers no clue as to the degree of temperature inside; it is quite possible for the outside to be cool to the touch while the insulation in the region of the hot spot is being nicely barbequed.

In conventional transformers, the hottest part of the windings is located close to the core. The heat generated in the resistance of the copper by the load current can escape only by flowing through the layers of insulation to the surface of the coil as shown in Fig. 4-15 (Chapter 4). This heat flow creates a temperature drop across the resistance of the insulation. The thicker the insulation, the greater is the temperature drop for a given watts loss.

In general, small transformers are designed to work at total temperatures of no more than 105°C. This, of course, is the hot spot temperature and is considered to be the maximum safe limit for standard insulation materials in long-term operation. This means that the maximum permissible temperature *rise* is determined by the ambient temperature in which the component is expected to work. For example, a transformer working in an ambient temperature of, say, 65°C can safely rise only 40°C, or if the ambient temperature is 45°C, the safe temperature rise of the transformer can be no more than 60°C.

It is quite possible to design safe transformers without calculating the temperature rise if one follows very conservative design practices using rules of thumb in which the safety factors are incorporated. However, the ability to design a specific temperature rise into a transformer not only ensures the safety of the component but gives the designer a high degree of flexibility in terms of achieving optimum designs.

In the following treatment, some basic assumptions are made that are not entirely true in practice. But they considerably simplify the problem without an unreasonably great sacrifice in accuracy. The first assumption is that in conventional transformers, any interchange of heat between the core and the coil is generally small enough to be negligible. This stems from the fact that the core is heated up by its own losses to a point approaching that of the coil and is heavily insulated from the coil by the bobbin and associated air spaces. Similarly, the cooling effects of the sides of the windings are considered as negligible because air trapped in the margins constitutes an effective thermal blanket. The heat then travels outwards to the surface of the coil, and the effective cooling surface is that indicated in Fig. 6-9A as C_s.

The Cooling Surface

Referring to Fig. 6-9B, inspection shows that the length of the perimeter of half of C_s is $E + 2B + \pi R$ (the two corners together make half of a circle of radius R, hence πR). Therefore, the area of half of C_s is $L(E + 2B + \pi R)$, and the total area C_s is

$$C_s = 2L(E + 2B + \pi R) \qquad (6\text{-}17)$$

NOTE: If the window space is filled, as it should be for maximum efficiency, R may be taken as the window depth.

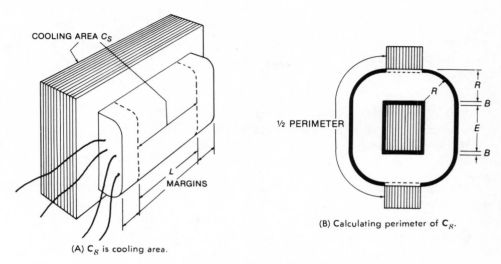

(A) C_S is cooling area.

(B) Calculating perimeter of C_S.

Fig. 6-9. Transformer cooling surfaces.

Temperature Gradients

A temperature gradient exists at the surface of the coil, where the heat is transferred to the air. This is given by

$$T_s = 20H \ \frac{W_T}{C_S} \qquad (6\text{-}18)$$

where

H is the height of the coil in inches and has the meaning illustrated in Fig. 6-10, W_T is the total watts lost in the windings.

The temperature gradient across the thickness of the winding must now be found. It is the sum of the gradients due to the watts lost in each winding and is found as follows.

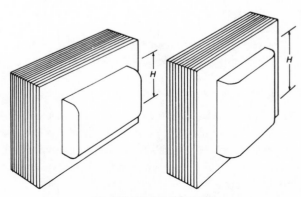

Fig. 6-10. Meaning of factor H.

The temperature gradient across the first winding (usually the primary) is

$$T_1 = 250t_1 \ \frac{W_1}{C_s} \qquad\qquad (6\text{-}19)$$

where
  t_1 is half the thickness of insulation in the primary plus the primary cover, plus all the insulation in the secondaries and the final cover,
  W_1 is the watts loss in the primary winding only,
  C_S is the outside area as already defined.

The same equation is used for the temperature drop across the next winding, but here t_2 is half the thickness of the insulation on that winding, plus all the insulation in the winding above, and W_2 is the watts loss in that winding. For the third winding, t_3 is half the thickness of insulation in the winding, plus the outside cover. If there are other windings above the third one, the same scheme is used to define t—that is to say, half the thickness of insulation on the winding plus all the insulation in the winding above. Note that the thickness of copper in the windings is not considered, because copper, having very low heat resistance, has a negligible temperature drop across it.

The total temperature rise T_T is then the sum of the temperature rises. Thus

$$T_T = T_1 + T_2 + \ \ldots \ + T_S \qquad\qquad (6\text{-}20)$$

This will be clear from the complete examples in later chapters.

Short Method

A shorter method that usually gives fairly good results is the following:

$$T_T = \frac{20W}{C_S} \ (9.4t \ + \ H) \qquad\qquad (6\text{-}21)$$

where
  W is the total copper loss in watts,
  C_s is the cooling surface area in square inches,
  t is half No. 1 winding plus the rest of the insulation,
  H is the height as already defined.

MINIMIZING STRAY CAPACITANCE AND LEAKAGE INDUCTANCE

In Chapter 4, I explained that stray capacitance is chiefly a shunt effect while leakage inductance is a series effect. Because the reactance of the first decreases and the second increases with frequency, their effects are apparent mostly in the higher frequency range. It should be understood, however, that they are *always* there and under some conditions should be considered even at lower frequencies. Here are a few comments that might be helpful.

Stray Capacitance

Capacitance is distributed throughout the windings—between turns, between layers, from windings to ground, and so on. At the usual power-line frequencies and in the low audio range, it is seldom a problem. But an exception to this occurs sometimes in very high ratio step-up transformers. Here the secondary winding might have so many turns that the capacitance is appreciable, even at low frequencies, and large capacitive currents could prove to be an embarrassment. Moreover, while capacitance can by itself constitute a problem, it also has a predilection for teaming up with that other electrical waif, stray (or leakage) inductance, to form resonant or partly resonant conditions that cause nasty humps and sometimes soaring peaks in the response curves of amplifiers.

Most of the stray capacitance occurs between layers and windings in multilayer coils. It can be reduced by dividing the windings in the manner shown in Fig. 6-11, but this generally has the effect of increasing the capacitance to ground and might not be a net improvement. Other more complex arrangements are possible but are rarely used.

Generally, more attention is paid to keeping the leakage inductance small by using other winding arrangements, but usually penalties have to be paid here in increased capacitance.

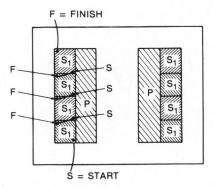

Fig. 6-11. Sectioned secondary reduces capacitance between layers and between windings but could increase it to ground.

Leakage Inductance

This phenomenon, as we have seen, is due to the fact that not all the magnetic flux links the windings. The effect of leakage flux is that of placing small inductances in series with the windings. At ordinary powerline frequencies, the effect of leakage inductance can usually be ignored, but not always. It depends on specific cases.

One method of minimizing leakage is to split and intermingle the primary and secondary windings as shown in Fig. 6-12A. Another variant, shown in Fig. 6-12B, achieves low leakage inductance and good balance in a fairly simple way. The bifilar winding shown in Fig. 6-12C is a frequently used method of obtaining reduced leakage inductance.

In transformers (or inductors) that have a gap in the magnetic circuit, it is desirable to place the windings over the gap as in Fig. 6-13, if this is possible. Some lamination

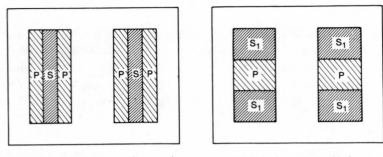

(A) By splitting and intermingling windings. (B) Method giving good balance.

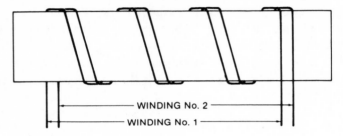

WINDING No. 2

WINDING No. 1

(C) Bifilar winding minimizes leakage inductance.

Fig. 6-12. Minimizing leakage inductance.

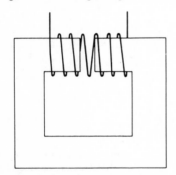

Fig. 6-13. Windings placed over gap minimize leakage inductance.

shapes have the gap placed in the center of the main limb, as in Fig. 5-11, which helps to achieve this. In cut cores, the gap occurs naturally at, or near, the center.

OTHER KINDS OF WINDINGS

Screened Windings

Sometimes it is an advantage to place an electrostatic screen between the primary and the secondary windings of power transformers. There are several reasons why this is done. One is to prevent relatively high frequency interference, carried on the power line, from entering the secondary winding circuit via capacitive coupling between the windings. Or, a screen may be used simply as a safety device to minimize the danger of primary-to-secondary short circuits. In some kinds of windings, a screen helps to reduce the capacitance between the primary and secondary windings by redistributing it from primary to secondary to primary and secondary to ground, the screen being grounded in most cases.

The screen can take one of several forms. A common type is a thin copper foil wrapped around the winding as in Fig. 6-14. Care must be taken to insulate the overlap. If this is not done, the screen becomes equivalent to a short-circuited, single-turn, secondary winding. The massive current that results can quickly burn out the transformer. A lead is usually soldered to the foil and brought out of the winding to ground.

Another common and simple screen is made by winding on one layer of insulated copper wire. One end of the coil is left open; the other end is brought out of the winding for connection, usually to ground. The gauge of wire is not critical and will usually be rather small. This type is shown in Fig. 6-15.

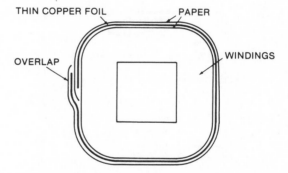

Fig. 6-14. Thin copper foil with insulated overlap makes electrostatic screen.

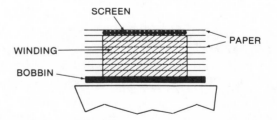

Fig. 6-15. Screen composed of a single-layer winding of thin wire.

Balanced Windings

In some applications, notably in push-pull amplifiers, it is an advantage to have the windings balanced with regard to resistance, capacitance, and numbers of turns, about a center tap as in Fig. 6-16. A simple but effective balance format is shown in Fig. 6-17. Note that the secondary is actually two parallel windings. This is also an example of a departure from the standard secondary-on-top-of-primary arrangement.

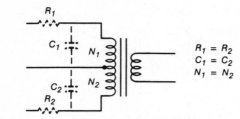

Fig. 6-16. Elements R_1, R_2, C_1, **and** C_2 **are the winding resistances and capacitances balanced about a center tap.**

$R_1 = R_2$
$C_1 = C_2$
$N_1 = N_2$

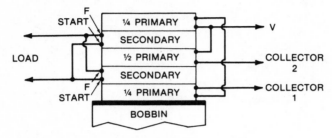

(A) Winding plan and connections.

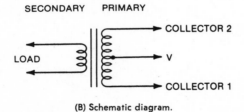

(B) Schematic diagram.

Fig. 6-17. Minimizing leakage inductance in a Class-B push-pull amplifier output transformer.

The bifilar winding mentioned earlier also achieves near-perfect balance about a center tap. The principle is shown in Fig. 6-18. Twin wires are wound simultaneously into the coil. In this way, both sides of the coil are precisely the same length and so have identical resistance and capacitance. The leads must be connected in the correct relationship; if they are not, the windings will oppose each other and reduce or cancel the inductance. If the center tap is opened, *two* balanced centertapped windings result.

Finally, returning to the "core" type of transformer, a winding format known as the *astatic winding* is sometimes used to advantage. Here the primary and secondary windings are divided with a half-primary and half-secondary on each leg as in Fig. 6-19. While this sacrifices one of the principal advantages of the core type of transformer—namely separation of the two windings in high-voltage situations—it improves leakage inductance and provides a valuable feature unobtainable in the shell type configuration.

It can also neutralize the effect of an interfering field. Referring again to Fig. 6-19, observe that although each half of a winding is in correct polarity to the half on the other

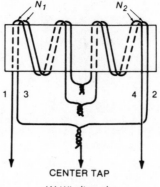

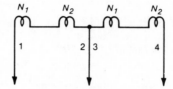

(A) Winding plan.

(B) Schematic diagram.

Fig. 6-18. Bifilar winding achieves low leakage inductance and excellent balance about a center tap.

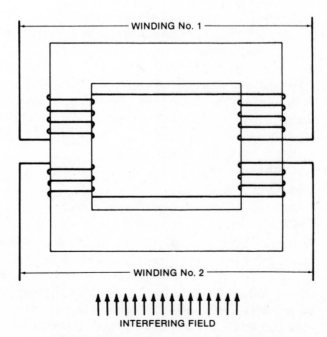

Fig. 6-19. Astatic winding in core type configuration.

leg as far as driving flux around the core is concerned—in other words they assist each other—this is not true for voltage induced in the windings by an external field cutting them in the direction shown in the figure. In this case, the direction of the windings is such that the induced interfering voltages in each half of the coil are in opposition and cancel out.

WINDINGS ON TAPE-WOUND CORES (TOROIDS)

The advantages of tape-wound cores were discussed in Chapter 5. There are also advantages in toroidal *windings*. The word "toroidal" is used here to describe the special shape of the toroid rather than just any old winding on a toroidal core, because not all windings on toroidal cores are toroidal in shape. If a winding occupies only a short length of the available 360-degree sweep of the core (and this is sometimes the case), it is not a toroidal winding. To be toroidal, the winding must occupy the entire length, the wire being spaced to achieve this if necessary.

A prime advantage of the toroidal winding is that it has a very small external field. The field is concentrated mainly inside the toroid. A second important feature is that leakage inductance is minimized because it is difficult for the flux lines to escape from the core without linking the windings. And third, the effects of external interfering fields, from whatever direction, tend to cancel in the windings in a manner similar to that in the astatic winding. These points are illustrated in Figs. 5-19A, 5-19B, and 5-19C in Chapter 5.

Considered as a whole, then, the toroid is altogether a handy configuration. As always, though, there are serpents in Eden. Because of its shape and the fact that the wire must be wound directly on to the core rather than on to a bobbin which is assembled later to the core, toroids present some problems.

It should be accepted right at the outset that to wind a large number of turns on to a toroidal core by hand is so incredibly tedious a task that few people in their right minds would attempt it; and of those who would, only people of immense strength of character or pronounced masochistic tendencies would complete the task. Fortunately, there are many applications for toroids that require relatively few turns, so they are always worth consideration by the experimenter when high performance is a requirement. in commercial practice where machines designed for the purpose are available, the problem of numbers of turns is not so acute.

Layers and Turns per Layer

A glance at the toroid in Fig. 6-20 reveals the following peculiarities. Unlike the standard coil, every layer is a different length; therefore for a given winding, every layer has a different number of turns. The length of the first layer is the length of the inside circumference, and the number of turns on it is determined by this length times the turns per inch figure for the wire gauge being used. The winding length available for the second layer is less than the first, and the third layer is shorter than the second, and so

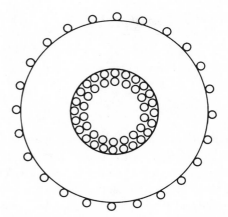

Fig. 6-20. First layer has wide spacing on outside of toroid (second layer not shown on outside).

on in an ever-diminishing pattern until eventually only a few turns are possible on the layer and the center hole of the core is nearly filled with wire.

Due to the magic of circles and their proportions, the number of turns on successive layers of a given winding diminishes at the rate of six turns per layer regardless of the wire gauge and provided that each layer is directly on top of the previous one without intervening material. When the number of turns has been established for the first layer, the others can be determined by simply deducting six turns from each succeeding layer.

For example, suppose that a core has an inner diameter of 2.13 inches. The winding length (internal circumference) is then 3.14 × 2.13 = 6.688 inches. Suppose the wire is No. 20 gauge. From the wire table (Table 6-1), the turns-per-inch figure for heavy insulation is 26, making a total of 26 × 6.688 = 173.9, say 174, turns on the first layer. The second layer will then be 168 turns, the third will have 162 turns, and the fourth 156 turns, and so on.

The usual case is that the number of turns on the winding is known, and you want to know how many layers are required. The easiest way of doing this is to simply deduct six turns layer by layer, totaling the turns as you go until the number of layers required is established. For example, if the number of turns on a winding is 250 and if 60 turns fill the first layer, then successive layers are 60 + 54 + 48 + 42 + 36 = 240 turns, leaving 10 turns to be wound as a partial sixth layer.

This is, of course, an arithmetic series. If many layers are involved it can be solved by the general equation

$$n = \frac{(a+3) - \sqrt{(a+3)^2 - 12N_T}}{6} \tag{6-22}$$

where

a is the number of turns in first layer,
n is the number of layers,
N_T is the total number of turns.

Using the same example as before where $a=60$ and $N_T=250$:

$$n = \frac{(60+3) \; - \; \sqrt{(60+3)^2 - 12(250)}}{6} = 5.3$$

say 6.0 layers.

Length of Wire

In toroids, as in other formats, the length-of-wire figure is needed to determine resistance, voltage drop, and power loss. It is also needed for obvious reason of knowing how long the wire is, a matter of some importance when winding toroids. Because in small transformers it is likely to be impossible to pass, say, a half-pound spool of wire through the window of the toroid in order to wind it, it is necessary to cut off the required length before starting to wind. If the winding is being done manually, the wire may be rewound on to a small spool, or, if it isn't too long, simply retained in a hank—but handled very carefully.

Manufacturer's design data sometimes gives wire length for each core in terms of minimum inches per turn for the first layer, and maximum inches per turn for the last layer. If the number of turns is known, an estimate can then be made of the total length. However, the length can be calculated in exactly the same way as was done earlier for conventional coils by figuring the mean length of turn (Fig. 6-21) and multiplying by the number of turns.

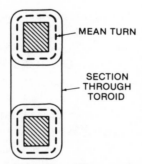

MEAN TURN

SECTION THROUGH TOROID

Fig. 6-21. Illustration of mean length of turn on a toroid.

Random Notes

At ordinary voltages, it is possible to get away with no interlayer insulation by using *heavy* insulation; see the wire table (Table 6-1). However, to use a phrase popularized by the undertaking industry, "peace of mind" can be obtained by winding thin paper cut into narrow strips over each layer if voltage conditions seem to justify it. Adhesive tape of the plastic sort used by electricians can stretch a little and is excellent as an overall cover for windings.

If a core is not of the plastic encased type, it will be necessary to protect the windings by wrapping tape around the metal core. Remember to take account of this addition when making calculations on build and winding space.

Don't forget also to allow for the inevitable bulge in the windings by choosing a slightly smaller K factor or adding, say, 10 percent to the calculated winding area. There is no reason why one should not work in terms of build thickness as was done earlier for standard windings if one wishes to do so or in square inches instead of circular mils. The only rule is to use whatever seems most comfortable to you.

Concerning build, a unique characteristic of toroidal windings is that the turns can be laid precisely side by side on the inside circumference but not on the outside. On the outside, there will always be spaces between the turns of a given layer (see Fig. 6-20). This means that the turns of the next layer tend to fall in between those of the one below. This makes sometimes for an untidy situation and possibly higher-than-normal voltage between turns.

It might not be possible to achieve a true layer-wound coil except with heavier-gauge wire. Generally speaking, it is best to first lay on the winding having the heaviest gauge wire, and then follow with the other windings in descending order of wire gauge diameter (which is ascending order of wire gauge number).

Toroidal *K* Factor

The need to design for a specific K factor has been emphasized, but what, in fact, should this be in a given case? It depends on the winding method—possibly a larger K factor will be possible for a hand-wound coil than for a machine-wound product because the experimenter's manual dexterity could be superior to that of the machine. In fact, it is possible to wind a toroid in the manner used for threading a needle by pushing the wire point first through the hole. If this can be done, then certainly the winding space can be better utilized.

To illustrate what a K factor of, say, 0.2 looks like, consider Fig. 6-22. The core illustrated has an inner diameter (ID) of 2.0 inches, which is 2000 mils and therefore its area is $2000^2 = 4 \times 10^6$ circular mils. A K factor of 0.2 means a winding area of $0.2 \times 4 \times 10^6 = 800,000$ circular mils. The unfilled portion of the window area is $0.8 \times 4 \times 10^6 = 3,200,000$ circular mils, or a circle of diameter equal to $\sqrt{3,200,000} = 1788.85$ mils. This is 1.788 inches. A K factor of 0.4 means that $0.6 \times 4 \times 10^6$ circular mils of window area is unfilled; in diameter, this is 1.55 inches, indicated by the dotted line in Fig. 6-22.

Build

The term "build" is defined in the same way as in standard cores and can be figured in exactly the same way. However, the maximum permissible build is set by the K factor. Assuming that a K factor has been selected, a convenient method of comparing a tentative winding design with the window area assigned to it is as follows.

Calculate the cross-sectional area of the winding by multiplying the area of a single strand of wire by the number of turns on the winding. If the wire area is in circular mils, the total area will then be in the same unit. Thus

$$w_1 = N_1 A_1$$

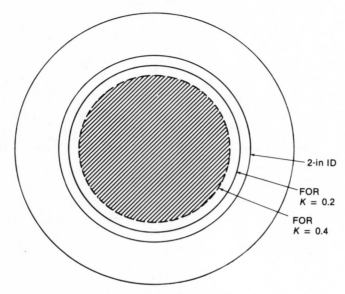

2-in ID

FOR
K = 0.2

FOR
K = 0.4

Fig. 6-22. Relative K factors to scale.

where

w_1 is the area of first winding in circular mils,

N_1 is the number of turns on the first winding,

A_1 is the area of single strand of wire on first winding in circular mils.

The area of second winding can be calculated similarly. The total winding area is then

$$w_T = w_1 + w_2 + \ldots + w_n$$

for n windings.

The K factor that the proposed design will yield is then

$$K = \frac{w_T}{W}$$

where W is the window area $(ID)^2$ in circular mils. Comparison of this figure with the K factor originally selected tells the story.

Wire Table

The wire table in Table 6-2 is constructed specifically for use with toroids. It lists only heavy-insulation wire, which is commonly used on toroids because interlayer insulation is not convenient to apply. The casual experimenter might have problems procuring this type of wire in small quantities. However, the single-insulation wire can be used with very thin paper that is wound in narrow strips between layers. An exacting job to be sure, but the experimenter generally winds by hand and usually has a little more time to spare for the job than does the professional.

Table 6-2. Wire Table Designed for Toroids

Wire Size AWG	Diameter With Heavy Insulation (Inches)	Area Circular Mils Nominal	Resistance Ohms Per 1000 Ft.	Weight Pounds Per 1000 Ft.	Layer Winding Turns Per Inch	Random Winding Turns Per Inch^2	Machine Minimum Wound I.D. (Inches)	Maximum Turns For Case I.D. of 0.5 In.	1.0 In.
8	.132	16510	.628	50.4	6	42	2.000	—	—
9	.118	13090	.793	40.0	7	57	1.750	—	—
10	.106	10380	.999	31.7	8	75	1.500	—	—
11	.094	8230	1.26	25.2	9	95	1.250	—	—
12	.084	6530	1.59	20.1	11	130	1.000	—	—
13	.075	5190	2.00	15.9	12	159	.875	—	—
14	.067	4110	2.52	12.6	13	193	.875	—	—
15	.060	3260	3.18	10.0	15	248	.875	—	—
16	.054	2580	4.02	7.95	17	316	.750	—	120
17	.048	2050	5.05	6.32	19	394	.750	—	180
18	.043	1620	6.39	5.02	21	487	.750	—	260
19	.039	1290	8.05	3.99	23	596	.500	60	360
20	.035	1020	10.13	3.16	26	792	.500	80	450
21	.031	812	12.77	2.51	29	982	.500	90	560
22	.028	640	16.20	1.99	32	1210	.438	120	680
23	.025	510	20.30	1.59	36	1260	.438	150	850
24	.022	404	25.67	1.26	40	1550	.313	180	1040
25	.020	320	32.37	1.01	45	1940	.313	250	1310
26	.018	253	41.02	.799	50	2700	.300	310	1560
27	.016	202	51.44	.634	55	3550	.300	370	1870
28	.014	159	65.31	.504	62	4180	.300	470	2500
29	.013	128	81.21	.401	68	5160	.300	620	3250
30	.012	100.0	103.7	.318	77	6560	.250	750	4000
31	.011	79.2	130.9	.254	85	8090	.250	920	5050
32	.010	64.0	162.0	.202	94	10000	.250	1250	6870
33	.009	50.4	205.7	.161	105	12500	.250	1510	8740
34	.008	39.7	261.3	.127	119	16250	.218	1920	10620
35	.007	31.4	330.7	.101	133	20600	.218	2440	13120
36	.0060	25.0	414.8	.0803	145	25000	.218	2930	16250
37	.0055	20.2	512.1	.0641	161	30900	.218	3500	19370
38	.0049	16.0	648.2	.0509	181	39300	.218	4300	23750
39	.0043	12.2	846.6	.0403	205	51500	.218	5300	30000
40	.0038	9.61	1079	.0319	226	72000	.218	7450	42500
41	.0034	7.84	1323	.0252	250	89800	.218	9950	58120
42	.0030	6.25	1659	.0199	283	116500	.218	12600	72500
43	.0027	4.84	2143	.0159	315	143000	.187	14900	85000
44	.0025	4.00	2593	.0127	340	168500	.187	17400	100000

Courtesy Magnetic Metals Corp.

SEVEN

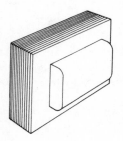

Summarized
Data and General
Design Considerations

Many people are not comfortable with formulas and might be a little dismayed that it seems necessary to use so many in the task of wrapping a few turns of wire around a hunk of iron. This chapter attempts to ease the pain with a summary of essential formulas and other design considerations.

SUMMARY OF FORMULAS

The truth must be told: Much of Table 7-1 is not essential. Using only one or two of the simplest equations one can obtain excellent results. In fact, some design examples are given in the next chapter just to demonstrate this point.

The collection of equations given in Table 7-1 can be very handy for those who want to use them, however. The collection should be viewed as a kind of designer's tool kit— mathematical power tools devised to reduce the time taken and allow one to accomplish sometimes tricky parts of the job with more accuracy and less effort. An important point to keep in mind is that only a few equations are needed for any given design.

In the design examples given in Chapter 8 and those given in the following, I have set out to demonstrate the use of as many of the formulas as possible. For this reason, the reader should expect to find variations in the methods used from one example and another. The original discussions on these formulas can be easily located for reference by the chapter numbers given in the first column of Table 7-1.

Table 7-1. Summary of Formulas

Chapter 1

$$V = IR \qquad R = \frac{V}{I} \qquad I = \frac{V}{R}$$

$$P = VI \qquad V = \frac{P}{I} \qquad I = \frac{P}{V}$$

$$P = I^2R \qquad I = \sqrt{\frac{P}{R}} \qquad R = \frac{P}{I^2}$$

$$P = \frac{V^2}{R} \qquad V = \sqrt{PR} \qquad R = \frac{V^2}{P}$$

Chapter 2

$$\Phi = aB$$

where a = core cross-sectional area in square centimeters, B = flux density in gauss, and Φ = total flux in maxwells.

$$\Phi = 6.45aB$$

where a = core cross-sectional area in square inches, B = flux density in gauss, and Φ = total flux in maxwells.

$$\Phi = \frac{V \times 10^8}{4FfN}$$

where F = form factor, f = frequency in hertz, N = number of turns, V = volts, and Φ = total flux in maxwells.

Chapter 3

$$\frac{V_1}{V_2} = \frac{N_1}{N_2} \qquad \frac{I_2}{I_1} = \frac{N_1}{N_2} \qquad \frac{V_1}{V_2} = \frac{I_2}{I_1}$$

$$\frac{N_1}{N_2} = \sqrt{\frac{R_L{}'}{R_L}} \qquad R_L{}' = R_L\left(\frac{N_1}{N_2}\right)^2$$

where N_1 = number of turns in primary, N_2 = number of turns in secondary, R_L = secondary load resistance, $R_L{}'$ = load resistance reflected to primary.

$$V = 25.8FfaNB \times 10^{-8} \qquad\qquad N = \frac{V \times 10^8}{25.8FfaB}$$

$$\frac{N}{V} = \frac{10^8}{25.8FfaB}$$

$$B = \frac{V \times 10^8}{25.8FfaN} \qquad\qquad P = \frac{fBWa}{17.26S}$$

$$a = \frac{V \times 10^8}{25.8FfNB} \qquad\qquad Wa = \frac{17.26SP}{fB}$$

$$D = \frac{17.26PS}{WEfB}$$

where V = volts, F = form factor, f = frequency in hertz, a = core cross-sectional area in square inches, N = number of turns, B = flux density in gauss, P = power in watts (or volt-amperes), W = core window in square inches, S = current density in circular mils per ampere, D = core stack in inches, E = core width in inches.

Table 7-1. Cont.

Chapter 4

With reference to Fig. 4-7:

$$I_P = I_1 + I_{Ro} \qquad V_1 = V_{in} - V_{d1}$$
$$V_1 = V_{in} - I_P R_1 \qquad V_2 = V_{d2} + V_o$$

With reference to Fig. 4-9:

$$V_1 = \sqrt{V_{in}^2 - (I_X R_1)^2} - I_1 R_1$$

With reference to Fig. 4-10:

$$V_1 = \sqrt{V_{in}^2 - (I_X R_1)^2} - R_1(I_1 + I_{Ro})$$

When iron-loss current is negligible:

$$I_P = \sqrt{I_X^2 + I_1^2}$$

With iron-loss current:

$$I_P = \sqrt{I_X^2 + (I_1 + I_{Ro})^2}$$

where V_{in} = supply volts, $V_1 = V_{in} - V_{d1}$, I_P = total primary current, I_X = magnetizing current, I_1 = primary load current, V_{d1} = primary voltage drop, R_1 = primary resistance, V_o = secondary output voltage, V_{d2} = secondary voltage drop, $V_2 = V_o + V_{d2}$.

$$\frac{N_1}{N_2} = \frac{V_{in} - V_{d1}}{V_o + V_{d2}}$$

$$N_1 = \frac{N_2(V_{in} - V_{d1})}{V_o + V_{d2}}$$

$$N_2 = \frac{N_1(V_o + V_{d2})}{V_{in} - V_{d1}}$$

$$V_o = \frac{N_2(V_{in} - V_{d1})}{N_1} - V_{d2}$$

where V_{in} = input volts, V_o = output volts, V_{d1} = primary voltage drop, V_{d2} = secondary voltage drop, $V_1 = V_{in} - V_{d1}$, $V_2 = V_o + V_{d2}$, N_1 = number of turns in primary, N_2 = number of turns in secondary.

$$R_T = R_2 + R_1 \left(\frac{N_2}{N_1}\right)^2$$

$$R_T = R_2 + R_1 \left(\frac{V_2(\text{unloaded})}{V_{in}}\right)^2$$

where R_T = transformer resistance seen by load looking into the secondary, R_2 = secondary resistance, R_1 = primary resistance, N_2 = number of turns in secondary, N_1 = number of turns in primary, V (unloaded) = open-circuit secondary voltage, V_{in} = input voltage.

Table 7-1. Cont.

Chapter 4

$$\eta = \frac{P_o}{P_{in}} \qquad \eta = 1.0 - \frac{P_L}{P_{in}}$$

$$\frac{V_o}{V_{in}} = \sqrt{\eta}\,\frac{N_2}{N_1}$$

$$V_2 = \frac{V_o}{\sqrt{\eta}}$$

$$\% \text{ Regulation} = 100\left(\frac{1}{\sqrt{\eta}} - 1\right)$$

where η = efficiency, P_o = power output, P_{in} = power input, P_L = power loss, V_o = secondary output voltage, V_{in} = primary input voltage, N_2 = number of secondary turns, N_1 = number of primary turns, V_2 = secondary no-load volts.

Chapter 5

$$I_M = \frac{\text{Wt} \times \text{VA/lb}}{V_1} + \frac{1.43Bgs}{N_1}$$

where I_M = magnetizing current in amperes, Wt = weight of core in pounds, VA/lb = apparent loss in volt-amperes per pound, V_1 = primary voltage, B = flux density in gauss, g = gap width in inches, s = stacking factor for laminations, N_1 = number of primary turns.

Chapter 6

$$K = \frac{w_T}{W}$$

$$K = \frac{(N_1A_1 + N_2A_2 + N_3A_3 + \ldots + N_nA_n)7854 \times 10^{-10}}{W}$$

$$K = \frac{N_1\left(A_1 + \dfrac{A_2}{r_1} + \dfrac{A_3}{r_2} + \ldots + \dfrac{A_n}{r_{n-1}}\right)7854 \times 10^{-10}}{W}$$

$$Wa = \frac{V_1\left(A_1 + \dfrac{A_2}{r_1} + \dfrac{A_3}{r_2} + \ldots + \dfrac{A_n}{r_{n-1}}\right)2.741}{KfB}$$

where K = ratio of window area to copper area, w_T = total copper area, W = window area, N_1 = number of primary turns, r_1 = turns ratio N_1/N_2, r_2 = turns ratio N_1/N_3, A_1 = area of N_1 wire (single strand), A_2 = area of N_2 wire (single strand), A_3 = area of N_3 wire (single strand), f = frequency in hertz, B = flux density in gauss, V_1 = primary voltage.

THE SPECIFICATION_____

The first step in the process is to clearly define the aims of the design. Any vagueness here will result in a vague component with a vague and unsatisfactory performance. The aim is not necessarily to achieve the highest possible performance but rather an adequate performance. For example, a cheap electric bell transformer might be inefficient and have poor regulation yet still be totally satisfactory because it is adequate for its purpose and meets one of the basic aims—in this case, low cost. All designs are a compromise between conflicting requirements—a choice between cost and performance or a tradeoff between performance in one area and performance in another.

The designer, then, must first know what he or she wants. To establish this you ask and answer a series of questions. For instance, for a power transformer, what will be the input voltage and frequency? What outputs are required in terms of volts and amperes? Is there a specific need for good regulation? Of what order? Is size or weight a critical factor? Will there be unusual ambient temperature conditions—extremes of heat or cold? What order of efficiency should be aimed for? How will external circuitry affect the insulation requirements or ratings of the windings? What sort of use will it have—continuous or intermittent, and for what periods? How important is cost? And so on.

Points of View

Of course, casual experimenters and professional circuit designers are apt to view specifications in rather different lights. Even within each group, opinions on what constitutes an adequate specification are likely to be conflicting.

For instance, the engineer designing a transformer for a mass-produced TV or radio will probably not work to such stringent specifications as one designing a component for a space vehicle destined to travel to Jupiter. But it is certain that whatever the guidelines, they will have been fully defined with regard to size, weight, performance, and cost. The need to cope with the vagaries of manufacturing will also be foremost in the engineers' minds, and to get everything just right they will be prepared to bench-test prototype designs and to redesign and rebuild the component until it meets the specification precisely.

On the other hand, the hobbyist, experimenter, amateur, or whoever designs for pleasure is independent. He or she is not usually tied to format, cost, or performance and often cannot afford to optimize the design on the work bench. He or she wants it to be right the first time and is usually prepared to yield a point here or there if it isn't. If wise, his or her initial specification will not be too tight, safety allowances will be generous, and the designer will underrun materials rather than teeter on a thin line between squeezing the maximum out of the component and disaster.

It is not possible here to deal with all the things that go into a specification under all conditions, but there are a number of criteria that are common to most situations. Some of the most important of these are given in the following.

MOUNTING POSITIONS

Cooling Considerations

How will the transformer be mounted and where will it be located? Depending on the shape and size, the choice of the mounting positions in Figs. 7-1A and 7-1B could

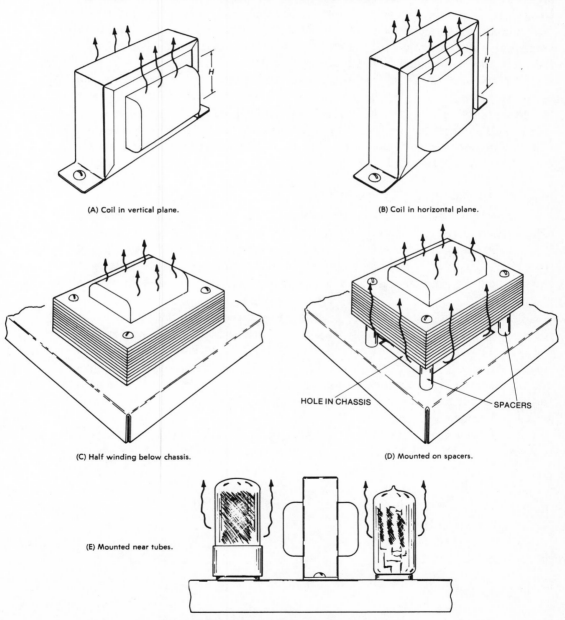

(A) Coil in vertical plane.

(B) Coil in horizontal plane.

(C) Half winding below chassis.

(D) Mounted on spacers.

HOLE IN CHASSIS

SPACERS

(E) Mounted near tubes.

Fig. 7-1. Transformer mounting affects its cooling.

have an effect on temperature rise, although with the shell shapes generally employed, the difference will not be great. If there is a choice, though, it is usually best to have the longest dimension of the *coil* placed horizontally. But what about the position in Fig. 7-1C? Here half of the winding surface is below a chassis and therefore contributes much less to the cooling of the component. Moreover, there could be heat-producing components below the chassis—resistors, for example. All the ¼-, ½-, and 1-watt dissipations add up to a fair amount of heat; not only has half of the cooling surface been neutralized, but it could actually be absorbing heat. The position in Fig. 7-1D, in which the transformer is mounted on spacers, improves this situation.

Then consider Fig. 7-1E. On either side of the transformer are rectifier vacuum tubes (there are still a few of these being used). This transformer position is similar to a slice of bread in a toaster and has the same effect. If this position must be used, then make allowances and design for a low temperature rise. Incidentally, the effects of heat might not be immediately apparent, but in a few months or a year the delicately crisped insulation could fail.

Minimizing Magnetic Coupling

A transformer has a magnetic field that intrudes into the space around it, sometimes for a considerable distance. For example, a power transformer in the vicinity of a cathode-ray tube can upset the operation of the tube. Or, it might interact with other types of components. There are a number of after-the-fact solutions to such problems, but it is advisable to think about them as part of the transformer design. The following will help you do this.

If it is important in a specific case to reduce magnetic coupling, the stray field can be reduced by reducing the flux density to, say, half of that depicted by the normal criteria. This, of course, means that the turns-per-volt figure must be modified. In a very critical application, one might consider the use of a toroid, which tends to have a small external field. (If the problem is encountered after the transformer has been built, keep in mind that unnecessarily large gaps in the core will increase the stray field, so the core assembly should be carefully checked.)

The way in which a transformer is placed relative to another iron-core component is important. In Fig. 7-2A, the positions of the windings encourage inductive coupling.

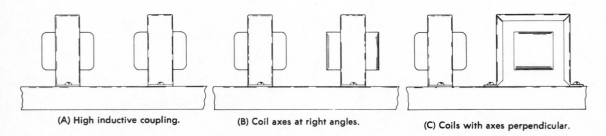

(A) High inductive coupling.   (B) Coil axes at right angles.   (C) Coils with axes perpendicular.

Fig. 7-2. Two transformers in close proximity.

Placing the axes of the coils at right angles to each other, as in Figs. 7-2B and 7-2C, is better but might not solve the problem entirely. If close proximity is unavoidable, it might be necessary to resort to one of the previous solutions in the design stage.

Keep in mind, too, that it is possible to shield a transformer from other components by enclosing the transformer, enclosing the other component, or both, in separate containers made with a high-permeability metal, such as SuperPerm 80 or Mumetal. This enclosure offers a low-reluctance path to the leakage flux and diverts it from the area to be screened. Shielding of this type is frequently used on cathode-ray tubes and to protect audio transformers from the effects of nearby power transformers.

VOLT-AMPERE RATINGS

The design examples given in Chapter 8 assume that the loads are of a purely resistive nature. That is, they have a power factor of 1.0. Some types of load, however, are reactive and have power factors less than 1.0. Such loads require transformers of larger volt-ampere capacity than that indicated by the load expressed in watts. Motors, for example, have power factors less than 1.0.

Other types of transformers require a greater volt-ampere capacity, because the currents in the windings are not sinusoidal and develop greater heating in the copper. Transformers with square-wave inputs, such as those used with vibrators or their electronic equivalents, come into this category, as do transformers which feed rectifiers. While the precise analysis of these effects can be complex, it is usually good enough to simply increase the volt-ampere rating according to rule of thumb. This is discussed later.

Time is also a factor in assessing ratings, and intermittent operation usually means that a transformer can carry a greater load than its volt-ampere rating indicates. Or, to put it another way, a transformer on intermittent load can be designed as though for a smaller volt-ampere rating. The extreme example of this is the domestic bell transformer, which might operate for only a few seconds in several hours. Less extreme examples are spot-welding transformers and certain kinds of radio transmitter operation.

A detailed analysis of the many conditions that affect ratings is outside the scope of this book, but the following rules of thumb are useful practical guides. The power capability, in watts, of a transformer is less than its volt-ampere rating in proportion to the amount that the power factor of the load is less than 1.0. The power factor is frequently marked on some kinds of loads; for instance, it is frequently specified on motors. In many cases, though, it is not known and if one does not possess the equipment to measure it—which is generally the case. The manufacturer of the device might help, and failing that, then "gee and by gosh" judgment is called for. In other words, you suspect that the load is substantially reactive, then add a guestimated percentage to the volt-ampere rating. Generally up to 20 percent is ample. A transformer rated at 300 volt-amperes can handle only 270 watts if the power factor is 0.9, or 240 watts if it is 0.8.

In intermittent operation where a transformer might have on-off periods of a few minutes each over several hours as is encountered in some kinds of applications, it is generally reasonable to add, say, 20 percent to the volt-ampere rating. But in an operation similar to that of the bell transformer, provision must be made for the fact that a faulty bell push could cause the circuit to operate continuously when no one is at home. It is better to be safe than have a hot transformer set the house on fire; so in this and comparable situations, it is best to design for continuous operation.

Rectifier transformers are specifically dealt with in Chapter 9.

SPECIFICATIONS FOR PUBLISHED PROJECTS

When a transformer is identified in a published project, the vendor's name and part number are usually given and quite likely also the transformer rating—the input and output voltages and load current. The experimenter must now put together his or her own specification using his basic knowledge of transformers and whatever information can be gleaned from the text with regard to the project's needs. But care is necessary in interpreting this data.

For example, a transformer for a project is specified by manufacturer's name and part number, and rated at 117-volt, 60-hertz input and 6.3-volt, 25-ampere output. This looks clear enough. However, although it is not specifically stated, the project actually takes only 5.0 amperes and the transformer delivers 8.5 volts at this current.

But if you design a transformer to deliver 6.3 volts at 25 amperes, you might obtain, say, 6.9 volts at 5.0 amperes instead of the 8.5 volts needed simply because you did a great design job in achieving high efficiency and small internal voltage drops. You should have designed for 8.5 volts at 5.0 amperes.

Why would an overrated transformer be chosen in the first place? There are lots of possible reasons—you happened to have one on hand; you got it in a surplus store; a friend gave it to you, and so on. The point is that before designing for a project, try to determine the real requirements of the circuit and design for those rather than the stated full-load rating.

Having decided what is really needed, the designer should now put together a total specification, and in so doing will be wise to lean towards the generous side so far as ratings are concerned. A point to remember in the design of a low-voltage transformer is that small adjustments of the secondary winding are usually relatively easy because it contains relatively few turns and they are on the outside. So do not gunk up the windings with varnish—at least not until the project has been checked.

THE DESIGN SHEET

Consider the function and form of the useful design tool shown in Fig. 7-3. Its purpose is to log the required string of small calculations in a logical sequence from beginning to end. It creates orderliness. It shows the designer where he or she is going and, just as important, where he or she has been.

At the top of the sheet, are spaces for recording the preliminary data: the numbers that are known or will be estimated or quickly calculated—input volts, hertz, efficiency, and so on. The core box appears under this. And below this again are the spaces in

TRANSFORMER DESIGN SHEET

Input volts		Hertz		Est. Efficiency		Turns/volt	
Lamination or core No.						B	
Window dimensions	G		F		W	M_P	
C.S.A. dimensions	E		D		a	v	
	Wt		Watts/lb		Iron loss		

Windings											
Coil											
Volts											
Amperes											
Turns											
Gauge											
Turns/inch											
Margins											
Winding length (L)											
Turns/layer											
Number of layers											

Build											Total
Copper											
Paper											
Cover											
Total											
Bulge Percent											
Total (R)											

Bobbin ____ clearance ____   Total bobbin (B)
Total depth

Losses											
Length mean turn (M)											
Total wire length (inches)											
Ohms/ft											
Resistance (hot)											
Voltage drop											
Copper loss (I^2R)											

Iron loss
Total loss
Efficiency

Temperature Rise

Coil	t	C_S	W	W/C_S	°C T_o
1					
2					
3					
4					

Temp. Rise $= T_o(\text{Total}) + 20\,HW/C_S =$

Fig. 7-3. Design sheet blank.

which to describe the windings and their voltages and currents and also the order in which they will be wound (the primary is not always first).

By the time the current line is entered, you have all the preliminary information needed to begin putting the design together. Tentative selections can now be made for the wire gauges and the gauge numbers entered. This clears the way for the calculation of the Wa product (if this is the design approach we are taking), and from this the tentative choice of a figure for the core cross-sectional area a. In turn, this leads directly to the choice of core or lamination size, the dimensions of which can now be entered in the core block. The basic transformer equation can now be used to obtain the numbers of turns needed for the windings. In quick succession, turns per inch, margins, layer length, turns per layer, and number of layers can now be entered for each winding.

In the "Build" block, the thicknesses of the windings are computed and totaled and then compared to the core dimensions to check the fit inside the window. The core size can be confirmed or rejected, or adjustments can be made to the stack thickness or wire gauges in order to obtain a good fit. When a satisfactory fit is achieved, the design process continues down the sheet through voltage drop and copper loss (I^2R). Then go back to the core box to complete the core details and calculate the core loss.

The result of the latter checks can be to readjust windings and/or the core until the original specifications are met (or possibly to adjust the specification to suit the results). If everything is fine, the temperature rise of the windings can be checked in the box for the temperature rise. If this is not satisfactory, the design is checked again to see whether appropriate changes can be made. Above all, checks are made for mistakes in arithmetic; nobody is perfect, even with a calculator. All of this should be perfectly clear in the worked-out examples.

If you wish, you can draw up your own format for the sheet. Some people like to have the relevant formulas right there on the sheet. Others have pictures of the windings and core with the important dimensions indicated for reference. Still others dispense with such refinements as design sheets, preferring to work on the backs of old envelopes and newspaper margins. In short, choose the method most comfortable for yourself.

GENERAL OBSERVATIONS

After formulating a specification outlining the required conditions, the first task is either to make a tentative core selection or to check whether a core on hand will do the job.

Choosing a Core Size

In either case, it can be done purely on the basis of successive trial designs, but it is much quicker to use the approximation formulas discussed earlier. This does not eliminate trial designs, because the equations use approximations of current and flux density, but it will probably reduce the number of trials needed. The estimates of current and flux density are initially based on what is "usual" for a transformer of the type and size and core material under consideration.

But there is nothing rigid about these choices. An initial figure of 1000 circular mils per ampere for current density could become 750 in the finished design. Similarly, the flux density initially chosen might not be the same as that used in the end; for that matter the core itself could be quite different from that originally selected.

Magnetizing Current

It is often asserted that the magnetizing current is usually so small relative to the load and loss currents that it can be ignored as a factor in the design of low-frequency power transformers.

In general, this is true enough, but why ignore it when it can be so easily checked? If it is checked as a matter of course, then it won't be forgotten when it really counts.

Adjustments

As the design proceeds, adjustments are continually being made to previously calculated figures, and you must keep in mind that an adjustment to one figure is likely to set off a chain reaction of adjustments throughout the entire design. This possibility should be considered and taken care of immediately.

For instance, it is usually necessary to adjust the numbers of turns on the windings to compensate for voltage drops. If this is done to the primary winding, the flux density is affected, as can be easily seen from the basic transformer equation. If the change is a large one, it is not impossible that the flux density could now be in the saturation region, a situation that could be serious if it is not detected. In any case, whether the change be large or small, the iron loss and the magnetizing volt-amperes will be changed; it can be seen from the watts per pound loss and the volt-amperes per pound curves that a relatively small change in flux density can produce a large change in these values, especially if the flux density is increased when the core is already close to saturation. These changes in turn produce changes in the current in the primary winding and in the copper loss in the primary.

Even before one gets to the copper loss stage, a change in the numbers of turns changes the build, which might require revising the core dimensions. Whatever the case, it will certainly affect the resistance of the windings, which again results in a change to the copper loss, and so on.

All of this does not mean that you are condemned to spend the rest of your days in an eternal round of adjustments while the fires of madness flare ever more brightly in your soul. It simply implies you should watch what you're doing and make considered judgments. If the design is following its proper course, a point is reached after a trial or two when the law of diminishing returns sets in and further adjustments are not worthwhile. Moreover, some of the data on which calculations are based, such as the loss and magnetizing curves, is likely to differ quite widely from actuality in the core that you will eventually use. In short, the accuracy obtained from working to several places of decimals, as we are apt to do, is largely illusory; if the secondary voltage works out on paper to 25.175 volts, full load, instead of 25 volts, be happy with it and hope that you are close to the mark in practice.

Selecting the Efficiency Figure

An important part of the design process is the choice of an efficiency figure. What this figure should be depends on a number of things, including the intended application and cost. For a given specification and core material, however, the mere choice of a figure does not guarantee that it can be achieved. In some cases, 70 percent efficiency can be difficult or impossible to obtain, especially if a design is being "forced" around a core on hand rather than on one to be selected and purchased. In other cases, a figure of 95 percent and higher is easily attainable. Nevertheless, this is a case where any number is better than none. Whatever number is chosen, its suitability will soon be apparent. Adjustments can then be made to achieve the figure, or perhaps a lower figure can be accepted.

Flux Density

Generally speaking, the amount of flux density in a power transformer is chosen to be as high as possible below the knee of the B-H curve and to give acceptable levels of loss and magnetization volt-amperes. At higher frequencies the loss figure rather than saturation may be the governing factor.

For power transformers, the suggested working figures in Table 8-4 (Chapter 8) are usually satisfactory. But keep in mind what was said earlier about field reduction in certain conditions.

In other types of transformers, the flux density might be deliberately kept very low in order to obtain maximum permeability and low losses.

Current Density

If good regulation and/or low temperature rise are desired, the current density figure selected should be at the low end of the "usual" range. If smallness is the main design criterion, regardless of regulation or temperature (within limits), the current density will be at the high end of the range.

Philosophy of the Examples: A Reminder

It should be realized that the performance of a transformer is governed to a large extent by the nature of the load, as was discussed previously. In all the examples that follow, purely resistive loads are used. Usually, other types of load do not affect the performance too much, except in the case of rectifiers, which have a chapter to themselves. However, keep the point in mind, and if highly reactive load conditions are anticipated, be rather generous in the design approach.

In working out the design examples, one of the objects was to demonstrate as many of the design techniques as possible, even to the extent of taking a circuitous route to the final solution, where a more direct one could have served. You will also note that in examples with the same or similar problems, different routes to the solutions have sometimes been used.

HIGH CURRENT AND HIGH VOLTAGE

The basic design procedures for high-current and high-voltage components are identical with those used for more conservative specifications. Questions of flux density, current density, and insulation requirements are handled in the same way; iron and copper losses bear the same relationship to efficiency, and so on.

At extremes (and sometimes not so extreme amounts) of current and voltage, however, certain practical considerations arise that can affect the direction of the design. A little foresight in the pencil and paper stage can often make things easier later.

High Current

High current means heavy-gauge conductors that, if heavy enough, can be the very devil to wind. If you are not familiar with heavy-gauge wires, try to get a "feel" for them before specifying them in your design. Check around the house and workshop for wire of a similar order of thickness—wire coat hangers are examples of heavy-gauge wire. Go down to a store and check material on the shelf; ask to see samples. It would be in your best interest to check these things rather than wind up being wound up about a winding that cannot be wound.

Consider using two (or even more) smaller-gauge wires in parallel as shown in Fig. 7-4. This method often has bonus advantages. For one thing, it can make better use of the winding space by filling the available layer length with thinner wire rather than partly filling it with thick wire.

In this connection, there is nothing more wasteful of space than to be forced to put one or two heavy-gauge turns on an additional layer. This can often be avoided by designing the heaviest-wire winding first and adjusting the turns per volt to give exactly even layers. Even if the thinner-wire windings do not exactly fit the layer lengths (which is likely), there still will have been a substantial saving of space.

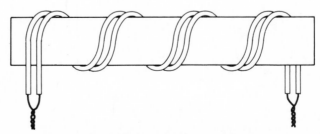

Fig. 7-4. Small-gauge wires wound in parallel replace one
heavier-gauge wire.

More often than not, when the current requirement is high, the voltage requirement is low. This means few turns in the winding, which leads to problems in adding or subtracting turns to compensate for voltage drop. For instance, the task of increasing the turns on a winding by 10 percent if it already has ten turns is a lot easier than if it has only three turns to start with. In the first case, one full turn does the job, but in the second case, 0.3 of a turn is needed, and partial turns like this are inconvenient. Here again it might be best to design the high-current winding first and apply the compensation to the primary.

Single-turn, heavy-current *primary* windings are commonly found in current transformers (Chapter 13). Because accuracy is usually a prime requirement in this type of transformer, compensation is nearly always applied to the secondary, which could consist of hundreds or even thousands of turns of relatively fine wire, thus making accuracy of adjustment easy.

Voltages and Insulation

All voltages, even as low or lower than the standard supply voltages of around 117 volts, present elements of risk. If the rules given in Chapter 6 with regard to insulation are applied with common sense, there should be no problems at moderately high voltages. But one must be clear as to where, inside the component, the maximum voltage differences will appear and what values they will have. These things depend on the kind of circuit with which the transformer is to be used. Examples of points to watch are given in Chapter 9, but the circuit should be carefully analyzed and adequate margins of safety built in.

Very high voltages are another subject and are not covered in this book.

EIGHT

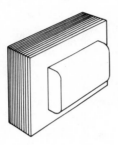

Power Transformers

Basically, there is no difference in the design methods for low- and high-power, or low-and high-voltage, or low- and high-current transformers. Therefore, no attempt has been made in the examples in this chapter to deal with a wide range of specifications. On the contrary, the first five examples are all based on the same low-voltage, medium-current specification. The purpose of this is to illustrate the effects on the performance of changing certain parameters in the design and to provide direct comparisons of the results. In the initial worked-out designs, the magnetizing current is considered to be negligible, a common practice in noncritical designs. In a later exercise, however, magnetizing current is discussed in relation to the same designs. The object of this is to illustrate the fact that magnetizing current cannot always be ignored. For this reason, the examples should not be used in a practical sense without first checking these comments.

FIVE VARIANTS OF A BASIC SPECIFICATION

The specification is as follows: A transformer is required to work from 110 volts, 60 hertz, and deliver 5.0 volts at 4.0 amperes in the simple circuit of Fig. 8-1A. The ambient temperature does not rise above about 40°C. The transformer runs continuously for hours at a time at a steady 4.0-ampere output and is mounted as in Fig. 8-1B.

Variant No. 1

The application for this design is deemed noncritical. Size is not an important factor, and regulation receives no special attention. In order to get the design started, an arbitrary

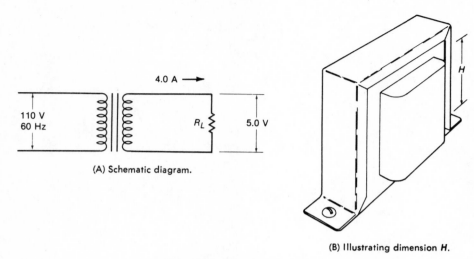

(A) Schematic diagram.

(B) Illustrating dimension **H.**

Fig. 8-1. Specification for variants 1 through 5.

figure of 85 percent is assigned to the efficiency. It is assumed that we have a choice of core-stamping patterns from a vendor's catalog rather than a make-do situation using stampings on hand. (It is a simple matter to transfer the data and methods described here to a make-do design.)

An initial outline of the design is easily obtained as follows. The required output power is 5 × 4 = 20 watts. At 85 percent efficiency, this means an input power of 20/0.85 = 23.5, say 24, watts. Because it is a small noncritical transformer, a current density of about 600 circular mils per ampere can probably be tolerated, but 700 is slightly more conservative. For this application, standard EI-type laminations can be used, and the material could be, say, low-carbon steel to be run at a conservative 12,000 gauss. The Wa product is

$$Wa = \frac{17.26 SP}{fB} = \frac{17.26 \times 24 \times 700}{60 \times 12,000} = 0.4 \text{ inch}^4$$

according to Equation 3-39.

A core catalog quickly yields candidate laminations for the job. Because the catalog usually gives both the window area W and the area a of a stack having a square cross section, selections can be made by simply multiplying the W and a numbers and comparing the result to 0.4 inch⁴. Another method is to divide 0.4 by the window area of a likely looking type, thus obtaining a number for the cross-sectional area. The cross-sectional area divided by the center limb width E and a stacking factor then gives the stack thickness required.

This exercise, carried out in a Magnetic Metals catalog, turned up a type 87 MH or 87 EI (which are identical except for the size and spacing of the clamping holes) with a window area of 0.575 square inch and a cross-sectional area a of 0.766 square inch to yield a *Wa* product of 0.44 inch^4. Using this size, a *Wa* of exactly 0.4 can be obtained with a stack 0.8 inch thick to obtain a cross-sectional area of 0.7 square inch. The vendor's sheet for this lamination is shown in Fig. 8-2. There are a number of other possible lamination sizes, but the nearly square cross-section obtained with this size makes for low winding resistance and therefore good regulation. Theoretically it looks good, so let's try it.

The turns needed can now be calculated

$$\frac{N}{V} = \frac{3.5 \times 10^6}{faB} = \frac{3.5 \times 10^6}{60 \times 0.7 \times 12,000} = 6.9$$

where *a* is in square inches and *B* in gauss. The number of turns in the primary is then $110 \times 6.9 = 759$ turns. In figuring the number of turns in the secondary, an approximate compensation for voltage drops in the winding can be built-in right away, if desired by rearranging Equation 4-29 as

$$N_2 = \frac{V_2 N_1}{V_1 \sqrt{\eta}} = \frac{5 \times 759}{110 \times 0.922} = 37.4 \text{ turns}$$

for the secondary. The number of uncompensated turns would have worked out to $6.9 \times 5 = 34.5$ turns. Therefore, the total transformer voltage drop in the secondary is about

$$V = \frac{37.4 - 34.5}{6.9} = \frac{2.9}{6.9} = 0.42 \text{ volt}$$

In other words, if 85 percent efficiency is achieved, the secondary no-load voltage will be 5.42 volts, and on load, it is hoped, exactly 5.0 volts.

If the design so far looks satisfactory for its proposed application, we can continue by selecting the wire gauges for the windings. The secondary winding current is specified at 4.0 amperes, and the primary winding current is given by input watts divided by input volts, or $24/110 = 0.218$ ampere. The wire gauge should be selected for the same current density figure used in calculating the *Wa* product—in this case, 700 circular mils per ampere. The secondary thus needs a wire with an area of $4 \times 700 = 2800$ circular mils, and the primary needs $0.218 \times 700 = 153$ circular mils.

From the wire tables, the nearest gauges are seen to be No. 16 at 2581 circular mils for the secondary and No. 28 at 159 circular mils for the primary.

The Details—If you like what you see so far, it is best now to start using the design sheet in Fig. 8-3. The following parameters can be specified in the appropriate spaces at the top of the sheet: input volts, hertz, estimated efficiency, and turns/volt. Also enter lamination number, window dimensions *G* and *F* (length and width), M_P (magnetic path length), *E* (center limb width), and *D* (stack). Note that the figure 0.8 obtained for the stack in the calculation must be divided by a stacking factor, say 0.9, to obtain the actual thickness; then *D* works out to $0.8/0.9 = 0.888$, say 0.9. Enter the cross-sectional area $a = 0.7$; this is the magnetic area for the electrical calculations, while 0.9 and 0.875 are the physical dimensions of the area.

$\mathbf{M}$AGNETIC $\mathbf{M}$ETALS

LAMINATION TYPE 87 EI

CHARACTERISTICS OF A CORE STACK HAVING A SQUARE CROSS SECTION

ALSO SEE
87 MH

VOLUME AND WEIGHT

VOLUME	– 4.03 in.³	– 66.2 cm.³
WINDOW AREA	– .575 in.²	– 3.71 cm.²
WT. SUPER Q 80	– 1.27 lb.	– 580 g.
WT. SUPERPERM "49"	– 1.18 lb.	– 539 g.
WT. SUPERFLUX	– 1.17 lb.	– 529 g.
WT. SILICON	– 1.06 lb.	– 481 g.

MAGNETIC DESIGN FORMULAE

Properties of Core Stack with Winding of "N" Turns

$$B_{max} = \frac{76.0 \times 10^3}{K_1 N} \text{ Gausses Per Volt at 60 Hertz}$$

$H_o = (.094 \times 10^{-3}) N$ Oersteds
(Gilberts per centimeter) per milliampere of direct current

$L_o. = (.4644 \times 10^{-8}) K_1 N^2 \mu_{ac}$ Henries

MAGNETIC PATH DIMENSIONS

$l = 5.25$ in. 13.34 cm.
$A = .766$ in.² 4.94 cm.²

K₁ (STACKING FACTOR)

Thickness	Butt Jointed	Interleaved one per layer
.004"	.90	.80
.006"	.90	.85
.014"	.95	.90
.0185"	.95	.90

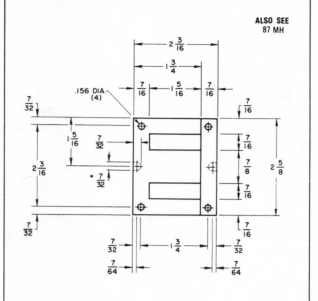

* Available with 7/32" slot as HS87EI.

PERFORMANCE DESIGNATION	MATERIAL TYPE	THICKNESS (Inches)	CATALOG NUMBER	WEIGHT AND COUNT	
				LBS. /M PCS.	PCS./ LB .
SUPERPERM 80	HyMu 80	.004	87EI8404	5.60	179
SUPERPERM 80	HyMu 80	.006	87EI8406	8.40	119
SUPERPERM 80	HyMu 80	.014	87EI8414	19.61	51
SUPER Q 80	HyMu 80	.004	87EI8004	5.60	179
SUPER Q 80	HyMu 80	.006	87EI8006	8.40	119
SUPER Q 80	HyMu 80	.014	87EI8014	19.61	51
SUPERTHERM 80	HyMu 80	.006	87EI7406	8.40	119
SUPERTHERM 80	HyMu 80	.014	87EI7414	19.61	51
SUPERPERM 49	49	.004	87EI4904	5.21	192
SUPERPERM 49	49	.006	87EI4906	7.81	128
SUPERPERM 49	49	.014	87EI4914	18.24	54.8
SUPERFLUX	PERMENDUR	.006	87EIVP06	7.72	130
SUPERFLUX	PERMENDUR	.010	87EIVP10	12.88	77.5
MICROSIL	Gr. Or. Silicon	.004	87EI3304	4.85	206
MICROSIL	Gr. Or. Silicon	.006	87EI3306	7.28	137
MICROSIL	Gr. Or. Silicon	.014	87EI3314	16.97	58.9
SILICON	Non Or. Silicon*	.014	87EI**14	16.97	58.9
SILICON	Non Or. Silicon*	.018	87EI**18	21.83	45.8
SILICON	Non Or. Silicon*	.025	87EI**25	30.32	33.0
HYPERTRAN	Low Carbon	.025	87EI2125	31.09	32.2

* Customer to designate AISI grade of material desired.
** See "How To Order Section" for Code Number.

Courtesy Magnetic Metals Corp.

Fig. 8-2. Lamination data for design variant No. 1.

TRANSFORMER DESIGN SHEET

Input volts	*110*	Hertz	*60*		Est. Efficiency	*85*	Turn/volt	*6.9*

Lamination or core No.	*87 EI LOW-CARBON STEEL*	B *12,000*

Window dimensions	G *1.312 (15/16)* F *0.437 (7/16)* W *0.575*	M_p *5.25*

C.S.A. dimensions	E *0.875 (7/8)* D *0.9* a *0.7*	v *4.3*

	Wt *1.2*	Watts/lb *3.83*	Iron loss *4.6*

Windings	W_1	W_2								
Coil	PRIM	SEC								
Volts	110	5								
Amperes	.218	4								
Turns	759	37								
Gauge	28	16								
Turns/inch	66	17								
Margins	.125	.188								
Winding length (L)	1.062	.936								
Turns/layer	70	16								
Number of layers	10.8	2.3								

Build										Total
Copper	.149	.157								
Paper	.015	.020								
Cover	.020	.020								
Total	.184	.197								
Bulge Percent *15*	.027	.029								
Total (R)	.211	.226								.437

Bobbin _____ clearance _____

Total bobbin (B) _____
Total depth

Losses										
Length mean turn (M)										
Total wire length (inches)										
Ohms/ft										
Resistance (hot)										
Voltage drop										
Copper loss (I^2R)										

Iron loss
Total loss
Efficiency

Temperature Rise

Coil	t	C_S	W	W/C_S	°C T_o
1					
2					
3					
4					

Temp. Rise = T_o(Total) + $20HW/C_S$ =

Fig. 8-3. Variant No. 1, design sheet No. 1.

Enter the figure for the windings. At "Coil" enter "PRIM" in column 1, and "SEC" in column 2, indicating that the primary is the winding nearest the core. Enter the volts, amperes, turns, and gauge for each winding. From the wire tables, enter turns/inch (the single-insulation column) and margins (called "edge distance" in the tables). From the core window dimension G, deduct the margins 2×0.125 for the primary and 2×0.188 for the secondary to obtain the winding lengths. Thus $1.312 - (2 \times 0.125) = 1.062$ for the primary, and $1.312 - (2 \times 0.188) = 0.936$ for the secondary. The turns/layer are then winding lengths times turns/inch; for the primary this is $1.062 \times 66 = 70$, and for the secondary it is $0.936 \times 17 = 16$. The number of layers is then $759/70 = 10.8$ and $37/16 = 2.3$.

In the "Build" box, the thickness of copper is the number of layers times the wire diameter. From the wire tables (Table 6-1), the diameter of No. 28 gauge is 0.0136 inch and No. 16 gauge is 0.0523 inch. The thickness of copper in the primary is therefore $11 \times 0.0136 = 0.149$ inch, and the secondary $3 \times 0.0524 = 0.1572$ inch. Note that the number of layers is taken to be the higher whole number in each case. The total paper thickness is the thickness of paper between each layer times the number of layers, less one. Assuming that the paper thicknesses given in Table 6-1 are used, then the primary is $0.0015 \times 10 = 0.015$ inch, and the secondary is $0.010 \times 2 = 0.020$ inch. The primary cover could be, say, 0.020 inch and the secondary cover 0.020 inch. Considering the small voltages involved, the covers could be much thinner, but the figures chosen are safe.

Totaling the build figures for the primary gives 0.184, of which 15 percent allowance for bulge is 0.027 inch, and for the secondary at 0.197 inch, of which 15 percent is 0.029. Enter the 15 percent figures and enter the totals of 0.211 inch for the primary, and 0.226 inch for the secondary. The total of these two figures, 0.437 inch, is entered in the total column.

Comparing this figure with the window depth of 0.437 inch, it can now be seen that the space left in the window, $0.437 - 0.437 = 0.000$ inch, is not enough to accommodate the bobbin. An additional 0.050 to 0.080 inch or thereabouts is needed for the bobbin.

There are several possible solutions. The core area could be increased by increasing the stack, thus reducing the turns per volt and therefore the total turns on the winding. The wire gauge could be made smaller. The flux density might be increased, thus reducing the number of turns in the windings. A slightly larger lamination size might be considered. Or a combination of several of these measures might be applied. For the purpose of this exercise, I chose to reduce the number of turns by increasing the core area slightly. The object of this is to reduce the number of turns on the secondary winding to 32, which will thus cut out one layer of thickness and save at least 0.052 inch. To preserve the required ratio, the primary winding will also have to be reduced and could result in cutting out one primary winding layer thickness for a further saving of 0.013 inch plus savings due to less paper and percentage of bulge.

Putting the number of turns on the secondary at 32 and plugging this number into another rearrangement of Equation 4-29 gives

$$N_1 = \frac{N_2 V_1 \sqrt{\eta}}{V_2} = \frac{32 \times 110 \times 0.922}{5} = 649 \text{ turns}$$

An arrangement of the basic transformer equation then gives the required core area:

$$a = \frac{V \times 3.5 \times 10^6}{N_1 fB} = \frac{110 \times 3.5 \times 10^6}{649 \times 60 \times 12,000} = 0.82 \text{ sq in}$$

The amended numbers stemming from these changes are entered for clarity in a new sheet (Fig. 8-4). Ordinarily, the old numbers would have been stroked out and the new ones entered in two fresh columns. The numbers affected are turns, number of layers, all the "Build" figures, and the core dimensions D and a; the new number for D is obtained by dividing 0.82 by 0.875 (the center limb width E) and then dividing the quotient by the stacking factor of 0.9 to give 1.0 inch.

The changes in the core box can be followed up immediately by entering the figures for v (core volume in cubic inches), Wt (core weight in pounds), and Iron loss. Now $v = M_P \times a = 5.25 \times 0.82 = 4.3$ cubic inches, and Wt is $4.3 \times 0.28 = 1.2$ lbs (0.28 is the pounds-per-cubic-inch figure for low-carbon steel obtained from Table 5-7). The loss for this material at the stated flux density can be put at 3.83 watts per pound for the present purpose, making the total iron loss $3.83 \times 1.2 = 4.6$ watts; this assumes a lamination thickness of 0.018 inch.

The "Losses" box can now be entered. The "length mean turn M" for each winding is given by Equations 6-3 and 6-4. So

$$M_1 = 2(E + D + 4B) + 6.28 \frac{W_1}{2}$$

and

$$M_2 = 2(E + D + 4B) + 6.28 \left(W_1 + \frac{W_2}{2}\right)$$

The numbers to be assigned to E and D are taken directly from the core section at the top of the design sheet. The term B is the total bobbin figure marked B on the right of the sheet, in the "Build" section (not to be confused with B for flux density at the top), W_1 and W_2 are the respective numbers for the totals just after the bulge percent number in the Build section. Thus,

$$M_1 = 2[0.875 + 1.0 + (4 \times 0.093)] + 6.28 \frac{0.194}{2}$$
$$= 5.100 \text{ inches}$$

and

$$M_2 = 2[0.875 + 1.0 + (4 \times 0.093)]$$
$$+ 6.28 \left(0.194 + \frac{0.155}{2}\right)$$
$$= 6.200 \text{ inches}$$

Enter these numbers in the appropriate columns. The total wire length is $M \times$ turns. For the primary winding, this is $5.100 \times 649 = 3310$ inches, and for the secondary $6.200 \times 32 = 198$ inches.

The ohms/ft figure is entered from the wire tables (Table 6-1). The "hot" resistance is then the product of the total wire length divided by 10 and the ohms per foot. For winding No. 1, this is $331 \times 0.0653 = 21.6$ ohms and for No. 2 it is $19.8 \times 0.0042 = 0.083$ ohm.

Voltage drop is calculated from Ohm's law: $V_d = I \times R$. For W_1 it is $0.218 \times 21.6 = 4.7$ volts, and for W_2 it is $4 \times 0.083 = 0.330$ volt. Copper loss is I^2R or $IV = 0.218 \times 4.7 = 1.022$ watts for the primary, and $4 \times 0.330 = 1.32$ watts for the secondary. The total copper loss is then 2.34 watts. Adding in the iron loss from the core box gives a total loss of 6.94 watts. The efficiency is then

$$\eta = \frac{\text{Output watts}}{\text{Input watts}} = \frac{20}{26.94} = 0.74 \text{ or } 74 \text{ percent}$$

which is considerably lower than the initial estimate. Because the difference between the efficiency originally estimated and that actually obtained is quite large, the input watts and primary winding current figures should be adjusted. The total loss of 6.94, say 7.0, watts then puts the input power at 27 watts and the primary winding current at 27/110 = 0.245 ampere. This in turn increases the voltage drop on the primary to 5.27 volts and the primary winding copper loss to 1.29 watts. These changes are shown in parenthesis on the design sheet.

Note that the new loss figure further decreases the efficiency and theoretically generates further adjustments to input watts and current. However, a point is very quickly reached where further adjustments create only very minor changes in the end result.

Adjusting the Turns—A check on the output voltage using Equation 4-10 gives

$$V_o = \frac{N_2(V_{in} - V_{d1})}{N_1} - V_{d2}$$
$$= \frac{32(110 - 5.27)}{649} - 0.330$$
$$= 4.83 \text{ volts}$$

TRANSFORMER DESIGN SHEET

Input volts	110	Hertz	60		Est. Efficiency	85		Turn/volt	

Lamination or core No.	87 EI	LOW-CARBON STEEL — .014 INCH	B 12,000 (12,460)

Window dimensions	G 1.313	F 0.437	W 0.575	M_p 5.25
C.S.A. dimensions	E 0.875	D 1.0	a 0.82	v 4.3
	Wt 1.2	Watts/lb 3.83	Iron loss 4.6	

Windings		W_1	W_2								
Coil		PRIM	SEC								
Volts		110	5								
Amperes	(.245)	.218	4								
Turns	(628)	649	32								
Gauge		28	16								
Turns/inch		66	17								
Margins		.125	.188								
Winding length (L)		1.062	.956								
Turns/layer		70	16								
Number of layers	(9.0)	9.2	2.0								

Build										Total
Copper		.136	.105							
Paper		.013	.010							
Cover		.020	.020							
Total		.169	.135							
Bulge Percent 15		.025	.020							
Total (R)		.194	.155							.349

Total bobbin (B) .088
Total depth .437

Bobbin .062 clearance .026

Losses										
Length mean turn (M)		5.087	6.167							
Total wire length (inches)		3310	198							
Ohms/ft		.0653	.0042							
Resistance (hot)		21.5	.083							
Voltage drop	(5.27)	4.69	.330							
Copper loss (I^2R)	(1.29)	1.022	1.320							

2.35
Iron loss 4.60
Total loss 6.95
Efficiency 74

Temperature Rise

Coil	t	C_s	w	W/C_s	°C T_o
1	.056	4.55	1.29	.284	3.97
2	.030	4.55	1.320	.290	2.18
3					
4					
				.574	6.15

$$T_s = 20H\,\frac{W_{TOTAL}}{C_s} = 20 \times 1.062 \times .574$$
$$= 12.19$$

Temp. Rise = T_o(Total) + $20HW/C_s$ = 18°C

Fig. 8-4. Variant No. 1, design sheet No. 2.

This full-load voltage is somewhat low due to the fact that while 85 percent efficiency was initially assumed, only 74 percent was achieved in the calculations. If necessary, it can be adjusted to give 5.0 volts by reducing the number of turns on the primary. Using Equation 4-13, a new figure can be obtained:

$$N_1 = \frac{N_2(V_{in} - V_{d1})}{V_o + V_{d2}} = \frac{32(110 - 5.27)}{5 + 0.330} = 628 \text{ turns}$$

This reduction in primary winding turns makes the number of layers of the primary almost exactly 9. These changes are entered in parentheses in Fig. 8-4.

Reviewing the Design—Looking over the design sheet with a critical eye, note the following points. When the transformer is supplying the specified load current, the output voltage will be 5 volts. But with no load connected and therefore very little voltage drop in the windings, the output voltage can be determined directly by

$$V_2 = \frac{N_2}{N_1}\left(V_{in} - \frac{W_o R_1}{V_{in}}\right) = \frac{32}{628}\left(110 - \frac{4.6 \times 21.5}{110}\right)$$
$$= 5.55 \text{ volts}$$

where
  W_o is the iron loss,
  R_1 is the resistance of the primary winding.

The regulation is therefore very approximately

$$\% \text{ Regulation} = \frac{100(5.55 - 5.0)}{5.0} = 11 \text{ percent}$$

For some purposes involving variable loading, this might not be considered a "good" figure. But there are many applications for which it would be perfectly adequate.

The distribution of copper loss in the primary and secondary windings is good, but the iron loss and copper loss are not well balanced. If trouble is taken to adjust the losses, some improvement in efficiency and regulation can be expected.

An interesting point to note for future reference is that the K factor—the ratio of copper area to window area—as given by Equation 6-11 is quite small in this transformer:

$$K = \frac{(N_1 A_1 + N_2 A_2)\, 7854 \times 10^{-10})}{W}$$
$$= \frac{[(628 \times 159) + (32 \times 2581)]\,(7854 \times 10^{-10})}{0.575}$$
$$= 0.25$$

where A_1 and A_2 are in circular mils. In other words in this transformer, a full 75 percent of the window space is lost to insulation and air. Any improvement to this figure—brought about perhaps by cutting down on the bobbin thickness or reducing the generous 15 percent bulge allowance—will improve overall efficiency and regulation, or alternatively will permit the use of a smaller core for the same efficiency.

Because the cross-sectional area was increased from an initial 0.7 square inch to 0.82 square inch, the actual Wa product is now $0.575 \times 0.082 = 0.47$ inch^4, somewhat greater than the original 0.4 figure. In terms of power capability, however, the initial 0.4 figure could probably have been retained by improving the K factor or even by using smaller-gauge wire; this latter solution would have reduced the efficiency and increased the voltage drop, but it is likely that the temperature rise in this small component would still be permissible.

Don't forget that in juggling with the numbers of turns on the primary winding, the flux density might change. If it does, the iron loss also changes together with efficiency, primary winding current, and to some degree, primary copper loss. Using the basic equation a check on this point gives

$$B = \frac{3.5V \times 10^6}{faN} = \frac{3.5 \times 110 \times 10^6}{60 \times 0.82 \times 628}$$
$$= 12{,}460 \text{ gauss}$$

This flux density is not so far removed from the original figure that you need worry about it.

Temperature Rise—So far no consideration has been given to this factor, which is probably the most important single parameter. Actually there should be no problem, but check it anyway, if only as an exercise. Referring to Chapter 6, enter figures for t in the "Temperature Rise" box of the design sheet. For winding No. 1, t is half the thickness of insulation on the primary winding plus the total thickness on insulation in the windings above. (Note that only insulation is considered here. The copper does not enter into these calculations.) From the "Build" box, observe that half the insulation on the primary winding is $0.013/2 = 0.006$ inch, and above this there is 0.020 inch for the primary winding cover plus 0.010 inch in the secondary plus 0.020 inch for the secondary winding cover, or a total t of 0.056 inch. Enter this at W_1. For coil No. 2, t is half the insulation in the secondary (because there is only one layer of paper in this case, take the 0.010 inch) plus the secondary winding cover of 0.020 inch for a total t of 0.030 inch. Enter this at W_2.

Equation 6-17 gives the area of the cooling surface, C_S,

$$C_S = 2L(E + 2B + \pi R)$$

The number required for each symbol is on the design sheet. For L, it is the winding length in the "Windings" section. Take the longer of the two values, which is 1.062. Also, E is 0.875 inch from the core section at the top of the sheet, and B is the bobbin thickness in the "Build" section (marked B). Finally, R is the total build figure—the number just before the total bobbin; it is marked R by "Total." Then

$$C_S = 2 \times 1.062 \, [0.875 + (2 \times 0.088)$$
$$+ \; (3.14 \times 0.349)]$$
$$= \; 4.55 \text{ sq in}$$

Enter this number under C_S for coils 1 and 2 in the "Temperature Rise" box. In the watts loss column, enter the respective *copper* losses from the "Losses" section, namely 1.29 watts for coil 1 and 1.320 for coil 2. W/C_S for coil 1 is then $1.29/4.55 = 0.284$ W/in^2, and for coil 2 it is $1.320/4.55 = 0.290$ W/in^2. Enter these figures.

Using Equation 6-19, solve and enter T_o for each winding as follows:
Coil No. 1:

$$T_o = 250 t_1 W_1 / C_S$$
$$= 250 \times 0.056 \times 0.284$$
$$= 3.97°C$$

and
Coil No. 2:

$$T_o = 250 \times 0.030 \times 0.290 = 2.18°C$$

Next, using Equation 6-18, calculate the temperature gradient at the surface:

$$T_S = 20 \, H \, \frac{W_{total}}{C_S}$$

The mounting position was defined in the initial specification. This puts $H = 1.062$ inch (winding length). Now W_{total}/C_S is the total of the W/C_S column. Thus

$$T_S = 20 \times 1.062 \times 0.574 = 12.19°C$$

The total temperature rise is the sum of the calculated temperature rises.

$$T_T = 3.97 + 2.18 + 12.19 = 18.34, \text{ say } 18.0°C$$

This is a low and usually very safe temperature. Because a maximum tolerable temperature (Chapter 4) is on the order of 105°C, this transformer could work in an ambient temperature up to $105 - 18 = 87°C$.

As a point of interest, the abbreviated method of calculation, Equation 6-21, also puts the temperature at 18°C.

The finished design, complete with all entries, is shown in Fig. 8-5.

Variant No. 2—Reducing the Size

Circuit designers are apt to regard power transformers as necessary evils—indispensable but ugly hunks of copper and iron that destroy the beautiful symmetry of their layouts and frustrate their attempts at miniaturization. Consequently, size reduction

TRANSFORMER DESIGN SHEET

Input volts 110		Hertz 60		Est. Efficiency 74		Turn/volt	

Lamination or core No.	87 EI LOW-CARBON STEEL — .014 INCH B 12,460

Window dimensions	G 1.313	F 0.437	W 0.575	M$_P$ 5.25

C.S.A. dimensions	E 0.875	D 1.0	a 0.82	v 4.3

Wt 1.2	Watts/lb 3.83	Iron loss 4.6

Windings		W$_1$	W$_2$									
Coil		PRIM	SEC									
Volts		110	5									
Amperes		.245	4									
Turns		628	32									
Gauge		28	16									
Turns/inch		66	17									
Margins		.125	.188									
Winding length (L)		1.062	.956									
Turns/layer		70	16									
Number of layers		9.0	2.0									

Build												Total
Copper		.136	.105									
Paper		.013	.010									
Cover		.020	.020									
Total		.169	.135									
Bulge Percent 15		.025	.020									
Total (R)		.194	.155									.351

Total bobbin (B) .086
Total depth .437

Bobbin .062 clearance .024

Losses												
Length mean turn (M)		5.087	6.167									
Total wire length (inches)		3300	19									
Ohms/ft		.0653	.0042									
Resistance (hot)		21.5	.083									
Voltage drop		5.27	.330									
Copper loss (I^2R)		1.29	1.320									2.6

Iron loss 4.6
Total loss 7.2
Efficiency 74

		Temperature Rise				
Coil	t	C$_S$	W	W/C$_S$	°C T$_o$	
1	.051	4.55	1.29	.284	3.97	
2	.030	4.55	1.323	.291	2.2	
3						
4						

Temp. Rise = T_o(Total) + 20 HW/C$_S$ = 18°C

Fig. 8-5. Variant No. 1, design sheet No. 3.

is often high on the list of their specification "wants" for a power transformer. This means making full use of the available window space and running the core at maximum flux density. Where the designer has control of the input as in converter and inverter circuits, it might mean increasing the input frequency to hundreds or even thousands of hertz. Sometimes efficiency and other normally desirable features are sacrificed and the design taken to the maximum limits of temperature rise.

If size reduction is important, a high-grade core material is likely to be chosen, but it is interesting to see what happens with the type used in the last example. Consider the result of attempting to reduce the size for the basic specification and the same core material—carbon steel.

Suppose the current density in the windings increases to about 500 CM/A and the flux density to 14,000 gauss. Put the efficiency at about 70 percent. (Because the last design worked out at 74 percent, expect this one to be less.) The figures are worked out in the same way as for the previous design. These are entered in sheet of Fig. 8-6.

The input watts figure is 20/0.70 = 28.6, say 29, watts. The primary winding current is then 29/110 = 0.264 ampere. The required Wa product is

$$Wa = \frac{17.6 \times 29 \times 500}{60 \times 14,000} = 0.3 \text{ inch}^4$$

A type 75 EI lamination (Fig. 8-7) looks close. With a window area of 0.422 square inch, it needs a core area of 0.3/0.422 = 0.71 square inch. The center limb width is 0.75 inch and a stacking factor of 0.9 is advised, making the stack thickness 0.71/(0.75 × 0.9) = 1.0 inch.

The number of turns per volt works out to

$$\frac{N}{V} = \frac{3.5 \times 10^6}{60 \times 0.71 \times 14,000} = 5.87 \text{ turns/volt}$$

The number of primary turns is then 110 × 5.87 = 646 turns. This time, calculate the number of secondary turns directly from the turns per volt figure: 5 × 5.87 = 29.35, say 30, turns. The only reason for this slight change in procedure is to demonstrate that there are other approaches to the design problem.

At 500 CM/A, the wire gauges work out to 500 × 0.264 = 132 CM for the primary and 500 × 4 = 2000 CM for the secondary. From the wire table (Table 6-1), the nearest gauges are seen to be No. 29 at 127.7 and No. 17 at 2052 CM for the primary and secondary windings, respectively. From Fig. 8-6, the numbers stemming from these gauges culminate in a build thickness—before adding the bobbin—of 0.193 + 0.199 = 0.392 inch. This is 0.017 inch more than the available window depth of 0.375 inch and therefore there is not sufficient room for the bobbin. One solution, which was used in the last example also, is to reduce the number of secondary turns to exactly 28, or 5.6 turns per volt. This will reduce the secondary winding thickness by one layer. In order to preserve the correct ratio, the number of turns in the primary must also be reduced

TRANSFORMER DESIGN SHEET

Input volts *110*		Hertz *60*			Est. Efficiency *70*		Turns/volt *5.87*	
Lamination or core No. *75 EI LOW-CARBON STEEL — .014 INCH* B *14,000*								
Window dimensions	G *1.125*		F *0.375*		W *0.422*		M_P *4.5*	
C.S.A. dimensions	E *0.75*		D *1.0*		a *0.71*		v *3.15*	
	Wt *0.85*		Watts/lb *8.0*		Iron loss *6.8*			

Windings		W_1	W_2	W_1	W_2							
Coil		PRIM	SEC									
Volts		110	5	110	5							
Amperes		.264	4	.259	4							
Turns		646	30	616	28							
Gauge		29	17	29	17							
Turns/inch		73	19	73	19							
Margins		.125	.188	.125	.188							
Winding length (L)		.875	.749	.875	.749							
Turns/layer		64	14	64	14							
Number of layers		10.68	2.21	9.6	2							

Build												Total
Copper		.134	.140	.122	.094							
Paper		.015	.014	.014	.007							
Cover		.020	.020	.020	.020							
Total		.169	.174	.156	.121							
Bulge Percent		.024	.025	.022	.017							
Total (R)		.193	.199	.178	.138							.316
											Total bobbin (B)	.059
Bobbin *.030* clearance *.029*			*.392*								Total depth	.375

Losses												
Length mean turn (M)				4.71	5.64							
Total wire length (inches)				2901	15.8							
Ohms/ft				.0812	.0051							
Resistance (hot)				23.55	.081							
Voltage drop				6.1	.324							
Copper loss (I^2R)				1.6	1.3							2.9
											Iron loss	
											Total loss	
											Efficiency	

Temperature Rise						
Coil	t	C_S	W	W/C_S	°C T_o	
1						
2						
3						
4						

Temp. Rise = T_o(Total) + $20HW/C_S$ = *31°C*

Fig. 8-6. Variant No. 2, design sheet No. 1.

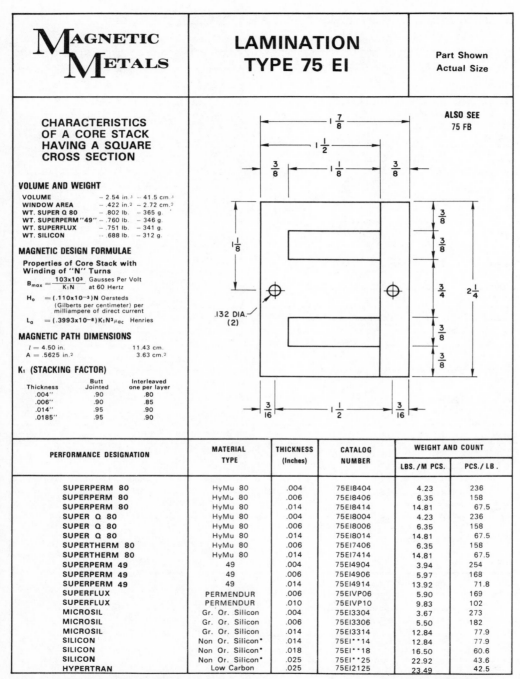

MAGNETIC METALS

LAMINATION TYPE 75 EI

Part Shown Actual Size

ALSO SEE
75 FB

CHARACTERISTICS OF A CORE STACK HAVING A SQUARE CROSS SECTION

VOLUME AND WEIGHT

VOLUME	– 2.54 in.³	– 41.5 cm.³
WINDOW AREA	– .422 in.²	– 2.72 cm.²
WT. SUPER Q 80	– .802 lb.	– 365 g.
WT. SUPERPERM "49"	– .760 lb.	– 346 g.
WT. SUPERFLUX	– .751 lb.	– 341 g.
WT. SILICON	– .688 lb.	– 312 g.

MAGNETIC DESIGN FORMULAE

Properties of Core Stack with Winding of "N" Turns

$$B_{max} = \frac{103 \times 10^3}{K_1 N} \text{ Gausses Per Volt at 60 Hertz}$$

$$H_o = (.110 \times 10^{-3})N \text{ Oersteds}$$
(Gilberts per centimeter) per milliampere of direct current

$$L_a = (.3993 \times 10^{-8})K_1 N^2 \mu_{ac} \text{ Henries}$$

MAGNETIC PATH DIMENSIONS

$l = 4.50$ in.   11.43 cm.
$A = .5625$ in.²   3.63 cm.²

K_1 (STACKING FACTOR)

Thickness	Butt Jointed	Interleaved one per layer
.004"	.90	.80
.006"	.90	.85
.014"	.95	.90
.0185"	.95	.90

.132 DIA. (2)

PERFORMANCE DESIGNATION	MATERIAL TYPE	THICKNESS (Inches)	CATALOG NUMBER	WEIGHT AND COUNT	
				LBS./M PCS.	PCS./LB.
SUPERPERM 80	HyMu 80	.004	75EI8404	4.23	236
SUPERPERM 80	HyMu 80	.006	75EI8406	6.35	158
SUPERPERM 80	HyMu 80	.014	75EI8414	14.81	67.5
SUPER Q 80	HyMu 80	.004	75EI8004	4.23	236
SUPER Q 80	HyMu 80	.006	75EI8006	6.35	158
SUPER Q 80	HyMu 80	.014	75EI8014	14.81	67.5
SUPERTHERM 80	HyMu 80	.006	75EI7406	6.35	158
SUPERTHERM 80	HyMu 80	.014	75EI7414	14.81	67.5
SUPERPERM 49	49	.004	75EI4904	3.94	254
SUPERPERM 49	49	.006	75EI4906	5.97	168
SUPERPERM 49	49	.014	75EI4914	13.92	71.8
SUPERFLUX	PERMENDUR	.006	75EIVP06	5.90	169
SUPERFLUX	PERMENDUR	.010	75EIVP10	9.83	102
MICROSIL	Gr. Or. Silicon	.004	75EI3304	3.67	273
MICROSIL	Gr. Or. Silicon	.006	75EI3306	5.50	182
MICROSIL	Gr. Or. Silicon	.014	75EI3314	12.84	77.9
SILICON	Non Or. Silicon*	.014	75EI**14	12.84	77.9
SILICON	Non Or. Silicon*	.018	75EI**18	16.50	60.6
SILICON	Non Or. Silicon*	.025	75EI**25	22.92	43.6
HYPERTRAN	Low Carbon	.025	75EI2125	23.49	42.5

*Customer to designate AISI grade of material desired.
**See "How To Order Section" for Code Number.

Courtesy Magnetic Metals Corp.

Fig. 8-7. Lamination data for variant No. 2.

in the same ratio, or if you like, to the same turns per volt figure, giving $110 \times 5.6 = 616$ turns.

With this change, the build now works out to 0.316 inch, leaving 0.059 inch for the bobbin and its fit, as shown in Fig. 8-6. This is reasonable. Continuing the design along these lines as shown, a total winding loss of 2.9 watts is obtained. The modified numbers up to this point are entered in the two adjacent columns in Fig. 8-6.

It is now necessary to compensate for voltage drop, which in turn creates changes in other factors. To keep this clear, the new numbers have been entered in a fresh design sheet (Fig. 8-8). Compensation for the voltage drops in the windings in this design is best achieved by reducing the number of primary winding turns by using Equation 4-13 thus:

$$N_1 = \frac{N_2(V_{in} - V_{d1})}{V_o + V_{d2}} = \frac{28(110 - 5.76)}{5 + 0.324}$$
$$= 548 \text{ turns}$$

This reduces the number of primary winding layers to 8.56 (9) and modifies the build to 0.30 inch, leaving 0.075 inch for the bobbin and its clearance. Because the primary winding dimensions and turns have been changed, the resistance and losses in both windings will be changed somewhat. In addition, the flux density and iron loss will be increased. The flux density now works out as

$$B = \frac{3.5 \times 110 \times 10^6}{60 \times 0.72 \times 548} = 16{,}492 \text{ gauss}$$

This flux density is substantially higher than before, but it is still within the capability of this material, which is a high-flux type. But its losses are quite high, and at this density, its loss will be relatively large. Using the figure of the last example as a guide, that is, 3.83 watts/lb at 12,000 gauss, refer the loss to the new flux density with Equations 5-3 and 5-4. Equation 5-3 gives

$$k = \frac{W/lb}{B^2} = \frac{3.8}{12{,}000^2} = 2.64 \times 10^{-8}$$

Then Equation 5-4 gives the loss at 16,492 gauss:

$$W/lb = 2.64 \times 10^{-8} \times 16{,}492^2 = 7.2 \text{ W/lb}$$

and the total core loss is $0.85 \times 7.2 = 6.12$ watts. It is now clear that the total transformer losses are 6.12 for the iron, plus 2.78 for the copper, or 8.9 watts total. Efficiency is on the order of $20/28.9 = 0.69$ or 69 percent.

A final check on the output voltage now works out at

$$V_2 = \frac{28(110 - 5.65)}{548} - 0.32 = 5.012 \text{ volts}$$

by means of Equation 4-10.

TRANSFORMER DESIGN SHEET

Input volts *110*	Hertz *60*		Est. Efficiency *69%*		Turn/volt	
Lamination or core No. *75 EI LOW-CARBON STEEL — .014 INCH*					B *16,492*	
Window dimensions	G *1.125*	F *0.375*		W *0.422*	M_p *4.5*	
C.S.A. dimensions	E *0.75*	D *1.0*		a *0.71*	v *0.316*	
	Wt *0.85*	Watts/lb *7.2*		Iron loss *6.12*		

Windings		W₁	W₂								Total
Coil		PRIM	SEC								
Volts		110	5								
Amperes		.263	4								
Turns		548	28								
Gauge		29	17								
Turns/inch		73	19								
Margins		.125	.188								
Winding length (L)		.875	.749								
Turns/layer		64	14								
Number of layers		8.56	2								

Build											Total
Copper		.110	.094								
Paper		.012	.007								
Cover		.020	.020								
Total		.142	.121								
Bulge Percent		.020	.017								
Total (R)		.162	.138								.300

Bobbin *.042* clearance *.033*

Total bobbin (B) *.075*

Total depth *.375*

Losses											
Length mean turn (M)		4.731	5.620								
Total wire length (inches)		2596	.157								
Ohms/ft		.0812	.0051								
Resistance (hot)		21.1	.08								
Voltage drop		5.65	.32								
Copper loss (I^2R)		1.5	1.28								2.78

Iron loss *6.12*

Total loss *8.90*

Efficiency *69%*

	Temperature Rise					
Coil	t	C_S	W	W/C_S	°C T_o
1					
2					
3					
4					

Temp. Rise = T_o(Total) + $20HW/C_S$ =

Fig. 8-8. Variant No. 2, design sheet No. 2.

Review of the Design—There is one important point to be wary of in all this. The formulas used to work out the iron loss (Equations 5-3 and 5-4) are based on the assumption that the flux density is not so high as to be beyond the knee of the *B-H* curve. Recall that above the knee, the core loss is much more than proportional to the square of the flux density. The figure of 16,500 gauss is a very high value, and one sample of material is not necessarily the same as another, so the loss might be higher than estimated. If accuracy is important, the manufacturer should be consulted. If the characteristics of the material are not known precisely, it is wisest to incline towards lower figures for flux density.

Again, it is of interest to note that the *K* factor is on the same order as that in the first example. The temperature rise is still quite low, indicating that a further reduction in size is possible. Assuming that the higher voltage drops do not present a problem, a smaller window with smaller-gauge wire would do the trick. Again, the *K* factor can be reduced by paying closer attention to insulation and bobbin thickness.

Before proceeding to the other variants of this design theme, it might be helpful to take a closer look at efficiency and regulation

IMPROVING EFFICIENCY AND REGULATION

In the designs just completed, rather low efficiency figures were initially selected to initiate the design procedure and low figures were obtained in the end result. These were allowed to stand because there was no stated need for anything better.

Remember, however, that the need for a given level of efficiency could be implied in a specification rather than stated directly. For instance, a low-percentage regulation may be called for; this implies a compatible level of efficiency.

It was shown earlier that when losses are correctly proportioned for maximum efficiency, the regulation is a function of the efficiency as shown by the approximation stated by Equation 4-31:

$$\% \text{ Regulation} = 100 \left(\frac{1}{\sqrt{\eta}} - 1 \right)$$

where η is expressed as a decimal. A rearrangement of this gives an approximate value for η to achieve a given regulation. If, for example, a regulation of 3 percent is required, then

$$\eta = 100 \left(\frac{1}{\text{Regulation} + 1} \right)^2 = 100 \left(\frac{1}{0.03 + 1} \right)^2$$

$$= 94.25 \text{ percent}$$

where regulation is expressed as a decimal fraction. Obviously, then, there would be little point in starting the design with an arbitrary figure of, say, 85 percent in this case. The sensible figure to start with would be 94 or 95 percent. But to state a figure does not mean that it will be achieved. At very low power ratings, such as that used in the two design examples, very high efficiencies are not always possible. For one thing, the poor K factors are against achieving high efficiency. Core loss, too, is a prime limiting factor, and if for any reason, only high-loss material is available, low efficiency will probably result. Nevertheless, the stated figure, achievable or not, is a starting point; whether or not it can be attained will be discovered in the course of the design.

Criteria for Maximum Efficiency

Recall from Chapter 4 that apart from such specifics as the kind of core material and K factor, certain general criteria must be met for maximum efficiency. In summary these are:

1. The total power loss should be distributed equally between the iron and the copper.
2. The copper loss should be distributed equally between the primary winding and the sum of the secondary windings.
3. The space occupied by the primary winding should equal that occupied by the sum of the secondary windings.
4. In general, the highest permissible flux density should be employed.
5. Insulation clearances and margins should be the smallest possible, compatible with meeting mechanical and voltage requirements.
6. The core cross section should be square.

A Useful Approach

Once an efficiency figure has been selected, it can be converted first to a watts loss figure for the transformer, and then to a watts loss figure for the core and used along with the Wa product to select a suitable core.

Suppose the efficiency has been calculated at an optimistic 94 percent based on a required regulation of 3 percent. Still adhering to the basic specification of the small 20-watt transformer, the watts input will be $20/0.94 = 21.3$ watts. The watts loss is thus 1.3 watts, which ideally will be divided between the core and the windings, making the required core loss $1.3/2 = 0.65$ watt.

If you consider using the same core material and a flux density of, say, 16,500 gauss, the loss is 7.2 watts per pound, and the required core weight using this material would be $0.65/7.2 = 0.09$ lb. The specific density of this material is given in Table 5-8 as 0.284 lb/cubic inch and therefore the required core volume is $v = 0.09/0.284 = 0.316$ cubic inch.

Compared to designs No. 1 and No. 2, this figure is ludicrously small. A trial design quickly reveals how impractical it would be. For example, Magnetic Metals lamination type 27 EI (Fig. 8-9) stacked to give a square cross-sectional area yields a core about equal to that calculated. The cross-sectional area is given as 0.14 square inch. From this, the number of turns on the primary works out to be 2612 turns! A very small gauge wire would be needed to get the turns into the window space, with a resulting massive copper loss. If the flux density is substantially increased in order to reduce the number of turns, the iron loss will increase alarmingly, resulting in a further reduction in the

MAGNETIC **M**ETALS	**LAMINATION TYPE 27 EI**	

CHARACTERISTICS OF A CORE STACK HAVING A SQUARE CROSS SECTION

VOLUME AND WEIGHT

VOLUME	− .316 in.³	− 5.18 cm.³
WINDOW AREA	− .125 in.²	− .806 cm.²
WT. SUPER Q 80	− .0998 lb.	− 45.2 g.
WT. SUPERPERM "49"	− .0942 lb.	− 42.8 g.
WT. SUPERFLUX	− .0931 lb.	− 42.2 g.
WT. SILICON	− .0866 lb.	− 39.3 g.

MAGNETIC DESIGN FORMULAE

Properties of Core Stack with Winding of "N" Turns

$$B_{max} = \frac{414 \times 10^3}{K_1 N} \quad \text{Gausses Per Volt at 60 Hertz}$$

$H_o = (.220 \times 10^{-3}) N$ Oersteds (Gilberts per centimeter) per milliampere of direct current

$L_o = (.1995 \times 10^{-8}) K_1 N^2 \mu_{ac}$ Henries

MAGNETIC PATH DIMENSIONS

l = 2.25 in.	5.72 cm.
A = .1406 in.²	.907 cm.²

K₁ (STACKING FACTOR)

Thickness	Butt Jointed	Interleaved one per layer
.004"	.90	.80
.006"	.90	.85
.014"	.95	.90
.0185"	.95	.90

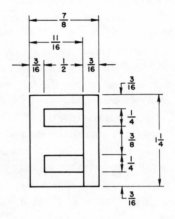

PERFORMANCE DESIGNATION	MATERIAL TYPE	THICKNESS (Inches)	CATALOG NUMBER	WEIGHT AND COUNT	
				LBS./M PCS.	PCS./LB.
SUPERPERM 80	HyMu 80	.004	27EI8404	1.066	938
SUPERPERM 80	HyMu 80	.006	27EI8406	1.598	626
SUPERPERM 80	HyMu 80	.014	27EI8414	3.730	268
SUPER Q 80	HyMu 80	.004	27EI8004	1.066	938
SUPER Q 80	HyMu 80	.006	27EI8006	1.598	626
SUPER Q 80	HyMu 80	.014	27EI8014	3.730	268
SUPERTHERM 80	HyMu 80	.006	27EI7406	1.598	626
SUPERTHERM 80	HyMu 80	.014	27EI7414	3.730	268
SUPERPERM 49	49	.004	27EI4904	.993	1,007
SUPERPERM 49	49	.006	27EI4906	1.487	673
SUPERPERM 49	49	.014	27EI4914	3.470	288
SUPERFLUX	PERMENDUR	.006	27EIVP06	1.468	680
SUPERFLUX	PERMENDUR	.010	27EIVP10	2.447	408
MICROSIL	Gr. Or. Silicon	.004	27EI3304	.924	1,082
MICROSIL	Gr. Or. Silicon	.006	27EI3306	1.386	722
SILICON	Non Or. Silicon*	.014	27EI**14	3.233	309

* Customer to designate AISI grade of material desired.
** See "How To Order Section" for Code Number.

Courtesy Magnetic Metals Corp.

Fig. 8-9. Lamination data for trial design.

core size required to meet the specified 0.65-watt iron loss. The reduction in turns is thus offset by having less space to accommodate them—which is an obviously self-defeating proposition.

The trick, it would seem, is to find a core material and format that has an inherently low loss per pound at high flux density. Several alternatives can be considered, such as gain-oriented silicon steel or vanadium-cobalt-iron stampings. Even better, the same materials in C-core style, and still better, in uncut toroidal form. But, of course, the additional cost of these fancy-dan cores can't be ignored.

Variant No. 3—A "C" Core Design

Consider grain-oriented silicon iron in C-core format made from 0.012-inch tape. The curves in Fig. 5-7B show a loss of about 0.9 watt per pound at 15,000 gauss, 60 hertz. The question is: Can an efficiency of around 94 percent be achieved in our small 20-watt transformer using this core? First of all, the chances of success will improve if a space factor K better than the 0.26 obtained in the previous examples can be achieved. By cutting down on bobbin thickness and fit allowance and not being so prodigal with the primary and secondary cover thicknesses, space can be saved. Nothing much can be done about the space-stealing winding margins; it is safest to retain the spacings given in the wire tables, and this applies also to the interlayer paper. By paying due attention to space saving, a K figure of about 0.3 or better should be possible.

The required iron loss is 0.65 watt. The catalog shows that several pairs will give this order of loss at 15,000 gauss. At a current density of, say, 1000 cm/A, which is a common figure for high-efficiency transformers, the power capability equation gives a Wa product figure of 0.408.

Therefore, look for a core having a loss of 0.325 watt and a Wa of 0.204 (half of the required figures because the catalog lists single C cores, two of which are needed to make one shell type core). Consulting Table 8-1, which is part of a Magnetic Metals listing, a core that is very close in watts loss at 0.322 watt and not too far off in Wa at 0.215, is No. MA128K (see arrow). A pair of these seem to be a good choice to start with. The larger Wa figure will require a slightly larger current density S as determined by a rearrangement of Equation 3-38 as follows:

$$S = \frac{WafB}{P \times 17.26} = \frac{0.215 \times 2 \times 60 \times 15,000}{20 \times 17.26}$$
$$= 1121 \text{ CM/A}$$

Note that what we refer to as Wa is identified in the tables as the $DEFG$ product (sixth column). Also note that the figure given for this must be multiplied by 2, as shown in the equation, for reasons already stated.

In the design sheet in Fig. 8-10, other factors that must be doubled are the dimension E, the weight Wt, and the iron loss.

The wire gauges work out at No. 27 and No. 14. However, in order to get a better fit for the secondary winding layers, the No. 14 is replaced with No. 15, which then

Table 8-1. Data on Core MA-128K (Arrowed)

SINGLE-PHASE CUT CORES

MATERIAL- MICROSIL 12 MIL, STACKING FACTOR .95

- NOTE, THIS LISTING SEQUENCED ON CATALOG NUMBER -

CORE LIMITS BASED ON 1.70 VA/LB & .90 WATTS/LB @ 60 HZ, 15.0 KILOGAUSS
AMPERE TURNS ASSUMES .001 IN. AIRGAP, EXCEPT .002 IN. WHERE STARRED*

MAGNETIC METALS CATALOG NUMBER	DIMENSIONS, INCHES				DEFG PRO-DUCT	...IRON... WEIGHT LBS.	LEG AREA SQ.IN.	CORE DESIGN ...LIMITS... WATTS LOSS	AMPERE TURNS	TURNS PER VOLT	MAGNETIC METALS CATALOG NUMBER
	D	E	F	G							
MA-94F	1.500	.625	.875	2.625	2.15	2.15	.891	1.94	36.3	4.35	MA-94F
MA-94P	1.500	.500	.825	3.000	1.86	1.77	.713	1.59	36.8	5.44	MA-94P
MA-95	.750	.750	1.000	3.000	1.69	1.50	.535	1.35	38.8	7.25	MA-95
MA-95M	1.500	.500	.875	2.875	1.89	1.74	.713	1.57	36.5	5.44	MA-95M
MA-95N	2.000	.875	.875	2.500	3.83	4.26	1.66	3.83	37.3	2.33	MA-95N
MA-97A	1.125	1.000	.875	3.625	3.57	3.52	1.07	3.17	62.5*	3.63	MA-97A
MA-98A	1.000	.406	.437	2.593	.460	.758	.386	.682	33.4	10.1	MA-98A
MA-100H	.437	.250	.312	1.437	.049	.120	.104	.108	28.0	37.5	MA-100H
MA-100L*	.500	.219	.312	1.062	.036	.096	.105	.086	26.5	37.1	MA-100L*
▫MA-102	1.500	1.312	1.375	3.000	8.12	6.53	1.87	5.88	63.8*	2.07	▫MA-102
MA-103	1.500	1.000	1.250	3.750	7.03	5.08	1.43	4.58	64.3*	2.72	MA-103
▫MA-111	1.500	1.000	1.000	3.000	4.50	4.30	1.43	3.87	60.7*	2.72	▫MA-111
MA-111F	1.750	.843	1.125	2.875	4.77	4.04	1.40	3.63	39.4	2.77	MA-111F
MA-111H	1.500	.900	1.200	3.200	5.18	4.04	1.28	3.64	41.2	3.02	MA-111H
MA-111K	1.750	.812	1.125	2.750	4.40	3.76	1.35	3.38	38.7	2.87	MA-111K
MA-112F	.750	.500	1.125	2.250	.949	.797	.356	.717	35.1	10.9	MA-112F
MA-112G	.625	.625	1.125	2.500	1.10	.922	.371	.830	36.8	10.4	MA-112G
MA-113A	2.500	1.000	1.000	3.250	8.13	7.49	2.38	6.74	61.6*	1.63	MA-113A
MA-113B	3.000	1.000	1.000	3.250	9.75	8.99	2.85	8.09	61.6*	1.36	MA-113B
MA-117B	1.750	.750	1.000	2.625	3.45	3.23	1.25	2.91	37.5	3.11	MA-117B
▫MA-119	2.500	1.000	1.375	3.000	10.3	7.65	2.38	6.89	62.0*	1.63	▫MA-119
MA-119A	2.500	.875	1.375	3.000	9.02	6.47	2.08	5.82	40.9	1.87	MA-119A
MA-119B	2.500	1.125	1.375	3.000	11.6	8.90	2.67	8.01	62.7*	1.45	MA-119B
MA-125	1.250	.750	.750	2.312	1.63	2.03	.891	1.83	35.4	4.35	MA-125
MA-125E	1.250	.750	.750	2.250	1.58	2.00	.891	1.80	35.2	4.35	MA-125E
MA-126D	1.250	.500	.750	1.937	.908	1.12	.594	1.01	32.8	6.53	MA-126D
MA-126F	.600	.350	1.359	1.550	.442	.375	.200	.338	32.8	19.4	MA-126F
MA-126G	1.250	.875	1.375	1.875	2.82	2.62	1.04	2.36	37.0	3.73	MA-126G
MA-126H	1.250	.250	.750	2.000	.469	.506	.297	.455	31.6	13.0	MA-126H
MA-127	.875	.750	.750	2.500	1.23	1.49	.623	1.34	36.1	6.22	MA-127
MA-128B	1.125	.375	.437	1.375	.253	.519	.401	.467	28.9	9.67	MA-128B
➤MA-128K	.500	.500	.625	1.375	.215	.358	.238	.322	30.3	16.3	MA-128K
MA-129B	.750	1.000	.625	1.937	.908	1.60	.713	1.44	55.6*	5.44	MA-129B
MA-129E	.750	.500	.625	1.937	.454	.648	.356	.583	32.4	10.9	MA-129E
MA-130	2.000	.875	1.000	3.000	5.25	4.83	1.66	4.35	39.6	2.33	MA-130
MA-130E	1.750	.625	1.000	3.250	3.56	2.94	1.04	2.65	39.0	3.73	MA-130E
MA-130F	1.250	.687	1.000	3.125	2.68	2.30	.816	2.07	38.9	4.75	MA-130F
MA-132	2.000	1.500	1.000	3.000	9.00	9.83	2.85	8.85	63.5*	1.36	MA-132
MA-132E	1.937	.437	.750	3.250	2.06	2.03	.804	1.83	37.1	4.82	MA-132E
MA-133	2.000	1.000	2.000	5.000	20.0	8.88	1.90	7.99	71.6*	2.04	MA-133
MA-133D	2.000	1.000	2.000	4.500	18.0	8.35	1.90	7.52	69.7*	2.04	MA-133D
MA-133F	1.250	.750	2.000	4.250	7.97	3.60	.891	3.24	47.0	4.35	MA-133F
MA-134	3.000	1.000	1.000	3.875	11.6	9.97	2.85	8.97	63.8*	1.36	MA-134
MA-134B	1.000	1.250	1.000	4.000	5.00	4.49	1.19	4.04	65.7*	3.27	MA-134B
MA-134C	.875	.562	1.000	4.000	1.97	1.49	.467	1.34	41.4	8.30	MA-134C
MA-135	1.250	1.250	1.000	3.000	4.69	4.80	1.49	4.32	62.1*	2.61	MA-135
▫MA-138	1.250	.500	.937	2.500	1.46	1.35	.594	1.21	35.4	6.53	▫MA-138
MA-140A	3.000	1.000	1.375	3.000	12.4	9.18	2.85	8.27	62.0*	1.36	MA-140A
MA-144	1.000	1.000	1.500	2.500	3.75	2.87	.950	2.58	60.7*	4.08	MA-144
MA-144C	.750	.500	1.500	2.500	1.41	.920	.356	.828	37.4	10.9	MA-144C

MATERIAL- MICROSIL 12 MIL
▫ - PREFERRED SIZE

TRANSFORMER DESIGN SHEET

Input volts	110	Hertz	60		Est. Efficiency	94%	Turn/volt	8.17

Lamination or core No. 128 K × 2   C CORE MICROSIL — .012 INCH B 15,000

Window dimensions G 1.375   F 0.625   W   M_p

C.S.A. dimensions E 1.00 (0.5×2) D 0.5   a 0.476 (.238×2) v

Wt 0.716 (.358×2) Watts/lb   Iron loss 0.644 (.322 × 2)

Windings		W₁	W₂							
Coil		PRIM	SEC							
Volts		110	5							
Amperes	(.207)	.193	4							
Turns		898	41							
Gauge		27	15							
Turns/inch		59	15							
Margins		.125	.188							
Winding length (L)		1.05	.924							
Turns/layer		62	14							
Number of layers		14.5	3							

Build										Total
Copper		.228	.176							
Paper		.028	.020							
Cover		.010	.010							
Total		.266	.206							
Bulge Percent		.040	.031							
Total (R)		.306	.237							.543

Total bobbin (B) .082

Bobbin .062 clearance .020   Total depth .625

Losses										
Length mean turn (M)		4.617	6.38							
Total wire length (inches)		4141	268							
Ohms/ft		.05164	.00318							
Resistance (hot)		21.3	.0852							
Voltage drop	(4.4)	4.11	.34							
Copper loss (I²R)	(0.9)	.795	1.36							2.16

Iron loss .64
Total loss 2.8
Efficiency 87%

Temperature Rise						
Coil	t	C_S	W	W/C_S	°C T_o	
1						
2						
3						
4						

Temp. Rise = T_o(Total) + 20 HW/C_S =

Fig. 8-10. Variant No. 3, design sheet.

works out at exactly 3.0 layers. The rest of the design follows through in the normal way, culminating in a total loss of 2.80 watts. This figure is used to increase the input watts to the more realistic value of 22.80 watts and the primary winding current to 0.207 ampere. These figures in turn are used to upgrade the primary winding loss figures, as indicated, which gives finally a total loss of 2.90 watts and 87 percent efficiency.

While this is an improvement on previous designs, it still does not meet the specification. Perhaps it can be brought closer by considering another core as in the next variant. Possibly, too, a better balance can be achieved with regard to losses.

Variant No. 4

Another search of the catalog reveals core No. 224B (Table 8-2), which is slightly lighter than MA-128K and has a smaller loss at 0.347 lb and 0.312 watt, respectively, but has a substantially larger Wa product of 0.371 inch^4. However, the advantage of the larger Wa figure is to an extent illusory because a for two cores is only 0.40 square inch or 16 percent less than for the MA-128K, which means that the number of turns must increase by 16 percent. Dimension FG (W) is a massive 1.767 square inches compared to 0.859 square inches for the MA-128K, more than 100 percent greater, and the Wa product for two such cores is 0.742 compared with 0.430 inch^4.

A ball-park figure for the current density S to be employed on this core can be obtained by using Equation 3-38 as before: Thus

$$S = \frac{WafB}{P \times 17.26} = \frac{0.742 \times 60 \times 15,000}{20 \times 17.26}$$
$$= 1934 \text{ CM/A}$$

At first sight, this current density that is much lower than typical might be expected to result in a very low copper loss, but the greater number of turns needed offsets this to some extent. The primary winding gauge works out to 386 CM—between No. 24 and No. 25 gauge, say No. 25—and the secondary winding gauge works out to 7736 CM, about No. 12 gauge. When the build is worked out (Fig. 8-11) there is 0.161 inch to spare. In view of the heavy wire being used on the secondary, however, leave this as a safety margin for excessive bulge until you see how the rest of the design works out.

In the end, the improvement is only about 2 percent and there is still a substantial imbalance in losses. But the window space was not completely filled. If this space can be filled by using, say, heavier-gauge wire, some improvement in efficiency can be expected.

There is a practical problem here, especially if the coil is hand wound, because the No. 12 gauge already designed into the coil is rather heavy for ease of winding, and a heavier gauge will make winding that much harder. A solution to this problem is to use two smaller gauge wires wound on bifilar style then connected in parallel as in Fig. 7-4, Chapter 7.

For example, from the wire table (Table 6-1), it can be seen that two No. 14s are about equivalent to a No. 11 in cross-sectional area. The turns-per-inch figure for *two*

Table 8-2. Data on Core MA-224B (Arrowed)

SINGLE-PHASE CUT CORES

MATERIAL- MICROSIL 12 MIL, STACKING FACTOR .95

- NOTE, THIS LISTING SEQUENCED ON CATALOG NUMBER -

CORE LIMITS BASED ON 1.70 VA/LB & .90 WATTS/LB @ 60 HZ, 15.0 KILOGAUSS
AMPERE TURNS ASSUMES .001 IN. AIRGAP, EXCEPT .002 IN. WHERE STARRED*

MAGNETIC METALS CATALOG NUMBER	DIMENSIONS, INCHES				DEFG PRO-DUCT	...IRON... WEIGHT LBS.	LEG AREA SQ.IN.	CORE DESIGN ...LIMITS... WATTS LOSS	AMPERE TURNS	TURNS PER VOLT	MAGNETIC METALS CATALOG NUMBER
	D	E	F	G							
MA-144D	1.500	1.500	1.500	2.312	7.80	7.15	2.14	6.44	62.8*	1.81	MA-144D
MA-144E	.500	.500	1.500	1.500	.563	.489	.238	.440	34.0	16.3	MA-144E
□MA-146	1.125	.500	.937	2.500	1.32	1.21	.535	1.09	35.4	7.25	□MA-146
MA-147A	.500	.500	1.000	.875	.219	.342	.238	.308	29.9	16.3	MA-147A
MA-151	1.625	1.125	1.625	4.187	12.4	7.16	1.74	6.45	68.0*	2.23	MA-151
MA-151A	.625	.500	.500	2.000	.313	.529	.297	.476	32.1	13.0	MA-151A
MA-151C	1.500	.500	.500	2.250	.844	1.35	.713	1.21	32.9	5.44	MA-151C
MA-156A	1.500	.750	3.343	4.875	18.3	5.48	1.07	4.93	54.2	3.63	MA-156A
MA-159J	1.000	.500	2.000	3.500	3.50	1.62	.475	1.46	42.9	8.16	MA-159J
MA-160	1.375	1.375	2.500	2.750	13.0	7.24	1.80	6.52	67.3*	2.16	MA-160
□MA-163	1.000	.437	.625	1.937	.529	.732	.415	.659	32.0	9.34	□MA-163
MA-164A	1.500	.625	.625	1.937	1.14	1.72	.891	1.54	33.1	4.35	MA-164A
MA-168C	1.500	.562	.875	1.625	1.20	1.47	.801	1.32	32.5	4.84	MA-168C
MA-170Q	2.500	1.062	.750	2.750	5.48	7.05	2.52	6.34	59.2*	1.54	MA-170Q
MA-174A	4.000	1.000	.750	3.750	11.3	12.5	3.80	11.3	62.5*	1.02	MA-174A
MA-174B	4.000	1.000	.750	1.500	4.50	7.90	3.80	7.11	54.5*	1.02	MA-174B
MA-174C	4.000	.625	.875	2.875	6.29	6.06	2.38	5.46	57.6*	1.63	MA-174C
MA-176A	.500	.500	1.000	3.000	.750	.613	.238	.552	37.4	16.3	MA-176A
MA-177	1.500	.750	1.000	3.000	3.38	2.99	1.07	2.69	38.8	3.63	MA-177
MA-179	2.000	.750	.937	2.500	3.51	3.55	1.43	3.19	36.8	2.72	MA-179
MA-181D	2.375	1.000	.937	4.750	10.6	8.90	2.26	8.01	66.8*	1.72	MA-181D
MA-182F	1.125	1.062	.937	3.500	3.92	3.76	1.14	3.38	62.6*	3.42	MA-182F
MA-184	.750	.750	1.375	3.000	2.32	1.61	.535	1.45	40.2	7.25	MA-184
□MA-185	1.375	.625	1.312	3.500	3.95	2.26	.816	2.31	41.1	4.75	□MA-185
□MA-187	2.500	1.250	1.000	3.000	9.38	9.60	2.97	8.64	62.1*	1.31	□MA-187
MA-187B	1.000	1.000	1.000	3.250	3.25	3.00	.950	2.70	61.6*	4.08	MA-187B
MA-187D	2.250	.625	1.000	3.000	4.22	3.59	1.34	3.24	38.1	2.90	MA-187D
□MA-196	2.250	1.125	1.375	3.000	10.4	8.01	2.40	7.21	62.7*	1.61	□MA-196
MA-196A	2.250	1.000	1.375	3.000	9.28	6.89	2.14	6.20	62.0*	1.81	MA-196A
MA-196B	2.218	1.000	1.375	3.000	9.15	6.79	2.11	6.11	62.0*	1.84	MA-196B
MA-200B	1.500	1.000	.937	2.500	3.51	3.85	1.43	3.47	58.6*	2.72	MA-200B
MA-202E	1.250	1.218	.937	3.250	4.64	4.78	1.45	4.31	62.6*	2.68	MA-202E
□MA-203	2.250	1.125	1.625	3.375	13.9	8.84	2.40	7.96	65.0*	1.61	□MA-203
MA-204A	1.000	.625	.750	1.562	.732	1.06	.594	.956	32.2	6.53	MA-204A
MA-204B	1.375	1.375	.750	1.750	2.48	4.57	1.80	4.11	57.5*	2.16	MA-204B
MA-204D	1.000	.750	.750	1.562	.879	1.35	.713	1.22	32.9	5.44	MA-204D
MA-222A	2.250	.906	.968	3.562	7.03	6.25	1.94	5.62	41.7	2.00	MA-222A
➤MA-224B	.600	.350	1.359	1.300	.371	.347	.200	.312	31.9	19.4	MA-224B
MA-224C	.600	.350	2.145	1.300	.586	.434	.200	.391	31.9	19.4	MA-224C
MA-227B	1.125	.500	.625	1.625	.571	.879	.535	.791	31.2	7.25	MA-227B
MA-227F	1.000	.625	.625	1.625	.635	1.04	.594	.937	32.0	6.53	MA-227F
MA-227G	1.375	.875	.500	3.250	1.96	3.17	1.14	2.85	38.7	3.39	MA-227G
MA-227H	1.500	.500	.500	2.750	1.03	1.55	.713	1.39	34.7	5.44	MA-227H
MA-228B	1.250	.750	.625	4.000	2.34	2.80	.891	2.52	41.1	4.35	MA-228B
MA-230A	1.500	.750	.500	5.000	2.81	3.88	1.07	3.49	44.3	3.63	MA-230A
MA-232F	.250	.500	.500	.875	.055	.138	.119	.124	28.1	32.6	MA-232F
MA-232H	.375	.375	.500	1.125	.079	.159	.134	.143	28.2	28.9	MA-232H
MA-232J	1.000	.500	.500	1.500	.375	.716	.475	.644	30.3	8.16	MA-232J
MA-234	1.625	.687	1.000	2.000	2.23	2.36	1.06	2.12	35.0	3.66	MA-234
MA-234F	1.968	.968	1.000	2.031	3.87	4.44	1.81	4.00	36.6	2.14	MA-234F

MATERIAL- MICROSIL 12 MIL
□ - PREFERRED SIZE

TRANSFORMER DESIGN SHEET

| Input volts | 110 | | Hertz | 60 | | | Est. Efficiency | | | | Turn/volt | 9.7 | |

| Lamination or core No. | 224 B×2 C CORE MICROSIL — .012 INCH | B | 15,000 |

Window dimensions   G 1.3   F 1.359   W M_p

C.S.A. dimensions   E 0.7 (0.35×2)   D 0.6   a 0.40 (0.2×2) v

Wt 0.694 (.347×2) Watts/lb   Iron loss 0.624 (.312×2)

Windings		W_1	W_2		W_2								
Coil					TRIAL								
Volts		110	5										
Amperes		.197	4										
Turns		1067	39										
Gauge		25	12		14×2								
Turns/inch		47	11										
Margins		.125	.25										
Winding length (L)		1.05	.8										
Turns/layer		49	9										
Number of layers		21.7	4.33										

Build													Total
Copper		.418	.42		.590								
Paper		.044	.050		.050								
Cover		.020	.020		.020								
Total		.472	.490		.660								
Bulge Percent		.070	.073		.099								
Total (R)		.542	.563		.753								1.105

Total bobbin (B) .093

Total depth 1.198

Bobbin .062 clearance .031

Losses													
Length mean turn (M)													
Total wire length (inches)													
Ohms/ft													
Resistance (hot)													
Voltage drop													
Copper loss (I^2R)													

Iron loss

Total loss

Efficiency

Temperature Rise

Coil	t	C_S	W	W/C_S	°C T_o
1					
2					
3					
4					

Temp. Rise = T_o(Total) + $20HW/C_S$ =

Fig. 8-11. Variant No. 4, design sheet.

No. 14s is half the number for one, that is to say, 14/2 = 7 turns per inch. The turns per layer figure is 5 to 6, say 6 by dint of squeezing a little, and the number of layers is 8.5 (9). This gives a copper thickness of 0.59 inch. A quick check of build thickness, however, shows that only 0.063 inch is left for the bobbin. This is rather skimpy, but if the bulge can be squeezed into 12 percent, then 0.083 inch will be left for the bobbin, which isn't bad. A gentle squeeze in the vise will do the job. (This would not be recommended for factory production, perhaps; nevertheless, the casual experimenter can get away with such contrivances.)

Suppose this is done. The secondary winding loss works out to be 0.937 watt and the efficiency increases slightly to 89.76 percent. Whether this improvement is worth the risk of not achieving a proper fit of the windings to the core is a question everyone must answer for himself.

Variant No. 5

It seems likely that if the losses can be brought into better balance, an efficiency of better than 90 percent can be achieved, although 94 percent seems unlikely.

Assuming, say, 92 percent, then the total permissible loss is 1.74 watts of which 0.87 watt should be tied up in the core. A pair of C cores having a 0.435 watt loss each is then required. From the catalog, an MA-432 core meets this requirement. (See Table 8-3.) It has a Wa product of 0.445, which gives 0.890 for the pair. Another possibility is MA-434, with exactly the same weight and loss but a Wa product of 0.293, or 0.586 for the pair. A somewhat lighter core with a correspondingly lower loss is MA-234L with 0.449 watt loss per core and a Wa product of 0.531, or 1.062 per pair. The statistics for these three cores are shown together for comparison in Table 8-3.

Again, the choice is best done on a cut-and-try basis although there are guidelines that can influence the choice. For instance, the turns-per-volt figures are 12.4, 10.9, and 16.3. The smaller turns-per-volt figure of MA-434 might be attractive to anyone winding by hand, although the squarer cross-sectional area of MA-432 will probably result in a somewhat lower winding resistance and loss.

Suppose you opt for the lower turns-per-volt figure of MA-434, which is 10.9/2 = 5.45 for a pair. The results are given in Fig. 8-12. To be strictly accurate, the input watts, amperes, and primary winding loss should be increased somewhat on the basis of the 90.8 percent efficiency figure, which is slightly less than the 92 percent originally postulated. This doesn't make much difference to the design, but for interest, the corrected figures are shown in brackets.

Notes

If the copper loss is substantially greater than the iron loss, the efficiency initially postulated cannot be obtained. However, the efficiency can be maximized at a lesser figure by increasing the core size and adjusting the windings and core until the iron and copper losses are equal.

Table 8-3. Comparison of Data on MA-234L, MA-432, and MA-434

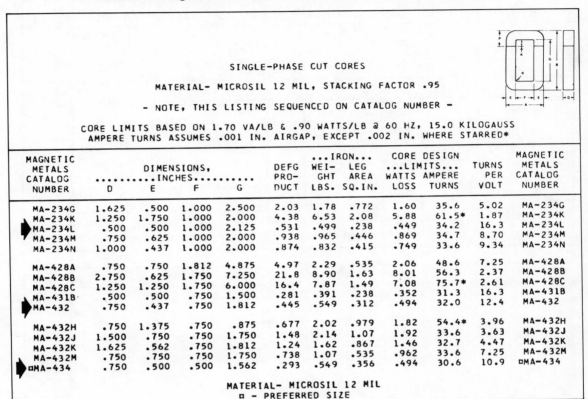

SINGLE-PHASE CUT CORES

MATERIAL- MICROSIL 12 MIL, STACKING FACTOR .95

- NOTE, THIS LISTING SEQUENCED ON CATALOG NUMBER -

CORE LIMITS BASED ON 1.70 VA/LB & .90 WATTS/LB @ 60 HZ, 15.0 KILOGAUSS
AMPERE TURNS ASSUMES .001 IN. AIRGAP, EXCEPT .002 IN. WHERE STARRED*

MAGNETIC METALS CATALOG NUMBER	DIMENSIONS, INCHES				DEFG PRO-DUCT	...IRON... WEI-GHT LBS.	LEG AREA SQ.IN.	CORE DESIGN ...LIMITS... WATTS LOSS	AMPERE TURNS	TURNS PER VOLT	MAGNETIC METALS CATALOG NUMBER
	D	E	F	G							
MA-234G	1.625	.500	1.000	2.500	2.03	1.78	.772	1.60	35.6	5.02	MA-234G
MA-234K	1.250	1.750	1.000	2.000	4.38	6.53	2.08	5.88	61.5*	1.87	MA-234K
▶MA-234L	.500	.500	1.000	2.125	.531	.499	.238	.449	34.2	16.3	MA-234L
MA-234M	.750	.625	1.000	2.000	.938	.965	.446	.869	34.7	8.70	MA-234M
MA-234N	1.000	.437	1.000	2.000	.874	.832	.415	.749	33.6	9.34	MA-234N
MA-428A	.750	.750	1.812	4.875	4.97	2.29	.535	2.06	48.6	7.25	MA-428A
MA-428B	2.750	.625	1.750	7.250	21.8	8.90	1.63	8.01	56.3	2.37	MA-428B
MA-428C	1.250	1.250	1.750	6.000	16.4	7.87	1.49	7.08	75.7*	2.61	MA-428C
▶MA-431B	.500	.500	.750	1.500	.281	.391	.238	.352	31.3	16.3	MA-431B
▶MA-432	.750	.437	.750	1.812	.445	.549	.312	.494	32.0	12.4	MA-432
MA-432H	.750	1.375	.750	.875	.677	2.02	.979	1.82	54.4*	3.96	MA-432H
MA-432J	1.500	.750	.750	1.750	1.48	2.14	1.07	1.92	33.6	3.63	MA-432J
MA-432K	1.625	.562	.750	1.812	1.24	1.62	.867	1.46	32.7	4.47	MA-432K
MA-432M	.750	.750	.750	1.750	.738	1.07	.535	.962	33.6	7.25	MA-432M
▶□MA-434	.750	.500	.500	1.562	.293	.549	.356	.494	30.6	10.9	□MA-434

MATERIAL- MICROSIL 12 MIL
□ - PREFERRED SIZE

To increase the efficiency further from this point, a core material having an inherently lower loss per pound must be chosen and the component redesigned along the same lines.

All this assumes that a square-section core area has been used in the design. If the core area departs considerably from the square format, the efficiency can be improved by choosing suitably proportioned laminations in order to achieve the square area. This should perhaps be considered before looking at lower-loss (and therefore more expensive) core materials.

LOW FREQUENCIES COMPARED WITH HIGH FREQUENCIES

As mentioned previously, in some power transformer applications the designer is not tied to a specific frequency—he or she can select his own as in converter circuits. Here he will often elect to work at higher frequencies in order to reduce the size and cost of transformers and associated components. In other situations, such as in the design of aircraft equipment, he will be presented with high-frequency inputs to work with. Often the equipment users are prepared to pay a premium for relatively high-performance

TRANSFORMER DESIGN SHEET

| Input volts | 110 | Hertz | 60 | Est. Efficiency | 92 | Turn/volt | |

Lamination or core No. MA-434×2 C CORE MICROSIL   B 15,000

| Window dimensions | G 1.562 | F 0.500 | W | M_P |

| C.S.A. dimensions | E 1.0 (.5×2) | D 0.75 | a 0.712 (.356×2) v |

Wt 1.098 (.549×2)   Watts/lb   Iron loss 0.988 (.494×2)

Windings			W_1	W_2							
Coil			PRIM	SEC							
Volts			110	5							
Amperes		(.200)	.197	4							
Turns			600	29							
Gauge			26	16×2							
Turns/inch			53	8.5							
Margins			.125	.188							
Winding length (L)			1.3	1.186							
Turns/layer			67	10							
Number of layers			8.96	2.9							

Build											Total
Copper			.153	.157							
Paper			.016	.020							
Cover			.020	.020							
Total			.189	.197							
Bulge Percent			.026	.028							
Total (R)			.215	.225							.440

Bobbin .040 clearance .020   Total bobbin (B) .060
                                     Total depth .500

Losses											
Length mean turn (M)			4.82	6.14							
Total wire length (inches)			2892	178							
Ohms/ft			.041	.002							
Resistance (hot)			11.85	.0359							
Voltage drop		(2.37)	2.3	.144							(1.049)
Copper loss (I^2R)		(.474)	.460	.575							1.035

Iron loss .988
Total loss 2.023
Efficiency 91%

Temperature Rise						
Coil	t	C_S	W	W/C_S	°C T_o	
1						
2						
3						
4						

Temp. Rise = T_o(Total) + 20 HW/C_S =

Fig. 8-12. Variant No. 5, design sheet.

core materials that not only reduce the size and weight of the transformer but substantially increase its efficiency and improve regulation.

It can be seen from the basic transformer equation that as f is increased, N is decreased for a given voltage, all other terms remaining constant. Again consider the difference in Wa product for two transformers having the same power capability and working at the same flux and current densities but at widely different frequencies. Suppose that the power requirement is 400 watts, the flux density 12,000 gauss, and the current density 850 CM/A in each case, but the frequency is 60 hertz in one, and 400 hertz in the other. Then, using Equation 3-39,

$$Wa = \frac{17.26PS}{fB}$$

we have for design No. 1:

$$Wa = \frac{17.26 \times 400 \times 850}{60 \times 12,000} = 8.15 \text{ inch}^4$$

and for design No. 2:

$$Wa = \frac{17.26 \times 400 \times 850}{400 \times 12,000} = 1.22 \text{ inch}^4$$

Unfortunately, the second result is not all gain, because the loss in a given core increases with frequency. For example, a grain-oriented silicon-steel cut core constructed from a 0.012-inch strip will have a loss on the order of 0.5 watt per pound at 12,000 gauss, 60 hertz, rising to about 8.0 watts per pound at 12,000 gauss, 400 hertz, and then to a monstrous 30 watts per pound at 1000 hertz.

The dramatic reduction in the Wa product, however, means fewer pounds and the net core loss is still low enough to give a marked design advantage, if the frequency is not too high. In comparing the effects of frequency, let's work at the 250-watt power level and at the same time introduce a little more variety into the design by using three windings, a higher voltage, and different configurations.

General Specifications

The input voltage is 115 volts, and for the outputs, one is an isolated 115 volts at 2.0 amperes, and one is a 26-volt output at 0.5 ampere. The circuit is in Fig. 8-13. The regulation is to be better than about 4 percent, and the temperature rise is to be no greater than 40°C to 50°C. From these figures, the total output power is 243 watts.

The 4 percent regulation stipulation suggests a required efficiency figure. Thus

$$\% \, \eta = 100 \left(\frac{1}{\text{Regulation} + 1} \right)^2 = 100 \left(\frac{1}{0.04 + 1} \right)^2$$

$$= 92.4 \text{ percent}$$

But let's shoot for an ambitious 95 percent just for kicks.

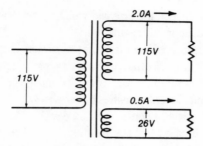

TEMPERATURE RISE = 40°C TO 50°C

Fig. 8-13. Specification for 60- and 400-Hz designs.

The input power is then 243/0.95 = 256 watts, making a loss of 13 watts, which for best efficiency will divide equally between the copper and the iron, or 6.5 watts each. As will be seen in the examples that follow, however, this desirable state of affairs, though fairly easily achieved at 60 hertz, is not so readily obtained at 400 hertz without resorting to more specialized core materials.

At 60 Hertz

In view of the high efficiency required, grain-oriented silicon steel in 0.014-inch-thick stampings is considered for the core first. For this material, a flux density of 15,000 gauss would be conservative, and a current density of 850 CM/A makes a good starting point for the wire. Then

$$Wa = \frac{256 \times 850 \times 17.26}{60 \times 15,000} = 4.17 \text{ inch}^4$$

is the Wa product.

A type 150 MH lamination (Fig. 8-14) looks reasonable. Its window area of 1.69 square inches means a core area of 4.17/1.69 = 2.46 square inches. The center limb width is 1.5 inches, and therefore the stack thickness must be 2.46/(1.5 × 0.9) = 1.8 inch. The magnetic path length is given on the sheet as 9.0 inches; therefore the volume v is $M_P \times a = 9 \times 2.46 = 22.14$ cubic inches. The density of silicon steel is 0.276 pounds per cubic inch and therefore the core weight is 22.14 × 0.276 = 6.1 pounds. The iron loss for this material is about 0.9 watt per pound, making the total core loss 6.1 × 0.9 = 5.5 watts. No design sheet entries were made for the following numbers, because changes were anticipated, but you might want to write them as an exercise.

The turns-per-volt figure is given by

$$\frac{N}{V} = \frac{3.5 \times 10^6}{60 \times 2.46 \times 15,000} = 1.58 \text{ turns/volt}$$

Magnetic Metals

LAMINATION TYPE 150 MH

CHARACTERISTICS OF A CORE STACK HAVING A SQUARE CROSS SECTION

VOLUME AND WEIGHT

VOLUME	– 20.3 in.³	– 332 cm.³
WINDOW AREA	– 1.69 in.²	– 10.9 cm.²
WT. SUPER Q 80	– 6.21 lb.	– 2900 g.
WT. SUPERPERM "49"	– 6.00 lb.	– 2710 g.
WT. SUPERFLUX	– 5.93 lb.	– 2689 g.
WT. SILICON	– 5.42 lb.	– 2457 g.

MAGNETIC DESIGN FORMULAE

Properties of Core Stack with Winding of "N" Turns

$$B_{max} = \frac{25.9 \times 10^3}{K_1 N} \text{ Gausses Per Volt at 60 Hertz}$$

$$H_o = (.055 \times 10^{-3})N \text{ Oersteds}$$
(Gilberts per centimeter) per milliampere of direct current

$$L_o = (.7981 \times 10^{-8})K_1 N^2 \mu_{ac} \text{ Henries}$$

MAGNETIC PATH DIMENSIONS

$l = 9.00$ in. 22.86 cm.
$A = 2.25$ in.² 14.51 cm.²

K₁ (STACKING FACTOR)

Thickness	Butt Jointed	Interleaved one per layer
.004"	.90	.80
.006"	.90	.85
.014"	.95	.90
.0185"	.95	.90

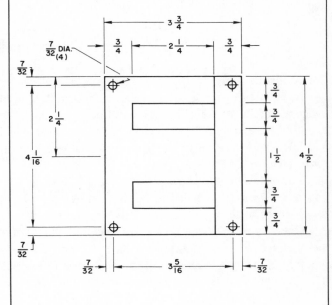

PERFORMANCE DESIGNATION	MATERIAL TYPE	THICKNESS (Inches)	CATALOG NUMBER	WEIGHT AND COUNT	
				LBS. / M PCS.	PCS. / LB.
SUPERPERM 80	HyMu 80	.004	150MH8404	16.7	59.9
SUPERPERM 80	HyMu 80	.006	150MH8406	25.0	40.0
SUPERPERM 80	HyMu 80	.014	150MH8414	58.4	17.1
SUPER Q 80	HyMu 80	.004	150MH8004	16.7	59.9
SUPER Q 80	HyMu 80	.006	150MH8006	25.0	40.0
SUPER Q 80	HyMu 80	.014	150MH8014	58.4	17.1
SUPERTHERM 80	HyMu 80	.006	150MH7406	25.0	40.0
SUPERTHERM 80	HyMu 80	.014	150MH7414	58.4	17.1
SUPERPERM 49	49	.004	150MH4904	15.5	64.4
SUPERPERM 49	49	.006	150MH4906	23.3	43.1
SUPERPERM 49	49	.014	150MH4914	54.3	18.4
SUPERFLUX	PERMENDUR	.006	150MHVP06	23.0	43.5
SUPERFLUX	PERMENDUR	.010	150MHVP10	38.4	26.0
MICROSIL	Gr. Or. Silicon	.004	150MH3304	14.4	69.2
MICROSIL	Gr. Or. Silicon	.006	150MH3306	21.7	46.1
MICROSIL	Gr. Or. Silicon	.014	150MH3314	50.5	19.8
SILICON	Non Or. Silicon*	.014	150MH**14	50.5	19.8
SILICON	Non Or. Silicon*	.018	150MH**18	65.0	15.4
SILICON	Non Or. Silicon*	.025	150MH**25	90.3	11.1

* Customer to designate AISI grade of material desired.
** See "How To Order Section" for Code Number.

Courtesy Magnetic Metals Corp.

Fig. 8-14. Lamination data for 60-Hz design.

Call it 1.6 turns per volt. Then the number of primary turns is $115 \times 1.6 = 184$, and the number of secondary turns is

$$N_2 = \frac{V_2 N_1}{V_1 \sqrt{\eta}} = \frac{115 \times 184}{115 \times 0.97} = 190 \text{ turns}$$

and

$$N_3 = \frac{V_3 N_1}{V_1 \sqrt{\eta}} = \frac{26 \times 184}{115 \times 0.97} = 42.8 \text{ turns}$$

Let N_3 be 43 turns.

At 850 CM/A, the wire gauges are: primary, $2.22 \times 850 = 1887$ CM, say No. 17 gauge at 2052 CM; the 115-volt secondary is $2 \times 850 = 1700$ CM, say No. 17 gauge at 2052 CM; and the 26-volt secondary is $0.5 \times 850 = 425$ CM, say No. 24 gauge at 404 CM. (With these figures, the real current densities in the windings are now 932, 1026, and 808 CM/A, respectively.)

The build before the bobbin works out at 0.789 inch. This is too great for the window space of 0.75 inch. It needs to be reduced by about 0.039 inch plus, say, 0.090 inch for the bobbin, for a total of 0.129 inch or thereabouts.

One way of achieving this is to reduce the numbers of turns so that the primary and first secondary windings are reduced by the thickness of one layer each. In other words, the 115-volt secondary winding becomes 175 turns, making exactly five layers. The number of turns of the primary winding then becomes

$$N_1 = \frac{N_2 V_1 \sqrt{\eta}}{V_2} = \frac{175 \times 115 \times 0.97}{115} = 169.75 \text{ turns}$$

Call it 170 turns. At 170 turns, the flux density is

$$B = \frac{115 \times 3.5 \times 10^6}{170 \times 60 \times 2.46} = 16,000 \text{ gauss}$$

This is within the capability of the core, so the 26-volt winding can be calculated:

$$N_3 = \frac{V_3 N_1}{V_1 \sqrt{\eta}} = \frac{26 \times 170}{115 \times 0.97} = 39.6 \text{ turns}$$

Call it 40 turns.

The figures up to this point are now entered on the sheet of Fig. 8-15. Now the build works out, leaving 0.057 inch for the bobbin. This should be satisfactory, but if necessary, a little could be trimmed from the insulation or a little less space allowed for bulge.

TRANSFORMER DESIGN SHEET

Input volts 115	Hertz 60	Est. Efficiency 95	Turn/volt

Lamination or core No. 150 MH	GRAIN-ORIENTED SILICON STEEL—.012 INCH B 15,000 (16,000)

Window dimensions	G 2.25	F 0.75	W	M_P 9.00

C.S.A. dimensions	E 1.5	D 1.8	a 2.46	v 22.14

	Wt 6.1	Watts/lb 0.9	Iron loss 5.5	

Windings		W_1	W_2	W_3							
Coil		PRIM	SEC	SEC							
Volts		115	115	26							
Amperes		2.22	2.0	0.5							
Turns		170	175	40							
Gauge		17	17	24							
Turns/inch		19	19	42							
Margins		.188	.188	.125							
Winding length (L)		1.87	1.87	2.00							
Turns/layer		35	35	84							
Number of layers		4.86	5.00	.48							

Build											Total
Copper		.234	.234	.021							
Paper		.028	.028								
Cover		.020	.020	.020							
Total		.282	.282	.041							
Bulge Percent		.041	.041	.006							
Total (R)		.323	.323	.047							.693
										Total bobbin (B)	.057
Bobbin .030 clearance .027										Total depth	.75

Losses											
Length mean turn (M)		8.134	10.132	11.274							
Total wire length (inches)		1383	1773	451							
Ohms/ft		.0051	.0051	.0257							
Resistance (hot)		0.70	0.90	1.16							
Voltage drop		1.554	1.80	.58							
Copper loss (I^2R)		3.45	3.60	.29							7.34
										Iron loss	5.50
										Total loss	12.84
										Efficiency	95%

Temperature Rise						
Coil	t	C_S	W	W/C_S	°C T_o	
1	.084	14.13	3.45	.244	5.124	
2	.041	14.13	3.60	.255	2.614	
3	.020	14.13	0.29	.020	0.100	
4						
				.519	7.84	

$$T_S = 20H \frac{W}{C_S} = 20 \times 1.87 \times .519 = 19.4°C$$

Temp. Rise = T_o(Total) + $20HW/C_S$ = 27°C

Fig. 8-15. Design sheet for 60-Hz transformer.

The losses work out as shown, and finally the efficiency works out at 95 percent, exactly as assumed in the beginning. A check on the secondary winding voltage gives

$$V_2 = \frac{N_2(V_1 - V_{d1})}{N_1} - V_{d2}$$

$$= \frac{175(115 - 1.554)}{170} - 1.8$$

$$= 114.98 \text{ volts}$$

and

$$V_3 = \frac{40(115 - 1.554)}{170} - 0.58 = 26.11 \text{ volts}$$

for the second secondary.

An approximate number for the regulation can be obtained by first calculating the no-load voltage for the 115-volt secondary winding thus:

$$V_{\text{no-load}} = \frac{N_2 V_1}{N_1} = \frac{175 \times 115}{170} = 118.38 \text{ volts}$$

Then approximate regulation for this secondary is

$$\% \text{ Regulation} = \frac{100(118.38 - 114.38)}{114.98}$$

$$= 2.96 \text{ percent}$$

Call it 3 percent. A similar calculation for the other secondary winding gives

$$\% \text{ Regulation} = 3.6 \text{ percent}$$

Call it 4 percent.

Review of the Design—Although the iron-copper loss distribution is not perfect, it is not too bad. Similarly, the winding space distribution is quite good. Further improvements in these areas would improve efficiency but only just perceptibly.

The winding space factor (K factor), which works out to be 0.336, is considerably better than in the smaller transformers. The temperature rise is a cool 27°C, as we would expect.

At 400 Hertz—Design No. 1

For the sake of making direct comparison with the 60-hertz case, grain-oriented silicon iron 0.014 inch thick is again considered initially, at the same flux density of 16,000 gauss. The current density of 850 CM/A is retained, and 95 percent efficiency is again (very optimistically) assumed. Thus

$$Wa = \frac{243 \times 850 \times 17.26}{400 \times 16,000} = 0.557 \text{ inch}^4$$

using Equation 3-39.

Try a No. 87 stamping (Fig. 8-2). The window area is 0.575 square inch, so the core area a needed to obtain a Wa of 0.557 inch^4 is 0.557/0.575 = 0.968 square inch. The center limb width is 0.875 inch and therefore the stack thickness is 0.968/(0.875 × 0.9) = 1.23 inch. (The 0.9 figure is, of course, the stacking factor.) This doesn't make the square core area so desirable for low copper resistance, but is seems closest to our requirement. The volume v works out at $M_P \times a$ = 0.968 × 5.25 = 5.082 cubic inches and the weight at 1.40 lbs. The core loss curve gives a figure of about 23 watts per pound at 400 hertz, 16,000 gauss; therefore the total core loss would be 23 × 1.40 = 32 watts!

Because the required total transformer loss has already been determined to be 13 watts for 95 percent efficiency, this core is obviously not going to do the job. In fact, even if the winding loss were negligible, the efficiency could not be better than 88 percent. The 95 percent efficiency figure was initially derived from a stated need to achieve better than 4 percent regulation. You might be able to relax the specification. This decision depends on the designer's situation—for instance, perhaps this is the only core at hand.

As an exercise, a design centered on this core is laid out in Fig. 8-16 using the same techniques as in the previous designs. Because 95 percent efficiency is not attainable, a more realistic figure of 85 percent is used.

Review of the Design—In a situation such as this where the balance of copper-iron losses is far from perfect, the formula

$$N_2 = \frac{V_o N_1}{V_1 \sqrt{\eta}}$$

cannot give close to the right answers for the number of secondary turns, because it is derived from an assumption of perfect balance for losses.

To demonstrate this point and also to show that regardless of how far out the initial assumptions may be, the cut-and-try process of transformer design leads inevitably to the right answers—the formula was used to start the ball rolling. The first turns numbers obtained, 71 and 16, are shown in the design sheet in columns 3 and 4. The amended numbers, together with other changes that stem from them, are given in columns 5 and 6.

Note that once again I resorted to the trick of using two paralleled conductors for the low-voltage secondary in order to reduce the build thickness. Even so, it is a rather tight squeeze with only 0.057 inch left for the bobbin, but this should be enough. Note that the bulge allowance is only 9 percent, which might be a little risky.

TRANSFORMER DESIGN SHEET

| Input volts | 115 | Hertz | 400 | Est. Efficiency | 85 | Turn/volt | |

Lamination or core No. 87 EI GRAIN-ORIENTED SILICON STEEL — .014 INCH B 16,000

| Window dimensions | G 1.313 | F .437 | W | M_P 5.25 |

| C.S.A. dimensions | E 0.875 | D 1.23 | a 0.968 | v 5.082 |

| Wt 1.4 | Watts/lb 23 | Iron loss 32 |

Windings		W_1	W_1	W_2	W_3	W_2	W_3					
Coil		PRIM	SEC	SEC								
Volts		115	115	26								
Amperes	(2.41)	2.48	2.0	0.5								
Turns		65	71	16	(66)	(15)						
Gauge		18	18	30x2								
Turns/inch		22	22	41								
Margins		.125	.125	.125								
Winding length (L)		1.05	1.05	1.05								
Turns/layer		23	24	43								
Number of layers		2.8	2.96	0.37	(2.75)	(0.346)						

Build												Total
Copper		.125	.125	.011								
Paper		.014	.014									
Cover		.020	.020	.020								
Total		.159	.159	.031								
Bulge Percent 9		.014	.014	.003								
Total (R)		.173	.173	.034								.380

Total bobbin (B) .057

Bobbin .030 clearance .027   Total depth .437

Losses												
Length mean turn (M)		5.23	6.34	6.98								
Total wire length (inches)		340	450	111	(418)	(105)						
Ohms/ft		.0064	.0064	.051								
Resistance (hot)		.217	.288	.566	(.267)	(.535)						
Voltage drop	(.523)	.539	.576	.283	(.534)	(.267)						(2.54)
Copper loss (I^2R)	(1.26)	1.338	1.152	.141	(1.068)	(.134)						2.63

Iron loss 32.00
Total loss 34.63
Efficiency 87.5%

(34.54)

Temperature Rise						
Coil	t	C_S	W	W/C_S	°C T_o	
1						
2						
3						
4						

Temp. Rise = T_o(Total) + $20 HW/C_S$ =

Fig. 8-16. Design sheet for 400-Hz transformer No. 1.

Note also the very high current density of around 400 CM/A for the primary and first secondary windings. This illustrates once again that what really counts is power dissipation; the selected density is simply a guide.

The relatively low level of efficiency of 87 percent is due entirely to the massive core loss. The copper loss is quite small, and regulation would probably be very good.

The low copper loss in this design should make for a low temperature rise, but the very high iron loss tends to invalidate the calculations that are based on no interchange of heat between the core and the windings. When the losses are balanced, this is a reasonable assumption, but in the present case, the core itself is likely to become rather hot—perhaps too hot—and contribute to the rise in temperature of the copper. The point is easily checked on the bench and the numbers of turns are so small that it is not a great chore to redesign the component if necessary.

At 400 Hertz—Design No. 2

It seems clear that an efficiency of 95 percent is probably not attainable with the previous core material, but it can be improved by using a thinner stamping of the same alloy. A 0.006-inch lamination, for instance, gives a loss of about 13 watts per pound at 16,000 gauss. To simply replace the core in the previous example with the thinner material makes the iron loss become $1.4 \times 13 = 18.2$ watts. The copper loss will still be roughly 2.5 watts, making a total loss of $18.2 + 2.5 + 20.7$ watts for a transformer efficiency of 91.7 percent.

The efficiency can be improved further to possibly 94 percent by bringing the losses more into balance. One way of doing this, still using the No.87 stamping, is simply to reduce the core area, thus reducing the core weight and iron loss and increasing the numbers of turns on the windings. This will result in thinner conductors, greater copper loss, and better all-round balance.

This technique is demonstrated in the design of Fig. 8-17 in which the core area is cut approximately in half to 0.5 square inch. The stack thickness is then $0.5/(0.875 \times 0.85) = 0.67$ inch, where 0.875 inch is the tongue width as before, and 0.85 is the stacking factor for the 0.006-inch lamination. The volume is 2.625 cubic inches and the weight is $0.276 \times 2.625 = 0.725$ lb and the core loss is $13 \times 0.725 = 9.4$ watts. The design then follows as shown.

The efficiency works out to be 93.8 percent, very close to the 94 percent assumed to initiate the design. The corrections to the numbers of secondary turns work out at 125 and 28, respectively. The number of layers were not changed, and the effects on other parameters are negligible.

Because the copper loss is now relatively high and the cooling area smaller, it would be wise to check the temperature rise. In this case, the losses are better balanced than in case No. 1 and the calculation will probably put the temperature figure in the ball park. The temperature should also be checked on the bench to be safe.

Finally, the substantial reduction in size from design No. 1 to design No. 2 should be noted. This prompts the question of whether further worthwhile reduction is possible.

TRANSFORMER DESIGN SHEET

Input volts 115	Hertz 400		Est. Efficiency 94		Turn/volt	

Lamination or core No. 87 EI   GRAIN-ORIENTED SILICON STEEL—.006 INCH B   16,000

Window dimensions	G 1.312	F .437	W		M_P 5.25

C.S.A. dimensions	E .875	D .672	a 0.5	v 2.625

Wt .724   Watts/lb 13.0   Iron loss 9.4

Windings		W_1	W_2	W_3							
Coil		PRIM	SEC	SEC							
Volts		115	115	26							
Amperes		2.25	2.0	0.5							
Turns		125	125	28							
Gauge		21	21	24							
Turns/inch		30	30	42							
Margins		.125	.125	.125							
Winding length (L)		1.05	1.05	1.05							
Turns/layer		32	32	44							
Number of layers		3.9	3.9	.636							

Build											Total
Copper		.119	.119	.021							
Paper		.015	.015								
Cover		.020	.020	.020							
Total		.154	.154	.041							
Bulge Percent 9		.014	.014	.004							
Total (R)		.168	.168	.045							.381

Total bobbin (B)   .056

Bobbin .032 clearance .024   Total depth   .437

Losses											
Length mean turn (M)		4.167	5.209	5.844							
Total wire length (inches)		521	651	164							
Ohms/ft		.0128	.0128	.0257							
Resistance (hot)		.667	.833	.421							
Voltage drop		1.500	1.666	.210							
Copper loss (I^2R)		3.37	3.332	.105							6.81

Iron loss   9.40
Total loss   16.21
Efficiency   93.8%

Temperature Rise					
Coil	t	C_S	W	W/C_S	°C T_o
1	.052	4.6	3.370	.732	9.50
2	.027	4.6	3.330	.724	4.89
3	.020	4.6	0.105	.023	0.12
4					
				1.479	14.51

$$T_S = 20HW/C_S = 20 \times 1.05 \times 1.479$$
$$= 31°C$$

Temp. Rise = T_o(Total) + 20HW/C_S = 45°C

Fig. 8-17. Design sheet for 400-Hz transformer No. 2.

Before answering this question in design No. 4, another method of core loss reduction and its effects is illustrated in design No. 3, which follows.

At 400 Hertz—Design No. 3

In this design (Fig. 8-18), the problem of excessive core loss has been tackled in two ways: first, by using thinner laminations as in design No. 2, and second, drastically reducing the flux density to 9500 gauss. In this case, the loss works out at about 5.5 watts per pound, or only 6.93 watts for the core.

With a No. 750 lamination (Fig. 8-19), an excellent loss distribution is achieved. The overall efficiency is improved to 95 percent, but the core is much larger. Because the numbers of layers in the windings are the same as for design No. 2, the copper loss is somewhat smaller, the cooling surface is larger, and therefore the temperature rise can be expected to be less and thus acceptable.

At 400 Hertz—Design No. 4

If one is really keen to cut the iron loss, then a vanadium-cobalt-iron alloy, such as Magnetic Metals' or Arnold Engineering's Supermendur (same name is used by both), in tape-wound toroidal format should fill the bill.

Because of the peculiarities of the toroidal shape, a detailed example of the design process is given in the following. Again the design is aimed at meeting the same specification as before so that direct comparisons of the results can be made. Reference should be made to Chapter 6, where the principles of designing windings for toroids were discussed.

The first thing to note about this material is the very high flux density at which it can be run—as high as 21,000 gauss (Fig. 8-20). This is obviously going to considerably reduce the turns required on the windings. Secondly, it can be seen from Fig. 5-6E, Chapter 5, that the loss in a core made from 0.004-inch material at this flux density and at 400 hertz is less than 10 watts per pound. You can expect, then, reduced loss in both copper and iron in comparison to grain-oriented silicon iron. Judged against the figures obtained in the silicon-iron case, 96 percent efficiency should not be too difficult to obtain. (If the losses are well balanced, a regulation of around 2 percent can then be expected together with a substantial reduction in size.)

In order to arrive at a suitable core, it is quite feasible to proceed as was done in the other designs and first work out the required Wa product, then select trial cores having appropriate Wa numbers from data such as that in the vendor's core tables. However, bearing in mind that in winding a toroid, especially in machine winding, it is necessary to ensure sufficient space to work in the center. The approach discussed in Chapter 6 of first estimating a K factor is a convenient one.

In any case, this is the method to be used if for no other reason than to provide an exercise in another aspect of the design process. Here a K factor of 0.28 is assumed. At an efficiency of 96 percent, the power in the primary winding works out at 243/0.96 = 253 watts, making the primary winding current 253/115 = 2.2 amperes. The primary

TRANSFORMER DESIGN SHEET

Input volts 115	Hertz 400	Est. Efficiency 95	Turn/volt

Lamination or core No. 750	GRAIN- ORIENTED SILICON STEEL −.006 INCH	B 9500

Window dimensions	G 2.25	F 0.625	W	M_p 7.98
C.S.A. dimensions	E 0.75	D 0.83	a 0.56	v 4.93
	Wt 1.26	Watts/lb 5.5	Iron loss 6.93	

Windings		W_1	W_2	W_3						
Coil		PRIM	SEC	SEC						
Volts		115	115	26						
Amperes		256	2.0	0.5						
Turns		190	196	44						
Gauge		18	18	24						
Turns/inch		22	22	42						
Margins		.125	.125	.125						
Winding length (L)		2.125	2.125	2.125						
Turns/layer		49	49	89						
Number of layers		3.88	4	0.49						

Build									Total
Copper		.167	.167	.0213					
Paper		.021	.021						
Cover		.020	.020	.020					
Total		.208	.208	.041					
Bulge Percent		.031	.031	.005					
Total (R)		.239	.239	.046					.524

Bobbin .062 clearance .039

Total bobbin (B) .101

Total depth .625

Losses									
Length mean turn (M)		4.8	6.28	7.15					
Total wire length (inches)		912	1231	315					
Ohms/ft		.00639	.00639	.0257					
Resistance (hot)		.58	.787	.80					
Voltage drop		1.276	1.574	.4					
Copper loss (I^2R)		2.81	3.148	.2					6.158

Iron loss 6.93

Total loss 13.088

Efficiency 95%

Temperature Rise

Coil	t	C_S	W	W/C_S	°C T_o
1					
2					
3					
4					

Temp. Rise = T_o(Total) + 20 HW/C_S =

Fig. 8-18. Design sheet for 400-Hz transformer No. 3.

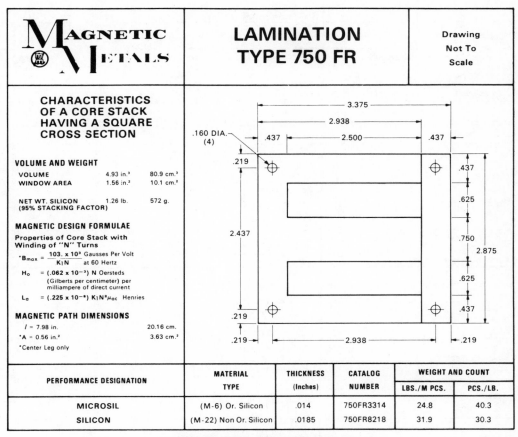

CHARACTERISTICS OF A CORE STACK HAVING A SQUARE CROSS SECTION

VOLUME AND WEIGHT

VOLUME	4.93 in.³	80.9 cm.³
WINDOW AREA	1.56 in.²	10.1 cm.²
NET WT. SILICON (95% STACKING FACTOR)	1.26 lb.	572 g.

MAGNETIC DESIGN FORMULAE

Properties of Core Stack with Winding of "N" Turns

$$*B_{max} = \frac{103. \times 10^3}{K_1 N} \text{ Gausses Per Volt at 60 Hertz}$$

$H_o = (.062 \times 10^{-3})$ N Oersteds (Gilberts per centimeter) per milliampere of direct current

$L_a = (.225 \times 10^{-8}) K_1 N^2 \mu_{ac}$ Henries

MAGNETIC PATH DIMENSIONS

l = 7.98 in.   20.16 cm.

*A = 0.56 in.²   3.63 cm.²

*Center Leg only

PERFORMANCE DESIGNATION	MATERIAL TYPE	THICKNESS (Inches)	CATALOG NUMBER	WEIGHT AND COUNT	
				LBS./M PCS.	PCS./LB.
MICROSIL	(M-6) Or. Silicon	.014	750FR3314	24.8	40.3
SILICON	(M-22) Non Or. Silicon	.0185	750FR8218	31.9	30.3

SHUNTS FOR 750 FR

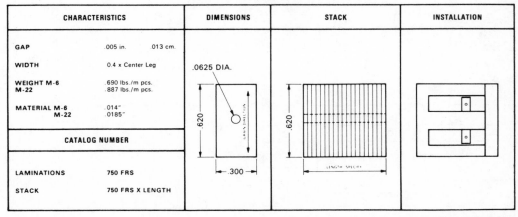

CHARACTERISTICS		DIMENSIONS	STACK	INSTALLATION
GAP	.005 in. .013 cm.			
WIDTH	0.4 x Center Leg			
WEIGHT M-6	.690 lbs./m pcs.			
M-22	.887 lbs./m pcs.			
MATERIAL M-6	.014"			
M-22	.0185"			

CATALOG NUMBER	
LAMINATIONS	750 FRS
STACK	750 FRS X LENGTH

Courtesy Magnetic Metals Corp.

Fig. 8-19. Lamination data for 400-Hz design No. 3.

TRANSFORMER DESIGN SHEET   *K FACTOR* = .28

Input volts 115	Hertz 400		Est. Efficiency 96		Turn/volt	

Lamination or core No. 493 TOROID   PERMENDUR — .006 INCH   B 21,000

Window dimensions   G 1.785 ID   F   W   M_P

C.S.A. dimensions   E $\frac{OD-ID}{2}$=.49 INCH D 0.631   a 1.089 cm²   v

Wt 0.395   Watts/lb   Iron loss 2.73

Windings		W_1	W_2	W_3						
Coil		PRIM	SEC	SEC						
Volts		115	115	26						
Amperes		2.2	2.0	0.5						
Turns		283	289	65						
Gauge		18	18	24						
Turns/inch		21	21	40						
Margins										
Winding length (L)-FIRST ONLY		5.542								
Turns/layer										
Number of layers		3	3	1						

Build										Total
Copper		.129	.129	.022						
Paper										
Cover		.020	.020	.020						
Total		.149	.149	.042						
Bulge Percent		.021	.021	.005						
Total (R)		.170	.170	.047						.387

Bobbin ____ clearance ____   Total bobbin (B)

Total depth   .387

Losses										
Length mean turn (M)		2.763	3.805	4.484						
Total wire length (inches)		782	1099	291						
Ohms/ft		.0064	.0064	.0257						
Resistance (hot)		.500	.703	.748						
Voltage drop		1.100	1.406	.37						
Copper loss (I^2R)		2.42	2.812	.187						5.42

Iron loss 2.73
Total loss 8.15
Efficiency 96.75

Temperature Rise						
Coil	t	C_S	W	W/C_S	°C T_o	
1						
2						
3						
4						

Temp. Rise = T_o(Total) + 20HW/C_S =

Fig. 8-20. Design sheet for 400-Hz transformer No. 4.

winding wire at 750 CM/A must then be 1650 CM, and the secondaries are 1500 and 375 CM, respectively. The nearest wire gauges for the primary and the two secondaries are No. 18, No. 18, and No. 24.

The actual circular mils for these gauges are 1624 CM for No. 18 gauges and 404 for No. 24 gauge (Table 6-1). Plugging these numbers into Equation 6-16 gives

$$Wa = \frac{115 \quad \left(1624 + \dfrac{1624}{1.0} + \dfrac{404}{4.42}\right) 2.741}{0.28 \times 400 \times 21,000}$$

$$= 0.446 \text{ inch}^4$$

From the core dimension numbers in Table 8-4, a No. 493 core is close at 0.469 inch^4. This core, made from 0.006-inch tape, has a cross-sectional area of 1.089 sq cm (iron area) and a core weight of $0.395 \times 0.988 = 0.39$ lb. The number 0.988 is a conversion factor supplied by the manufacturer to convert the weight of Square 50 type of material given in the table to the weight for permendur. The iron loss from the curves (Fig. 5-6E) is about 7 watts per pound, giving a core loss of 2.73 watts.

Because a is given in the data in square centimeters and B is in gauss, Equation 3-19 is used to determine the number of primary turns:

$$N_1 = \frac{115 \times 10^8}{4.44 \times 400 \times 1.089 \times 21,000} = 283 \text{ turns}$$

The turns on the secondaries are

$$N_2 = \frac{V_o N_1}{V_{\text{in}} \sqrt{\eta}} = \frac{115 \times 283}{115 \sqrt{0.96}}$$

$$= 288.77, \text{ say } 289, \text{ turns}$$

and

$$N_3 = \frac{26 \times 283}{115 \sqrt{0.96}} = 65.29, \text{ say } 65, \text{ turns}$$

The case inside diameter, plastic type, is 1.765 inches. Therefore the winding length is $3.14 \times 1.765 = 5.542$ inches. From the *toroid* wire table (Table 6-2), the turns per inch figure for *heavy* insulation wire is given as No. 18, 21, and 24, 40 turns per inch, layer wound. Recall that in a toroid, the number of turns per layer reduces by six turns per layer. The first layer of No. 18 gauge works out at 116 turns; therefore the second layer is 110, and 57 turns are left for the third layer. The balance of the space in the third layer can be filled with the start of the 115-volt secondary. Suppose that 40 turns of the secondary winding are filled into the third layer; then the fourth layer will have 98 turns, followed by 92 in the fifth layer, and so on. Thus, 59 turns will be left for the sixth layer. This leaves about 1.25 inches of winding space free on the sixth

Table 8-4. Core Dimensions

CORE SIZE	CORE DIMENSIONS IN			CASE DIMENSIONS IN						NET CORE AREA CM²			MEAN PATH LENGTH		CASE WINDOW AREA IN²		RATIO ID/OD	GR. CORE WEIGHT SQUARE 50		PRODUCT Wa X Ca IN⁴
	ID	OD	HT	ID		OD		HT		.001	.002	.004 .006	IN	CM	Metal	Plastic		Pounds	Grams	
				Metal	Plastic	Metal	Plastic	Metal	Plastic	SF=.75	SF=.85	SF=.90								
28	.875	1.250	.375	.800	.795	1.340	1.343	.452	.465	.340	.386	.408	3.338	8.48	.503	.496	.700	.0699	31.72	.0353
44	1.000	1.375	.375	.925	.925	1.455	1.455	.452	.445	.340	.386	.408	3.731	9.48	.672	.672	.727	.0782	35.46	.0473
12	1.125	1.500	.375	1.050	1.050	1.570	1.570	.452	.445	.340	.386	.408	4.123	10.47	.866	.866	.750	.0864	39.19	.0609
486	1.375	1.750	.375	1.295	1.275	1.840	1.860	.452	.467	.340	.386	.408	4.909	12.47	1.317	1.277	.786	.1029	46.65	.0926
6	.650	1.150	.375	.585	.585	1.240	1.240	.452	.455	.454	.514	.544	2.827	7.18	.269	.269	.565	.0790	35.83	.0252
52	.750	1.250	.375	.665	.665	1.340	1.335	.452	.455	.454	.514	.544	3.142	7.98	.347	.347	.600	.0878	39.81	.0326
484	.875	1.375	.375	.810	.790	1.450	1.470	.452	.467	.454	.514	.544	3.534	8.98	.515	.490	.636	.0987	44.79	.0483
11	1.000	1.500	.375	.925	.920	1.570	1.593	.452	.465	.454	.514	.544	3.927	9.97	.672	.665	.667	.1097	49.76	.0630
87	1.250	1.750	.375	1.170	1.170	1.820	1.822	.452	.445	.454	.514	.544	4.712	11.97	1.075	1.075	.714	.1317	59.72	.1008
42	1.500	2.000	.375	1.400	1.400	2.110	2 110	.452	.455	.454	.514	.544	5.498	13.96	1.539	1.539	.750	.1536	69.67	.1443
362	.750	1.250	.500	.685	.665	1.350	1.345	.607	.630	.605	.685	.726	3.142	7.98	.369	.347	.600	.1170	53.08	.0461
62	1.000	1.500	.500	.925	.920	1.570	1.593	.607	.610	.605	.685	.726	3.927	9.97	.672	.665	.667	.1463	66.35	.0840
29	1.250	1.750	.500	1.050	1.035	1.705	1.725	.607	.610	.605	.685	.726	4.712	11.97	1.075	1.057	.692	.1609	72.99	.1082
85	1.125	1.625	.500	1.170	1.160	1.820	1.850	.607	.610	.605	.685	.726	4.320	10.97	.866	.841	.714	.1755	79.62	.1344
487	1.375	1.875	.500	1.295	1.275	1.965	1.985	.607	.631	.605	.585	.726	5.105	12.97	1.317	1.277	.733	.1902	86.26	.1646
14	1.500	2.000	.500	1.400	1.400	2.110	2.110	.607	.620	.605	.685	.726	5.498	13.96	1.539	1.539	.750	.205	92.89	.1924
84	1.750	2.250	.500	1.650	1.650	2.350	2.350	.607	.600	.605	.685	.726	6.283	15.96	2.138	2.138	.778	.234	106.16	.267
17	2.000	2.500	.500	1.860	1.860	2.652	2.652	.607	.610	.605	.685	.726	7.069	17.95	2.717	2.717	.800	.263	119.43	.340
18	2.500	3.000	.500	2.360	2.360	3.152	3.152	.607	.620	.605	.685	.726	8.639	21.94	4.373	4.374	.833	.322	145.97	.547
151	.750	1.500	.375	.685	.665	1.575	1.595	.452	.475	.680	.771	.817	3.534	8.98	.369	.347	.500	.1481	67.18	.0518
75	1.250	2.000	.375	1.170	1.160	2.110	2.110	.452	.465	.680	.771	.817	5.105	12.97	1.075	1.057	.625	.214	97.04	.1512
243	2.000	2.750	.375	1.910	1.890	2.850	2.870	.452	.485	.680	.771	.817	7.461	18.95	2.865	2.806	.727	.313	141.83	.403
385	3.000	3.750	.375	2.900	2.880	3.860	3.880	.452	.467	.680	.771	.817	10.603	26.93	6.605	6.514	.800	.444	201.54	.929
125	1.000	1.750	.500	.920	.900	1.840	1.860	.607	.610	.907	1.028	1.089	4.320	10.97	.665	.636	.571	.241	109.48	.1246
60	1.250	2.000	.500	1.170	1.170	2.110	2.110	.607	.620	.907	1.028	1.089	5.105	12.97	1.075	1.075	.625	.285	129.39	.202
488	1.500	2.250	.500	1.420	1.400	2.340	2.360	.607	.631	.907	1.028	1.089	5.890	14.96	1.584	1.539	.667	.329	149.29	.297
490	1.625	2.375	.500	1.535	1.515	2.475	2.495	.607	.631	.907	1.028	1.089	6.283	15.96	1.851	1.803	.684	.351	159.24	.347
83	1.750	2.500	.500	1.650	1.650	2.600	2.600	.607	.600	.907	1.028	1.089	6.676	16.96	2.138	2.138	.700	.373	169.20	.401
493	1.875	2.625	.500	1.785	1.765	2.725	2.745	.607	.631	.907	1.028	1.089	7.069	17.95	2.502	2.447	.714	.395	179.15	.469
475	2.250	3.000	.500	2.160	2.140	3.100	3.120	.607	.631	.907	1.028	1.089	8.247	20.95	3.664	3.597	.750	.461	209.01	.687
498	2.500	3.250	.500	2.400	2.380	3.360	3.380	.607	.631	.907	1.028	1.089	9.032	22.94	4.524	4.449	.769	.505	228.91	.848
381	1.250	2.250	.500	1.170	1.150	2.340	2.360	.607	.631	1.210	1.371	1.452	5.498	13.96	1.075	1.039	.556	.410	185.78	.269
15	1.500	2.500	.500	1.400	1.395	2.600	2.615	.607	.610	1.210	1.371	1.452	6.283	15.96	1.539	1.528	.600	.468	212.33	.385
335	1.750	2.750	.500	1.660	1.640	2.850	2.870	.607	.631	1.210	1.371	1.452	7.069	17.95	2.164	2.112	.636	.527	238.87	.541
76	2.000	3.000	.500	1.910	1.895	3.100	3.115	.607	.610	1.210	1.371	1.452	7.854	19.95	2.865	2.820	.667	.585	265.41	.716
19	2.500	3.500	.500	2.313	2.380	3.688	3.630	.607	.610	1.210	1.371	1.452	9.425	23.94	4.202	4.449	.714	.702	318.49	1.050
334	1.000	1.500	1.000	.935	.915	1.575	1.595	1.135	1.112	1.210	1.371	1.452	3.927	9.97	.687	.658	.667	.293	132.70	.1717
55	1.250	1.750	1.000	1.170	1.170	1.820	1.822	1.135	1.105	1.210	1.371	1.452	4.712	11.97	1.075	1.075	.714	.351	159.24	.269
95	1.500	2.000	1.000	1.400	1.400	2.110	2.110	1.135	1.110	1.210	1.371	1.452	5.498	13.96	1.539	1.539	.750	.410	185.78	.385
128	1.750	2.250	1.000	1.670	1.650	2.340	2.360	1.135	1.110	1.210	1.371	1.452	6.283	15.96	2.190	2.138	.778	.468	212.33	.548

Courtesy Magnetic Metals Corp.

layer, which is not enough to accommodate the 26-volt secondary, so it might as well be put on as a complete seventh layer.

The build is thus 0.387 inch, as shown in the sheet. The diameter of the center hole in the winding is the window diameter minus (0.387 × 2), or 1.785 − 0.774 = 1.011 inch. If this is judged to be sufficient for your winding method, the design is continued as shown, culminating in total copper loss of 5.42 watts against an iron loss of 2.73 watts. The efficiency works out to be 96.75 percent, almost the same as that initially selected.

The losses are not very well balanced. But in this case, where efficiency is already very high, we cannot expect to obtain more than a decimal fraction of a percent of improvement by redesigning the transformer. As a matter of interest, note that the K factor actually obtained works out at 0.27—very close to the original figure.

Note

The winding resistance is figured in the same way as for standard coils except that the factor $8B$ is deleted from the expression $2(E + D) + 8B$ because there is no bobbin. The dimensions E and D are from the vendor's data table. Here D is the same as case dimensions H in the table, and E is derived from the case outside diameter (OD) and inside diameter (ID) figures; it is $E = \text{OD} - \text{ID}$. The core weight, as stated earlier, is the table figure multiplied by a conversion number.

In working out the Wa product using Equation 6-16 the respective ratios r_1 and r_2 are: $r_1 = 115/115 = 1.0$ and $r_2 = 115/26 = 4.42$ as used.

ADJUSTMENTS FOR MAGNETIZING CURRENT

At the beginning of the chapter, it was stated that magnetizing current would not be considered in the initial design examples. This is a normal approach, and most of the time it works out fine. However, as a matter of interest, let's see how *not* doing so has affected our results.

The method of determining the magnitude of the magnetizing current and how to add this vectorially to the load and loss currents was discussed in Chapter 5. Applying this method to the examples just completed should be instructive. The examples are identified by the design sheet figure numbers.

Fig. 8-5

For low-carbon steel in 0.018-inch stampings at 60 hertz, 12,000 gauss, the magnetizing volt-amperes are on the order of 18 VA per pound of core, giving 1.2 × 18 = 21.6 VA. The magnetizing current is then given by Equation 5-6. The equivalent air gap g can be taken as 0.003 inch for carefully assembled core. Thus

$$
\begin{aligned}
I_m &= \frac{\text{VA/lb} \times \text{Wt}}{V_1} + \frac{1.43\, Bgs}{N_1} \\
&= \frac{21.6}{110} + \frac{1.43 \times 12{,}000 \times 0.003 \times 0.9}{632} \\
&= 0.269 \text{ ampere}
\end{aligned}
$$

This, added vectorially to the primary winding current of 0.245 ampere, gives

$$I_p = \sqrt{0.245^2 + 0.269^2} = 0.364 \text{ ampere}$$

for the total current in the primary.

The primary winding voltage drop is then $0.364 \times 21.5 = 7.826$ volts and the primary winding copper loss is $7.826 \times 0.364 = 2.85$ watts. Added to the secondary winding copper loss previously calculated, this gives a total copper loss of 4.173 watts. This copper loss is, in turn, added to the previously calculated iron loss, giving a total transformer loss of 8.773 watts. The efficiency is now 20/28.773 = 0.69 or 69 percent.

If the turns ratio is left as is, the voltage on the secondary will be

$$V_2 = \frac{N_2(V_{\text{in}} - V_{\text{d1}})}{N_1} - V_{\text{d2}}$$

$$= \frac{32(110 - 7.826)}{632} - 0.331$$

$$= 4.84 \text{ volts}$$

by Equation 4-10. The adjustment applicable to the primary winding turns to bring the output voltage back to 5.0 volts is

$$N_1 = \frac{32(110 - 7.826)}{5 + 0.331} = 613 \text{ turns}$$

by Equation 4-13.

In this case, the effect of the magnetizing current is evidently not negligible, a fact that would have been obvious right at the start of the design from the relative magnitudes of the magnetizing and load-loss currents. However, whether it need be considered in the design is a matter of judgment, or one can simply take it into account anyway, negligible or not.

Fig. 8-8

Here again carbon-steel stampings are involved at about 16,500 gauss and 60 hertz. At this flux density, the magnetizing volt-amperes are likely to be around 70 VA per pound for this type of material, making the core loss $0.85 \times 70 = 59$ or 60 VA!

Judging from the effects of magnetizing current in the last case, there is no question that this current is going to be far from negligible. The magnetizing current is

$$I_M = \frac{60}{110} + \frac{1.43 \times 16{,}500 \times 0.003 \times 0.9}{548}$$

$$= 0.66 \text{ ampere}$$

The effective primary winding current is a massive

$$I_p = \sqrt{0.263^2 + 0.66^2} = 0.71 \text{ ampere}$$

The primary voltage drop is now $21.1 \times 0.7 = 14.77$ volts, and the loss is $14.77 \times 0.71 = 10.48$ watts. The primary winding copper loss plus secondary winding copper loss is now $10.48 + 1.28 = 11.76$ watts, which, added to the iron loss of 3.2 watts, gives a total transformer loss of 14.96 watts. The efficiency is now 0.57 or 57 percent—a far cry indeed from the 78 percent we thought we had. Moreover, the large primary winding copper loss is quite likely to play havoc with the temperature rise and bring the copper to a nice shade of cherry red, with a smoke or even flame accompaniment. It would be an excellent idea to check the temperature rise if a practical use is being considered for the design.

Figs. 8-10, 8-11, and 8-12

These three designs are based on C cores of grain-oriented silicon iron, all at 15,000 gauss, 60 hertz. The magnetizing volt-amperes from the curves of Fig. 5-7A are 1.7 VA per pound. The magnetizing current of the design of Fig. 8-10 works out to 0.03 ampere. Added vectorially to the existing load-loss current of 0.207 ampere, this is so small as to make virtually no difference. Thus

$$I_P = \sqrt{0.207^2 + 0.03^2} = .209$$

is the primary winding current.

Similarly, in the designs of Fig. 8-11 and 8-12, the effects of the magnetizing volt-amperes are negligible.

Fig. 8-15

In this case, the core has a magnetizing volt-ampere figure of about 4 VA per pound at 16,000 gauss, 60 hertz. (Note the very sharp increase in volt-amperes per pound from 15,000 gauss to 16,000 gauss). Because the core weighs 6.1 lbs, the magnetizing current is quite large at 0.571 ampere. Even so, it is still much smaller than the load-loss current of 2.22 amperes. With the two currents added vectorially, the effective primary winding works out to be 2.292 amperes. The increase over the previous figure is a little more than 4 percent. This would be considered negligible in most cases.

Fig. 8-16

This is a 0.014-inch grain-oriented silicon-steel lamination at 400 hertz with a magnetization volt-amperes per pound in excess of 200 at a flux density of 16,000 gauss. The core, however, is quite small at this frequency, and the magnetizing current is

$$I_m = \frac{1.40 \times 200}{115} + \frac{1.43 \times 16{,}000 \times 0.003 \times 0.9}{65}$$

$$= 3.38 \text{ amperes}$$

The total current in the primary is then

$$I_p = \sqrt{2.41^2 + 3.38^2} = 4.15 \text{ amperes}$$

something less than twice the previous figure. The primary winding copper loss now works out at 3.73 watts, bringing the total copper loss up to 4.9 watts, while the primary voltage drop is increased to 0.9 volt.

The output voltages work out to 115.3 volts and 26.06 volts for the respective secondaries. For most purposes, these results would be adequate and in any case are very likely well within the accuracy to which the transformer can be built.

Figs. 8-17, 8-18, and 8-20

The same methods applied to these three designs show that the effect of the magnetizing current is small enough to be ignored in every case.

Conclusion

The preceding discussion indicates that the original contention that magnetizing current can be considered negligible in most (but not all) cases is valid. Magnetizing current tends to become a potent factor at unusually high flux densities and frequencies (for power transformers, that is), but at what could be termed conservative ratings, it is usually negligible.

A RETURN TO THE SIMPLE LIFE

So far we have worked via vendors' catalogs and indulged in a lot of fancy manipulations involving Wa products, winding K factors, vectorial addition, and so on. A return to simpler methods might be instructive at this point; therefore two less esoteric formulas are called back into service from Chapter 1. It is also proposed to work with salvaged laminations of uncertain characteristics, but which are most likely good quality silicon steel—the type usually encountered in TV or radio power transformers.

The two formulas are

$$a = 0.16 \sqrt{P} \tag{8-1}$$

and

$$T = \frac{N}{V} = \frac{5}{a} \tag{8-2}$$

where

a is the cross-sectional area of core in square inches,

P is the power output in watts,

T is the turns per volt (N/V).

The formulas, if you recall, are based on silicon steel at approximately 12,000 gauss and at an input frequency of 60 hertz. They will work very well also for low-carbon steel if one unwittingly obtains it, although the losses will be higher.

A high-current, low-voltage specification has been selected for the exercise. The core dimensions are obtained by direct measurement, and volume, weight, magnetic path length, and so on are derived using the methods given in Chapter 5. The specification is as follows: input, 117 volts at 60 hertz; output 6.3 volts at 25 amperes.

A First Design

A tentative efficiency of 85 percent is chosen to get the ball rolling. The output power is therefore $25 \times 6.3 = 158$ watts, and the input power is $158/0.85 = 186$ watts. Fig. 8-21 illustrates the circuit.

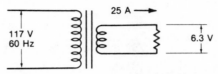

117 V
60 Hz

25 A →

6.3 V

Fig. 8-21. High-current, low-voltage specification.

The core cross section needed is then

$$a = 0.16 \sqrt{186} = 2.18 \text{ square inches}$$

and the turns per volt figure is

$$T = \frac{5}{2.18} = 2.29, \text{ say } 2.3, \text{ turns per volt}$$

This puts the number of primary winding turns at $117 \times 2.3 = 269$ and the number of secondary winding turns at $6.3 \times 2.3 = 14.49$, say 15.

Wire gauges can be chosen right away. A current density of, say, 800 CM/A will give No. 7 gauge for the secondary and No. 19 for the primary. The turns per inch are 24 for the primary and 6 for the secondary, and margins are 0.125 and 0.156 inch, respectively.

These figures are entered in the design sheet of Fig. 8-22. The lamination in Fig. 8-23 is tried for no better reason than that an old transformer with those laminations is on hand and they look about right. After fitting the wire to the laminations in the usual way, the primary winding build is found to be 0.483 inch and the secondary is 0.437 inch as shown. The total build is thus 0.92 inch, but the lamination window is only 0.73 inch, which is 0.19 inch too small, and the bobbin still has to be fitted.

One attractive thing about the build up to this point is that the space is about evenly divided between primary and secondary windings; and then we have those laminations on hand. So try again. If the core area is doubled, the number of turns will be halved and the build will be approximately halved to 0.451 inch. Add, say, 0.092 inch for the bobbin and the fit comes to 0.543 inch. A quick check shows that there are enough laminations to do this, but the windings are obviously too *small* for the window, leaving 0.187 inch to spare.

This arrangement would probably work, but it is by no means ideal. A problem here is that the heavy wire of the secondary winding and the few turns involved give very

TRANSFORMER DESIGN SHEET

Input volts 117	Hertz 60		Est. Efficiency 85		Turn/volt	
Lamination or core No.	SALVAGED — PROBABLY SILICON STEEL				B 12,000	
Window dimensions	G 1.77	F 0.73	W		M_P 6.947	
C.S.A. dimensions	E 1.26	D 1.73	a 2.18		v 15.18	
	Wt 4.25	Watts/lb 1.2	Iron loss 5.0			

Windings		W_1	W_2								
Coil		PRIM	SEC								
Volts		117	6.3								
Amperes		1.59	25								
Turns		269	15								
Gauge		19	7								
Turns/inch		24	6								
Margins		.125	.156								
Winding length (L)		1.52	1.458								
Turns/layer		37	9								
Number of layers		7.3	1.6								

Build											Total
Copper		.330	.330								
Paper		.070	.020								
Cover		.020	.030								
Total		.420	.380								
Bulge Percent		.063	.051								
Total (R)		.483	.437								0.92

Bobbin ____ clearance ____   Total bobbin (B) ____ Total depth ____

Losses											
Length mean turn (M)											
Total wire length (inches)											
Ohms/ft											
Resistance (hot)											
Voltage drop											
Copper loss (I^2R)											

Iron loss ____ Total loss ____ Efficiency ____

Temperature Rise

Coil	t	C_S	W	W/C_S	°C T_o
1					
2					
3					
4					

Temp. Rise = T_o(Total) + $20HW/C_S$ =

Fig. 8-22. Salvaged-part transformer design sheet No. 1.

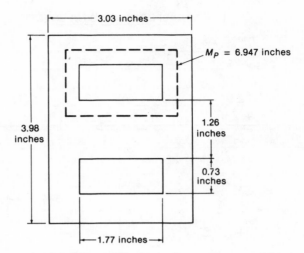

Fig. 8-23. Salvaged lamination for design sheet of Fig. 8-22.

little flexibility for adjustment to an exact fit; this is a characteristic of low-voltage, heavy-current windings, so the possibility of using even heavier-gauge wire to fill the space is not so attractive. Hence, go back to the original area and turns per volt and try thinner wire.

After some trial and error, gauges No. 21 and No. 9 provide a reasonable fit, although it's still not perfect. By continuing the design, the results shown in the design sheet of Fig. 8-24 are obtained. The turns have not yet been adjusted for voltage drop. These would be as follows.

The primary winding voltage drop is 4.2 volts. Because the turns-per-volt figure is 2.3, it can be compensated for by removing $4.2 \times 2.3 = 9.66$, say 10, turns from the primary. The secondary winding voltage drop is 0.275 volt, thus $0.275 \times 2.3 = 0.63$, say 0.7, turn can be added to the secondary. Because the secondary is already calculated at 14.5 turns, it could be made exactly 15, which is not the precise compensation calculated but probably good enough. These two corrections are shown in parenthesis. This method of calculating the compensation for voltage drop enables the adjustments to be applied to both windings, which is often convenient.

Finally, moving to the Temperature Rise box, 69.2°C is obtained. Using the temperature guideline figure of 105°C maximum permissible temperature, this transformer can be used in an ambient temperature of not more than 105°C − 70°C = 35°C.

A Second Design

If the core happens to be a mite bigger in the window depth, such as the one pictured in Fig. 8-25, the design data shown in Fig. 8-26 can be achieved. Overall efficiency has been increased to 91 percent. Voltage drops in both the primary and secondary windings

TRANSFORMER DESIGN SHEET

Input volts *117*	Hertz *60*		Est. Efficiency *85*		Turn/volt *2.3*
Lamination or core No.	*SALVAGED — SILICON STEEL*			B	*12,000*
Window dimensions	G *1.77*	F *0.73*	W	M_P	*6.947*
C.S.A. dimensions	E *1.26*	D *1.73*	a *2.185*	v	*15.18*
	Wt *4.25*		Watts/lb *1.2*	Iron loss *5.0*	

Windings		W_1	W_2									
Coil		PRIM	SEC									
Volts		117	6.3									
Amperes		1.59	25									
Turns	(259)	269	14.5	(15)								
Gauge		21	9									
Turns/inch		30	8.7									
Margins		.125	.156									
Winding length (L)		1.52	1.458									
Turns/layer		46	13									
Number of layers		6	2									

Build												Total
Copper		.200	.230									
Paper		.050	.020									
Cover		.020	.030									
Total		.270	.280									
Bulge Percent *15*		.040	.040									
Total (R)		.310	.320									.630
										Total bobbin (B)		.100
Bobbin *.62* clearance *.038*										Total depth		.730

Losses												
Length mean turn (M)		7.7	9.65									
Total wire length (inches)		2071	140									
Ohms/ft		.01277	.00079									
Resistance (hot)		2.65	.011									
Voltage drop		4.2	.275									
Copper loss (I^2R)		6.7	6.875									13.575
											Iron loss	5.000
											Total loss	18.575
											Efficiency	89%

Temperature Rise					
Coil	t	C_S	W	W/C_S	°C T_o
1	.095	10.39	6.7	.645	15.32
2	.050	10.39	6.875	.662	8.275
3					
4					
				1.307	23.595

$H = 1.77$

THEN $T_2 = 20 \times 1.77 \times 1.3 = 46°C$

TEMP. RISE = $23.595 + 46 = 69°C$

Temp. Rise = T_o(Total) + $20 HW/C_S =$

Fig. 8-24. Salvaged-part transformer design sheet No. 2.

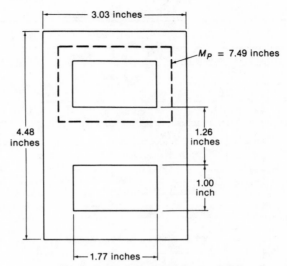

Fig. 8-25. Lamination for design sheet of Fig. 8-26.

are much smaller, thus contributing to improved regulation. Finally, the temperature rise is a low 38°C.

But—and this is a large "but"—No. 7 gauge wire is very thick. It is difficult to wind. The simple solution, as we have seen in past designs, is to use thinner wire paralleled. And what does this do to the efficiency, voltage drop, regulation, and so on? Transformer design is a compromise between conflicting requirements, so let's find out what we would have to live with in this case.

Figure 8-27 shows the voltage drops, copper loss, and temperature rise to expect if No. 7 gauge is replaced with No. 9 or two No. 12 gauges in parallel. The results of using one No. 9 or two No. 12s are the same from the point of view of resistance and therefore of voltage drops and copper loss. The secondary winding copper loss has increased about 50 percent from the No. 7 gauge case, the voltage drop has gone up by the same amount, and the temperature has increased to 58°C although the total efficiency is still good.

THE INSULATION

In the low-voltage transformers discussed so far, the choice of insulation has been rather arbitrary, but in practice, a more studied approach should be made, especially if higher voltages are involved.

For higher voltages, the general design procedure is identical with that used in the low-voltage designs. Considerations of flux density in the core, current density in the wire, and temperature rise in the windings are the same. Insulation between, and on top of, the windings may be greater than in the low-voltage case, but not necessarily so. As was discussed in Chapter 6, the amount of insulation depends on both electrical and mechanical requirements. Similarly, the winding margins might be no different because they are related to a great extent to the gauge of wire.

TRANSFORMER DESIGN SHEET

Input volts 117	Hertz 60		Est. Efficiency 85	Turn/volt 2.3

Lamination or core No.	SALVAGED — SILICON STEEL			B	$12,000$

Window dimensions	G 1.77	F 1.00	W	M_p 7.49

C.S.A. dimensions	E 1.26	D 1.73	a 2.185	v 16.36

	Wt 4.58	Watts/lb 1.2	Iron loss 5.5

Windings		W_1	W_2								
Coil		PRIM	SEC								
Volts		117	6.3								
Amperes		1.59	25								
Turns		269	14.5								
Gauge		19	7								
Turns/inch		24	6								
Margins		.125	.156								
Winding length (L)		1.52	1.458								
Turns/layer		37	9								
Number of layers		8	2								

Build											Total
Copper		.330	.330								
Paper		.070	.020								
Cover		.020	.030								
Total		.42	.380								
Bulge Percent		.063	.057								
Total (R)		.483	.437								.920

Total bobbin (B) · $.080$
Total depth 1.000

Bobbin $.050$ clearance $.030$

Losses											
Length mean turn (M)		8.23	11.12								
Total wire length (inches)		2214	161								
Ohms/ft		.008	.0004%								
Resistance (hot)		1.77	.008								
Voltage drop		2.8	.2								
Copper loss (I^2R)		4.47	5.00								9.47

Iron loss 5.50
Total loss 14.47
Efficiency 91%

Temperature Rise						
Coil	t	C_s	W	W/C_s	°C T_o	
1	.105	13.145	4.47	.34	8.93	
2	.040	13.145	5.00	.38	3.8	
3						
4						
					12.73	

$T_2 = 25.5°C$

TOTAL TEMP. $= 25.5 + 12.73 = 38°C$

Temp. Rise $= T_o$(Total) $+ 20HW/C_s =$

Fig. 8-26. Salvaged-part transformer design sheet No. 3.

TRANSFORMER DESIGN SHEET

Input volts		Hertz			Est. Efficiency		Turn/volt			
Lamination or core No.		SALVAGED CORE — SILICON IRON					B			
Window dimensions		G		F		W		M_p		
C.S.A. dimensions		E		D		a		v		
		Wt		Watts/lb		Iron loss				

Windings		W_1	W_2							
Coil										
Volts										
Amperes										
Turns										
Gauge		19	9	OR 2 × No. 12 IN PARALLEL						
Turns/inch										
Margins										
Winding length (L)										
Turns/layer										
Number of layers										

Build										Total
Copper										
Paper										
Cover										
Total										
Bulge Percent										
Total (R)										

Total bobbin (B)
Total depth

Bobbin _____ clearance _____

Losses										
Length mean turn (M)										
Total wire length (inches)		2214	161							
Ohms/ft		.008	.00079							
Resistance (hot)		1.77	.0127							
Voltage drop		2.8	.318							
Copper loss (I^2R)		4.47	7.96							12.43

Iron loss 5.50
Total loss 17.93
Efficiency

Temperature Rise

Coil	t	C_S	W	W/C_S	°C T_o
1	.105	10.87	4.47	.41	10.7
2	.040	10.87	7.96	.73	7.3
3					
4					
				1.14	18.00

$T_2 = 40.356$

TEMP. RISE = 40.356 + 18 = __58°C__

Temp. Rise = T_o(Total) + 20HW/C_S =

Fig. 8-27. Salvaged-part transformer design sheet No. 4.

The following simple case illustrates these points. A transformer is required to work from 117 volts, 60 hertz and deliver 300 volts. The circuit is shown in Fig. 8-28. In this circuit, the core is at ground potential, a common situation. Because one side of the supply is usually at, or near, ground potential, 117 volts rms (or 165 volts peak) exists between one side of the primary winding and the core—but you don't know which side unless you intend to ensure that one specific lead is always connected to the grounded side of the supply. Because this is rarely done, it is necessary to make sure that the insulation is good for 117 volts between the primary winding and core at all points.

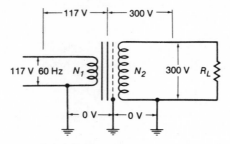

(A) Voltages between various points.

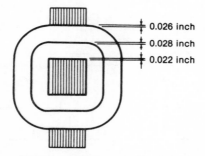

(B) Minimum thicknesses of insulation.

Fig. 8-28. Insulation in a high-voltage transformer.

If one side of the secondary winding happens to also be grounded, by design or accident, then 417 volts can exist between one end of the primary and one end of the secondary as shown in the illustration. Again, you don't know which side of the secondary might be grounded, so for safety, the insulation between the windings should be made good for at least 417 volts at all points, and between the secondary and the core it should be good for at least 300 volts.

In addition, the insulation itself should be designed to withstand much more than its bare working voltage. Referring back to Chapter 6, the insulation for 117 volts should, for safety, withstand a test voltage of 1117 volts! This calls for 0.002 inch of paper for each 100 test volts, or around 0.022 inch in total. Therefore, a bobbin constructed of material at least this thick should be satisfactory from the electrical point of view. Don't forget, though, that it has to take the mechanical stress of winding, so something a little stouter should probably be used.

Between the windings, there are 417 volts. The insulation must then be good for 1417 volts. This calls for at least 0.028 inch of paper on top of the first winding. There are 300 volts between the secondary and the core; therefore the secondary-to-core insulation will be fore 1300 volts, or 0.026 inch of paper.

For the winding margins, use the same test voltage and allow a minimum of 1/16 inch per kilovolt. In this case, 0.080 inch is fine for 1300 volts. Therefore, for wire gauges up to No. 37, the margins in the table are good, but above No. 37 the calculation rather than the table figure should be used.

THE AUTOTRANSFORMER

The remarkable reductions in cost, size, and weight coupled with the high apparent efficiency often possible with an autotransformer should not be overlooked in the search for miniaturization and economy. To be sure, a penalty must be paid as always for all these advantages, but perhaps you might not mind paying it in a given situation. The penalty, of course, is the lack of isolation of the supply line from the load, because they are directly connected to each other through the windings.

In this design, the advantages are greatest when the ratio of input and output voltage is closest to unity, a fact that makes the autotransformer popular as a device for adjusting small differences between a supply line voltage and that required for a load. For example, it might be necessary to boost a supply voltage from, say, 103 volts to a standard 115 volts, or to reduce an overly high supply voltage of, say, 120 volts to 115 volts. If 103 and 120 seem like odd numbers for line voltages, don't forget that so-called standard voltages are what the power company hopes for and usually tries to give you. What actually appears at the outlets, however, depends on local, and sometimes not-so-local, conditions of load in relation to impedances in the system.

Design Example

Suppose a 60-hertz supply voltage is constantly low at 105 volts and it is required to drive various 115-volt loads up to a maximum of 5.0 amperes as in Fig. 8-29. The

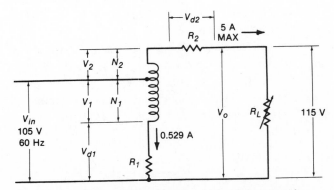

Fig. 8-29. Step-up autotransformer circuit for design example.

maximum output is therefore $115 \times 5 = 575$ VA. Because the load current is variable, reasonably good regulation is required. Therefore, winding resistance should be kept small by using a small current density in the windings, say 1000 CM/A for a start.

To achieve the 10-volt increase, section N_2 of the winding must have 10 volts across it and will carry a maximum of 5.0 amperes. Its required output rating is therefore $10 \times 5 = 50$ VA. Assuming 90 percent efficiency for the transformer, the input needed for 50 VA output is $50/0.9 = 55.56$ VA. The current in N_1, including losses, is thus $55.56/105 = 0.529$ ampere.

On the basis of 1000 CM/A, the wire for section N_2 must have an area (single strand) of about $1000 \times 5 = 5000$ CM. From the wire tables (Table 6-1), gauge No. 13 is about right. The wire for section N_1 needs $0.529 \times 1000 = 529$ CM, or about No. 23 gauge.

Using standard silicon-iron stampings, a flux density of about 14,000 gauss would be appropriate. The Wa factor can then be estimated from the power capability equation at

$$Wa = \frac{50 \times 1000 \times 17.26}{60 \times 14,000} = 1.027 \text{ inch}^4$$

A check of the catalog quickly shows that a 1.5-inch stack of type 100 MH laminations (Fig. 8-30) should be chosen. This was arrived at by dividing 1.027 by the window area of 0.75 square inch to get the required core area of 1.369 inch. The turns-per-volt figure is then

$$\frac{N}{V} = \frac{3.5 \times 10^6}{60 \times 1.369 \times 14,000} = 3.0 \text{ turns per volt}$$

Thus, N_1 is then $105 \times 3 = 315$ turns and N_2 is $10 \times 3 = 30$ turns. The window length is 1.5 inches. Gauge No. 23 for N_1 requires 0.125-inch margins, leaving $1.5 - (2 \times 0.125) = 1.25$ inches for the winding length. The turns per inch are 38, making the turns per layer $1.25 \times 38 = 47.5$, say 47, and the number of layers needed is $315/47 = 6.7$ (7) layers. Gauge No. 13 for N_2 needs 0.25-inch margins, leaving $1.5 - (2 \times 0.25) = 1.0$ inch for the winding length. The turns per inch, and in this case per layer, are 12; therefore 2.5 layers (3) are needed. A rough check on the build shows that it will be too large for the window depth by about 0.14 inch.

If the number of layers for N_2 are reduced to exactly 2.0 by reducing the turns to 24 and increasing the core area to compensate, the N_1 turns will reduce in the same proportion to 252 turns or 5.4, say 6, layers. In this case, you should probably compensate for voltage drop in the windings by adjusting the primary winding turns downwards, which takes the number of layers to less than 6. The core area required to retain B at 14,000 gauss is determined thus:

$$a = \frac{105 \times 3.5 \times 10^6}{252 \times 60 \times 14,000} = 1.736 \text{ square inches}$$

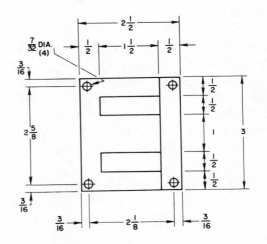

MAGNETIC METALS

LAMINATION TYPE 100 MH

CHARACTERISTICS OF A CORE STACK HAVING A SQUARE CROSS SECTION

VOLUME AND WEIGHT

VOLUME	– 6.00 in.^{3}	– 98.2 cm.^{3}
WINDOW AREA	– .750 in.^{2}	– 4.83 cm.^{2}
WT. SUPER Q 80	– 1.90 lb.	– 863 g.
WT. SUPERPERM "49"	– 1.80 lb.	– 818 g.
WT. SUPERFLUX	– 1.78 lb.	– 807 g.
WT. SILICON	– 1.57 lb.	– 712 g.

MAGNETIC DESIGN FORMULAE

Properties of Core Stack with Winding of "N" Turns

$$B_{max} = \frac{58.2 \times 10^3}{K_1 N} \text{ Gausses Per Volt at 60 Hertz}$$

$H_o = (.082 \times 10^{-3}) N$ Oersteds
(Gilberts per centimeter) per milliampere of direct current

$L_a = (.5289 \times 10^{-8}) K_1 N^2 \mu_{ac}$ Henries

MAGNETIC PATH DIMENSIONS

l = 6.00 in.	15.24 cm.
A = 1.00 in.^{2}	6.45 cm.^{2}

K$_1$ (STACKING FACTOR)

Thickness	Butt Jointed	Interleaved one per layer
.004"	.90	.80
.006"	.90	.85
.014"	.95	.90
.0185"	.95	.90

PERFORMANCE DESIGNATION	MATERIAL TYPE	THICKNESS (Inches)	CATALOG NUMBER	WEIGHT AND COUNT	
				LBS. /M PCS.	PCS./ LB.
SUPERPERM 80	HyMu 80	.004	100MH8404	7.25	138.
SUPERPERM 80	HyMu 80	.006	100MH8406	10.88	91.9
SUPERPERM 80	HyMu 80	.014	100MH8414	25.39	39.4
SUPER Q 80	HyMu 80	.004	100MH8004	7.25	138.
SUPER Q 80	HyMu 80	.006	100MH8006	10.88	91.9
SUPER Q 80	HyMu 80	.014	100MH8014	25.39	39.4
SUPERTHERM 80	HyMu 80	.006	100MH7406	10.88	91.9
SUPERTHERM 80	HyMu 80	.014	100MH7414	25.39	39.4
SUPERPERM 49	49	.004	100MH4904	6.75	148.
SUPERPERM 49	49	.006	100MH4906	10.12	98.8
SUPERPERM 49	49	.014	100MH4914	23.62	42.3
SUPERFLUX	PERMENDUR	.006	100MHVP06	9.98	100.
SUPERFLUX	PERMENDUR	.010	100MHVP10	16.66	60.2
MICROSIL	Gr. Or. Silicon	.004	100MH3304	6.28	159.
MICROSIL	Gr. Or. Silicon	.006	100MH3306	9.43	106.
MICROSIL	Gr. Or. Silicon	.014	100MH3314	21.98	45.5
SILICON	Non Or. Silicon*	.014	100MH**14	21.98	45.5
SILICON	Non Or. Silicon*	.018	100MH**18	28.27	35.4
SILICON	Non Or. Silicon*	.025	100MH**25	39.26	25.5
HYPERTRAN	Low Carbon	.025	100MH2125	40.32	24.9

* Customer to designate AISI grade of material desired.
** See "How To Order Section" for Code Number.

Courtesy Magnetic Metals Corp.

Fig. 8-30. Data sheet for lamination used in autotransformer design example.

The numbers up to this point are entered in the design sheet shown in Fig. 8-31.

The output voltage is given by the following calculation:

$$V_o = \frac{N_2(V_{in} - V_{d1})}{N_1} - V_{d2} + V_{in}$$

$$= \frac{24(105 - 1.96)}{252} - 0.205 + 105$$

$$= 114.61 \text{ volts}$$

where the terms are given in Fig. 8-31. The output is low by $115 - 114.61 = 0.39$ volt. This can be compensated for by removing turns from the primary winding as follows: Because the core area was changed, the turns-per-volt figure is now $252/105 = 2.4$. The voltage drop, referring to the primary winding, is $0.39 \times N_1/N_2 = 0.39 \times 252/24 = 4.095$ volts, and the turns to be removed are $4.095 \times 2.4 = 9.8$ (10) turns, making the total number of primary turns 242.

The result could also have been achieved from the following:

$$N_1 = \frac{N_2(V_{in} - V_{d1})}{V_o + V_{d2} - V_{in}}$$

$$= \frac{24(105 - 1.96)}{115 + 0.205 - 105} = 242 \text{ turns}$$

for the primary winding.

The core volume v is $a \times M_p = 1.736 \times 6 = 10.416$ cubic inches. The specific weight for silicon iron is 0.276 lb/inch^3; therefore the core weight is 2.875 lbs. At about 1.2 watts per pound, the iron loss is $2.875 \times 1.2 = 3.45$ watts. The total copper plus iron loss is $3.45 + 2.06 = 5.5$ watts. The transformer efficiency is thus $50/55.5 = 0.9$ or 90 percent. The magnetizing current in this example will probably be negligible but, if necessary, can be calculated as in previous examples.

Review of Design

The most significant point in all this is that a 50-VA transformer is all that is necessary to handle 575 VA output. Consequently, the transformer is relatively small and inexpensive. And the most dramatic point is the amazing efficiency of the *circuit* when considered from input to output. Because the losses amount to only 5.5 watts, the circuit input is 580.5 watts and the total efficiency is $575/580.5 = 0.99$ or 99 percent!

The efficiency of the *transformer* (as distinct from the circuit efficiency) is a respectable 90 percent. This stems from a good balance of secondary-to-primary copper loss while the copper-to-iron loss balance is not too bad. Note that the core cross-sectional area is far from the ideal square shape because it has a length-to-width ratio (*D/E*) of

TRANSFORMER DESIGN SHEET

Input volts 105	Hertz 60		Est. Efficiency 90 %		Turn/volt	

Lamination or core No.	100 MH SILICON STEEL				B	14,000
Window dimensions	G 1.5	F 0.5		W	M_P	6.0
C.S.A. dimensions	E 1.0	D 1.736÷0.9=1.93	a 1.736		v	10.416
	Wt 2.875	Watts/lb 1.2	Iron loss 3.45			

Windings		W_1	W_2							
Coil		N_1	N_2							
Volts		105	10							
Amperes			5							
Turns		252	24							
Gauge		23	13							
Turns/inch		38	12							
Margins		.125	.25							
Winding length (L)		1.25	1.00							
Turns/layer		47	12							
Number of layers		6	2							

Build										Total
Copper		.143	.150							
Paper		.025	.010							
Cover		.010	.010							
Total		.178	.170							
Bulge Percent		.027	.025							
Total (R)		.205	.195							.400
								Total bobbin (B)		.100
Bobbin ____ clearance ____								Total depth		.500

Losses										
Length mean turn (M)		7.25	8.54							
Total wire length (inches)		1827	205							
Ohms/ft		.0203	.002							
Resistance (hot)		3.709	.041							
Voltage drop		1.96	.205							
Copper loss (I^2R)		1.036	1.025							2.06
								Iron loss		3.45
								Total loss		5.51
								Efficiency		90 %

Temperature Rise						
Coil	t	C_S	W	W/C_S	°C T_o	
1						
2						
3						
4						
						Temp. Rise = T_o(Total) + 20 HW/C_S =

Fig. 8-31. Design sheet for autotransformer.

nearly 2:1. Improved copper loss and possibly better all-round efficiency would likely result from a squarer cross-sectional area, but with the efficiency already at 90 percent, the improvement if any would not be great.

Tapped Windings

Recall from Chapter 3 that autotransformers are frequently designed with tapped windings and can be reversed as shown in Figs. 8-32A and 8-32B. In Fig. 8-32A, the output is tapped above and below the input tap, and in Fig. 8-32B the input is tapped above and below the output tap. At first sight, it might appear that these two arrangements will give identical end results, but this is not so.

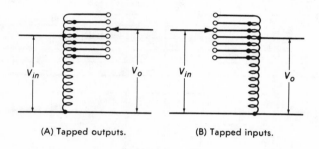

(A) Tapped outputs.   (B) Tapped inputs.

Fig. 8-32. Two methods of connecting a tapped autotransformer to permit adjustment of output voltage.

If in Fig. 8-32B, the input voltage is constant and the tapping point is selected to give higher and lower values of output volts, the flux density will change with the tapping point, because the number of primary turns is being changed for a given voltage. This is evident from the basic equation:

$$B = \frac{V \times 3.5 \times 10^6}{Nfa}$$

if N varies, then B varies, and if B varies, the iron loss varies in proportion to B^2. If, however, in the case of Fig. 8-32A, the input voltage is constant and the tapping point is varied to give higher or lower output voltages, the flux density remains substantially constant because both the input voltage and the number of primary turns are fixed.

On the other hand, if in Fig. 8-32A the input voltage is variable and the tapping point is varied in order to maintain a given output voltage, the flux density will vary with the input voltage. But in Fig. 8-32B, if the input is variable and the tapping point is changed to maintain a given output voltage, the flux density will be constant because both the input voltage and the turns are changing in the same proportion.

Ideally, the aim of the design should be to keep the flux reasonably constant, or at least precautions should be taken to avoid inadvertently saturating the core. The position can be restated as follows.

1. If the object is to achieve a given fixed level of output voltage with a varying input, then the case of Fig. 8-32B will give a constant flux density.
2. If the object is to achieve a variable output voltage with a fixed input, then the case of Fig. 8-32A will give a constant flux density.
3. If these functions are interchanged—the case of Fig. 8-32B used with a fixed input and adjusted to provide various output voltages—the flux density must be arranged to be below saturation at the lowest tapping point. The selection of higher taps will then result in lower flux densities so that the core can never saturate.
4. If the circuit of Fig. 8-32A is used with a variable input voltage, the flux density must be arranged to be below saturation at the highest expected input voltage. Lower voltages will then result in lower densities.

NINE

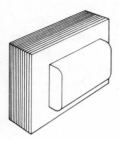

Designing
for Rectifiers

So far, it is assumed that these transformers supply "straight" resistive loads. Rectifier circuits, however, are among the most common types of transformer loads and require special consideration.

A basic problem with these circuits (Figs. 9-1 through 9-12) is that in converting ac power to dc, they create severe distortions of the current waveforms in the windings that then heat up to a greater extent than would be the case with pure sine waves. There are other side effects, too, but mainly these mean that transformers designed for rectifiers must have greater volt-ampere ratings than for straight loads.

To complicate matters, each of the various types of rectifier circuit has unique characteristics that must be considered in the transformer design. Further, in most circuits, a "smoothing" filter (that can take one of several forms) is interposed between the rectifier and the load, and each type of filter imposes a different condition on the transformer.

All of this sounds somewhat complicated, and so it is if a high level of theoretical treatment is demanded. Fortunately, the convoluted logic of higher mathematics can be bypassed in favor of rules of thumb, handy charts, and simple arithmetic to arrive at approximate but eminently practical solutions. It is also helpful that a great deal of modern circuitry is served by only a few forms of rectifier-filter circuits so that much of what would be of chiefly academic interest can be excluded.

Nevertheless, a good textbook knowledge of rectifier connections and associated circuitry is very helpful in designing transformers for them. In addition to the usual textbooks, excellent sources of information include transistor-diode data manuals put out by manufacturers at very low cost. When studying rectifier data from these sources, it is important to understand that, unless otherwise stated, voltage and current figures relating dc output to ac rms values usually refer to the performance without smoothing filters connected.

THE HALF-WAVE CIRCUIT

Consider the simplest rectifier circuit, shown in Fig. 9-1A. The alternating voltage across the secondary winding changes polarity at the frequency of the supply. When the top end of the winding is positive, the diode conducts current. When the top end is negative, the diode is nonconducting, thus cutting out the negative half-cycle. The voltage appearing across R_L is then a series of positive-going half-cycles as shown in Fig. 9-1B. Because the half-cycles are all in the same direction, this is a dc voltage although it varies from zero to maximum.

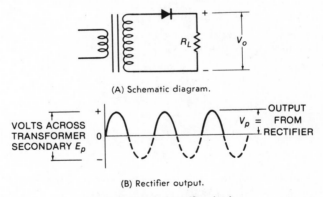

(A) Schematic diagram.

(B) Rectifier output.

Fig. 9-1. Half-wave rectifier circuit.

In order to avoid confusion between ac input and dc output in the following analysis, the symbols E and V are used as follows:

E_p	peak ac secondary volts
E_{av}	average ac secondary volts
E_{rms}	rms ac secondary volts
V_p	peak dc output volts
V_{av}	average dc output volts

This circuit uses only half the available sine wave across the transformer secondary. But, ignoring for the moment any resistance that might be in the circuit, the peak dc output voltage V_p is the same as the peak ac voltage E_p across the secondary. The average value of a half-cycle sine wave (Chapter 1, ''Kinds of Voltage'') is $E_{av} = E_p/\pi$;

therefore $V_{av} = V_p/\pi$ or $V_p = V_{av}\pi$. Obviously the *peak* dc output, V_p, is equal to the *peak* full-cycle voltage E_p across the secondary; therefore $E_p = V_{av}\pi$ also.

The rms value of the full-cycle secondary voltage E_{rms} is

$$E_{rms} = \frac{E_p}{\sqrt{2}}$$

or, because $E_p = V_p = V_{av}\pi$,

$$E_{rms} = \frac{V_{av}\pi}{\sqrt{2}} = V_{av}\frac{3.14}{1.414} = 2.22\ V_{av}$$

or

$$V_{av} = \frac{E_{rms}}{2.22} = 0.45 E_{rms} \tag{9-1}$$

In words, the average output dc voltage of the circuit is only 0.45 of the rms ac secondary voltage. Otherwise stated, for a given average output dc voltage, V_{av}, the secondary rms ac voltage, E_{rms}, must be 2.22 times as great.

The dc power output of this circuit divided by the ac power input, in other words the efficiency, is only about 0.4. In algebraic form

$$\frac{\text{dc load power}}{\text{ac power in}} = 0.4 \text{ approximately} \tag{9-2}$$

or

$$P_{ac} = \frac{P_{dc}}{0.4} = 2.5 P_{dc}$$

This takes no account of the losses in the copper and iron which reduce the circuit efficiency still further. Moreover, the unidirectional current in the windings tends to magnetically saturate the iron.

Because of the varying nature of its output, a circuit like this has few applications (although now and then a requirement for it arises). Generally, a steady, unvarying dc is needed, and to get this, it is necessary to insert a filter between the rectifier and the load. Various kinds of filters will be discussed, but basically the filter can be (and often is) simply a capacitor connected across the load as shown in Fig. 9-2A.

The effect of the capacitor is to smooth out the variations—to fill in, as it were, between the humps, as shown in Fig. 9-2B. The capacitor charges during the conducting period of the rectifier and discharges through the load during the cutoff period. It has a further important effect that the smoothed dc voltage can be higher than the secondary rms voltage by an amount that depends on the values of the capacitor C and the load R_L, together with the impedance of the transformer and any other series resistance

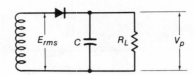

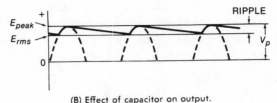

(A) Circuit with capacitor across load.

(B) Effect of capacitor on output.

Fig. 9-2. Rectifier circuit with filter.

that might be in the circuit. The apparent gain in voltage is due to the fact that the capacitor tends to charge up towards the *peak* value of the applied ac, which is 1.414 times the rms value.

Suppose in Fig. 9-2A that the secondary voltage is 7.0 volts rms. If the load R_L is disconnected, the capacitor will charge up to the full peak value of the voltage, which is $1.414 \times 7.0 = 9.89$ volts. With the load connected, current is drawn from the capacitor and the voltage is reduced, but if the current is not too great, the dc voltage can still be substantially higher than the secondary rms voltage. In designing a transformer for this circuit, then, it is necessary to take this effect into account.

In Fig. 9-3, R_T represents the total transformer resistance as seen by the load "looking into" the secondary. As discussed in Chapter 4, it is possible to show the resistance connected to one winding as a reflected resistance in the other. This applies not only to resistance connected to the winding, but also to the resistances of the windings themselves. Thus, in Fig. 9-3, R_T is the sum of the primary winding resistance R_1 reflected into the secondary winding as

$$R_1 \left(\frac{N_2}{N_1} \right)^2$$

and the secondary winding resistance R_2, or, as stated by Equation 4-16,

$$R_T = R_2 + R_1 \left(\frac{N_2}{N_1} \right)^2$$

The ac resistance of the rectifier is shown as r_{ac}. Generally, this can be considered to be negligible in solid-state rectifiers, which is fortunate because r_{ac} varies with forward current and temperature, among other things. Incidentally, r_{ac} is not to be confused with

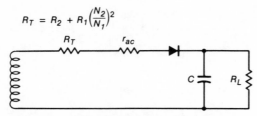

$$R_T = R_2 + R_1\left(\frac{N_2}{N_1}\right)^2$$

Fig. 9-3. What load R_L "sees" looking back into the secondary.

r_{dc}, which is a different element. If you are finicky, a close approximation of r_{ac} in ohms is given by

$$r_{ac} = \frac{24 + 0.0875T}{I} \tag{9-3}$$

where

    T is the temperature in degrees Celsius,
    I is the current in milliamperes.

Thus, if the temperature is 25°C and I is 4.0 amperes,

$$r_{ac} = \frac{24 + (0.0875 \times 25)}{4 \times 1000} = 0.0065 \text{ ohm}$$

If the temperature were 100°C, the ac resistance of the rectifier would still be only 0.00818 ohm just before it burns out (that is, if it is not rated for this temperature).

The curves of Fig. 9-4 relate secondary winding rms volts and dc volts across the load to values of C, R_T, and R_L. A high $C \times R_L$ factor together with low R_T/R_L ratio results in a high V_{dc}/V_{rms} ratio, which in turn means high dc output volts. Of course, this ratio can never be higher than 1.414 (maximum dc volts is 1.414 × V_{rms}). Note that this limitation is reflected in the top limit of the vertical axis of the graph. The graph is used as follows.

Example 1—A transformer is required for the circuit shown in Fig. 9-5. The dc output is to be 8.0 volts across a load resistance R_L of 16 ohms, making the load current 0.5 ampere. The capacitor C is to be 1000 μF. What ac voltage is required from the secondary winding at full load?

If the impedance of the transformer is not known (and how can it be since it hasn't been designed yet), then, in line with previously discussed design strategy, when a number is not known it must be estimated. But even a guess must have a basis in reality, and if the designer has little previous experience to call upon, a guess for R_T might not appear easy. This point is dealt with in more detail later.

In the meantime, to avoid getting too uptight about it, simply opt for an R_T/R_L ratio that is fairly low. This means that copper loss will be fairly low and the regulation relatively good. Suppose the ratio is tentatively assigned a value of 0.04. Then R_T is 0.04 × 16 = 0.64 ohm.

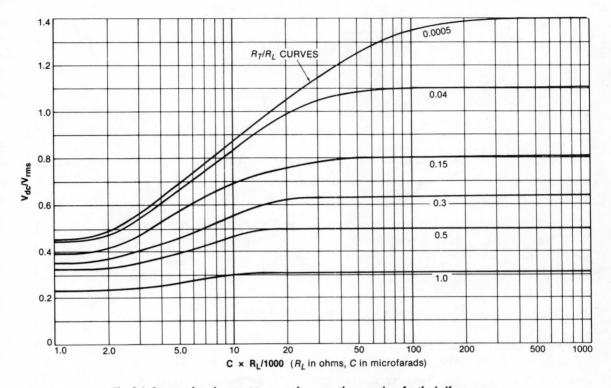

Fig. 9-4. Curves relate dc output to rms volts across the secondary for the half-wave case.

The product $C \times R_L/1000$ is $1000 \times 16/1000 = 16$. Find 16 on the bottom axis of the graph. Move vertically to the 0.04 curve, then horizontally to the V_{dc}/V_{rms} axis and read off the ratio 0.925. The required full-load terminal voltage of the transformer is then

$$V_o = \frac{V_{dc} + IR_T}{0.925} = \frac{8 + (0.5 \times 0.64)}{0.925} = 8.99 \text{ volts} \qquad (9\text{-}4)$$

where I and R_T are as shown in Fig. 9-5.

Example 2—A study of the curves reveals the effects of using different numbers for the various parameters. If the load current is put at 4.0 amperes and R_T at 0.2 ohm, then R_L is $8/4 = 2$ ohms and R_T/R_L becomes $0.2/2 = 0.1$. The product $C \times R_L/1000$ becomes $1000 \times 2/1000 = 2$. This puts V_{dc}/V_{rms} at about 0.42, or

$$V_o = \frac{8 + (4 \times 0.2)}{0.42} = 20.95 \text{ volts}$$

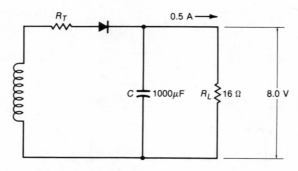

Fig. 9-5. Circuit for the first example.

On the other hand, try reducing the load current to 0.1 ampere and increasing R_T to 1.2 ohms. This puts the load resistance R_L at 8/0.1 = 80 ohms and R_T/R_L is 1.2/80 = 0.015. The product $CR_L/1000$ is $1000 \times 80/1000 = 80$. This puts V_{dc}/V_{rms} at about 1.13 and

$$V_o = \frac{8 + (0.1 \times 1.2)}{1.13} = 7.19 \text{ volts}$$

is the full-load transformer secondary voltage.

On a graph of this type, it is necessary to interpolate between curves—that is, to judge where a point may be between existing curves—for instance to place the value of $R_T/R_L = 0.015$. This can never be done with great precision, and it emphasizes the approximate nature of the graph when used in this manner. Nevertheless, the point picked is accurate enough for most purposes.

From this point, the design would proceed in the normal way with the selection of wire gauges, core area, numbers of turns, and so on. Once the actual winding resistances are determined, R_T would be calculated and compared to the figure used initially. Appropriate adjustments would then be made based on the new R_T figure and the process repeated and then repeated again if necessary. This process is demonstrated in examples later.

The half-wave circuit is usually confined to low-power operation but not often as low as the 0.8 watt implied in the last example. Nevertheless, there is no practical reason prohibiting its use for such low powers if it is needed.

THE CENTER-TAPPED FULL-WAVE CIRCUIT

Figure 9-6A shows the popular center-tapped full-wave rectifier circuit. Here two half-wave arrangements are in effect combined to use both half cycles of the secondary output voltage. The combination is in fact more in the nature of a two-phase half-wave circuit than a full-wave circuit although the end result is that of full-wave rectification.

Each rectifier conducts alternately, and as the waveform in Fig. 9-6B shows, both half-cycles appear in the output but going in the same direction. There is therefore a

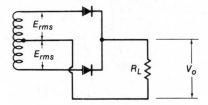

(A) With center-tapped secondary.

(B) Output voltage across R_L.

Fig. 9-6. Full-wave rectifier circuit.

series of humps, or a ripple, at twice the frequency of the half-wave circuit, making 120 per second if the supply is 60 Hz. In this circuit

$$E_{rms} = \frac{V_{av}\pi}{2\sqrt{2}} = 1.11\ V_{av} \tag{9-5}$$

or

$$V_{av} = \frac{E_{rms}}{1.11} = 0.9E_{rms} \tag{9-6}$$

where

V_{av} is the average dc output voltage,

E_{rms} is the rms ac secondary voltage.

Once more a filter is needed (Fig. 9-7A) to smooth out the voltage, a task that is somewhat easier here than in the half-wave case. Here I simply added a capacitor, with the effect to the waveform shown in Fig. 9-7B. This circuit is much more efficient than the half-wave arrangement. In addition to utilizing both half-cycles, there is effectively no core saturation due to unidirectional dc currents, as in the half-wave case, because the currents in the secondary halves flow in opposite directions.

The nature of the secondary winding current waveform is illustrated in Fig. 9-8. This shows the short-duration high-current surges that charge the capacitor. These peaks are a number of times greater than the steady dc current.

For design purposes, curves similar to those used in the half-wave case are used, as shown in Fig. 9-9. The appropriate quantities to be considered are identified in Fig. 9-10.

It is important to note that R_T in this case is the resistance as seen looking into *only half* of the secondary winding; therefore, the turns ratio used to compute R_T is *half* of the total secondary-to-primary turns ratio. By the same token, the V_{rms} value

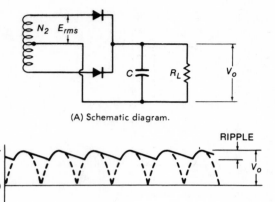

(A) Schematic diagram.

(B) Smoothed output waveform.

Fig. 9-7. Full-wave rectifier with smoothing capacitor.

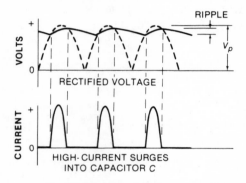

Fig. 9-8. Waveforms of full-wave rectifier.

used in the voltage ratio V_{dc}/V_{rms} is only half of the total secondary voltage. The rectifier ac resistance is considered negligible and is not shown.

Example 3—In the circuit of Fig. 9-11, the values used in the first case of Example 2 have been repeated for the purpose of comparing results. The output is 8 volts dc at 4 amperes, and R_L is 2 ohms. Each R_T is 0.2 ohm and therefore $R_T/R_L = 0.1$. Capacitor C is 1000 μF and therefore $C \times R_L/1000$ is $1000 \times 2/1000 = 2$. Again you must interpolate to place the value of $R_T/R_L = 0.1$ as shown by the dotted curve. The V_{dc}/V_{rms} ratio is thus about 0.84.

The required transformer terminal voltage is then obtained from the equation

$$V_o = \frac{V_{dc} + (I_{dc} \times R_T)}{0.84} = \frac{8 + (4 \times 0.2)}{0.84} = 10.476 \text{ volts} \qquad (9\text{-}7)$$

This is the voltage across *half* the winding. The total voltage is $10.476 \times 2 = 20.95$, say 21, volts.

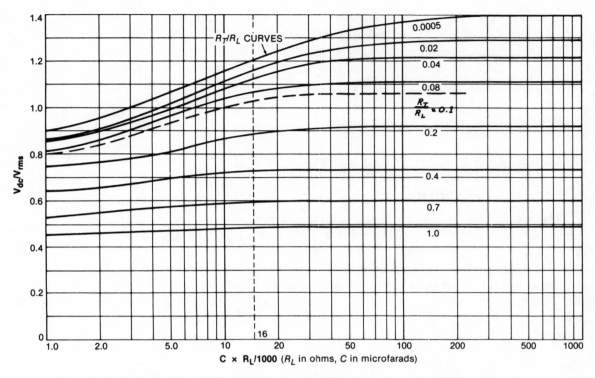

Fig. 9-9. Curves relate dc volts output to rms volts on transformer secondary for full-wave case.

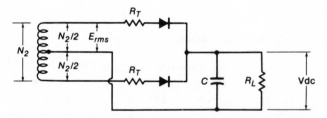

Fig. 9-10. Quantities related to curves in Fig. 9-9.

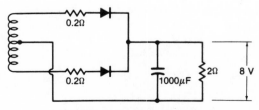

Fig. 9-11. Values for Example 3.

In practice, though, a much higher capacitance than 1000 μF would probably be used for a current of 4 amperes. A common rule of thumb for this is to use from 1000 to 2000 μF *per ampere*. For 2000 μF per ampere, then C would be $4 \times 2000 = 8000$ μF. And $C \times R_L/1000$ then becomes 16. This figure on the horizontal axis together with $R_T/R_L = 0.1$ leads to $V_{dc}/V_{rms} = 1.04$. The required terminal voltage in this case is then

$$V_0 = \frac{8 + (4 \times 0.2)}{1.04} = 8.46 \text{ volts}$$

or 16.9 volts across the entire winding.

THE FULL-WAVE BRIDGE

The circuit of Fig. 9-12 is a widely used arrangement in solid-state electronics. It is rather more efficient than the center-tapped full-wave circuit. The four rectifier elements are obtainable most often in neatly encapsulated units, but separate elements can be used. No center tap is required on the transformer secondary, and for a given dc voltage, the ac voltage across the secondary winding is approximately half that across the total winding in the center-tapped circuit. Again, there is no dc saturation of the core, and the entire secondary winding is in operation for both half-cycles.

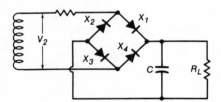

Fig. 9-12. Full-wave bridge rectifier.

In this case, when the top of the winding is positive, current flows from the bottom terminal through X_3, R_L, and X_1 back to the positive side. Figure 9-13 illustrates this action. When the top end swings to negative, X_3 and X_1 are cut off, and the other rectifiers conduct. Current then flows through X_2, R_L, and X_4. As in the previous circuits, using capacitive-input filters, the current flows through each rectifier for only that part of the voltage half-cycle which exceeds the voltage across the capacitor. For this short time, however, the current is considerably higher than the dc load current.

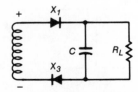

Fig. 9-13. Equivalent circuit of Fig. 9-12 when top end of winding is positive.

The output voltage and current waveforms are the same as for the center-tapped case.

As with other rectifier circuits, the design philosophy must be generous in order to cater for the larger volt-ampere requirements of rectifier circuits.

HYBRID CIRCUITS

The circuit shown in Fig. 9-14 is a combination of the center-tapped and the bridge type full-wave circuits, and it provides three different outputs from a single winding. If B is chosen as a reference point for A and C, the voltages between B and A and B and C are center-tapped full-wave outputs, equal in voltage, but of opposite polarity. Each voltage, as in other center-tapped circuits, will be very approximately the same as that across each half of the winding. The third output appears between C and A and is the bridge circuit output equal to double that from B to A or from B to C.

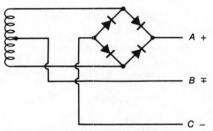

Fig. 9-14. A three voltage output circuit.

The relative polarities are as indicated. This circuit could be used for a dual polarity output (B to A, B to C) having equal voltages of opposite polarity as required by many integrated circuits. Or it could be used to supply two different voltages of the same polarity (C to B, C to A).

Another type of circuit, a voltage doubler, is shown in Fig. 9-15. In this circuit, the output voltage (V_{dc}) on small load currents can approach twice the *peak* voltage across the secondary winding. Here C_1 charges through R_1 (a surge limiter) and X_1, which conducts during one-half of the cycle. On the next half-cycle X_1 shuts off and X_2 conducts and charges C_2.

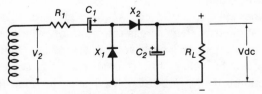

Fig. 9-15. Voltage doubler circuit.

VACUUM-TUBE RECTIFIERS

Although we tend to think of vacuum tubes as relics of yesteryear superseded by solid-state devices, there are still plenty of them around. Moribund, perhaps, but not yet dead, they still appear in a variety of applications. From the experimenter's point of view, the collecting of old radios is thriving and gaining ground all the time. If you collect them, it will be useful for you to be able to make your own transformers as replacements for those which have burned out.

A point of interest here is that most small vacuum-tube radios of 20 to 30 years ago had no power transformers in them. They operated directly from the supply line using the half-wave circuit shown in Fig. 9-16. The isolating safety feature of the transformer was lost, and these little radios were responsible for littering the domestic scene with more than a few electrocuted citizens, especially in the bath, which offers perfect grounding conditions for the human body. The scheme is not recommended.

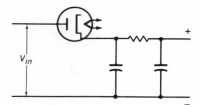

Fig. 9-16. Vacuum-tube half-wave rectifier circuit.

The three basic rectifier circuits in vacuum-tube format are shown in Figs. 9-17A, 9-17B, and 9-17C. The tubes pictured here are all of the so-called indirectly heated types. That is to say, the cathode consists of a coated tube enclosing a heater which brings it to a temperature high enough to "boil" off electrons. The electrons are attracted to the positive anode and constitute a current across the vacuum. The heaters are supplied by a separate heater winding, or windings, on the transformer.

Basically, the transformer design process is identical with that for the transformer of the solid-state rectifier. One difference is the need for a heater winding on the transformer, but this presents no special design problems. Another marked difference is in the value of r_{ac} for the rectifier; it is very much higher in vacuum tubes than in solid-state components. But the figure is easily obtainable from manufacturers' data.

If the tube is of the filament type, as shown in Fig. 9-18, it might have a marked effect on the insulation requirements for the transformer as compared with the heater type. The voltage distribution is shown in the figure.

Note that the tube has a voltage across it that can be as high as 2.8 V_{rms}. This voltage is composed of the dc voltage across the capacitor, which on low-current loads can approach 1.414 V_{rms}, plus the peak voltage across the winding, which is also 1.414 V_{rms}. In Fig. 9-18, this voltage exists between the two secondary windings, and the practical implications of it will be discussed later in the section on insulation requirements. It is called the *piv*, or peak inverse voltage (sometimes prv, the peak reverse voltage). It appears across rectifiers of all types, in every kind of circuit, solid-state and vacuum-tube alike. It is a most important figure in the selection of rectifiers, but our immediate interest is in what it does to the transformer. The actual value reached by the piv depends on the kind of circuit and the circuit constants.

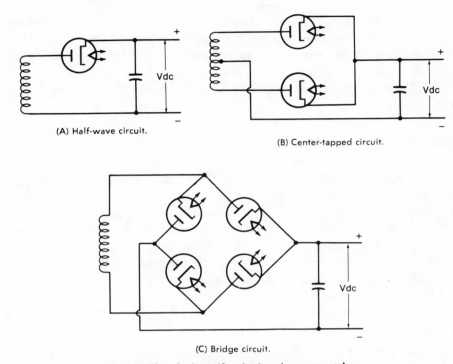

(A) Half-wave circuit.

(B) Center-tapped circuit.

(C) Bridge circuit.

Fig. 9-17. Three basic rectifier circuits using vacuum tubes.

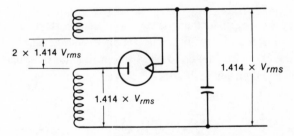

Fig. 9-18. Illustrating the electrical stress between circuit windings.

EFFECTS OF FILTERS

The purpose of the notes in this section is to put into perspective the effects of external circuitry on the design of transformers for rectifier circuits. The notes will, it is hoped, prove interesting, perhaps even useful, in specific situations. Certainly they should underline the need to understand the effects of the circuits attached to a transformer (and not only rectifier circuits) on the design of the transformer.

In the rectifier circuits just discussed, a single capacitor is used to act as a smoothing filter between the rectifier and the load. This is the simplest filter, widely used in solid-state equipment where smoothing requirements are not usually as stringent as in their vacuum-tube counterparts. In any case, the much lower voltages make the use of high-capacitance capacitors much more practical in solid-state design. There are, however, many other possible arrangements of capacitors, inductors, and resistors, all designed to improve filtering, and all affecting the transformer design in various ways.

Basically, these circuits can be grouped under two headings—capacitive input and inductive input.

Capacitive Input

Three other capacitive input types of the same ilk as those discussed previously (but a little more complex) are shown in Fig. 9-19. Although these employ inductors and resistors, they are still capacitive types because the first element is a capacitor. (Sometimes they are referred to as "shunt capacitance input filters.")

The first arrangement (Fig. 9-19A), using inductors, is not often encountered today. A more typical circuit in modern solid-state usage is shown in Fig. 9-19B and consists of a capacitor followed by a transistor or IC regulator to "hold" the voltage at a predetermined level. The circuit of Fig. 9-19C is typically found in the output of a vacuum-tube full-wave rectifier feeding a circuit of relatively low power.

Inductive Input

In the inductive input filter of which two examples are shown in Fig. 9-20, the first element is a series inductor. This element makes the circuit behave in a manner markedly different from the capacitive input arrangement, and it creates different conditions for the transformer.

Provided that the first inductance is no less than a certain critical value, this type of filter draws a constant current from the rectifier system when a load is connected to the output of the filter. For the duration of an entire half-cycle, one rectifier conducts at a constant level and then suddenly cuts off. At that instant, the other rectifier suddenly

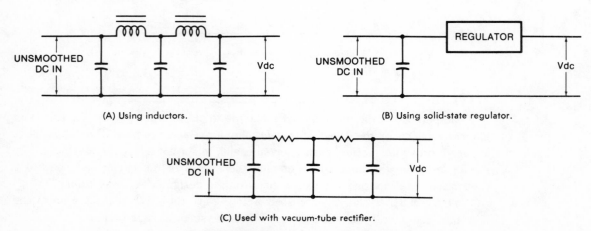

(A) Using inductors.

(B) Using solid-state regulator.

(C) Used with vacuum-tube rectifier.

Fig. 9-19. Three forms of the capacitive input filter.

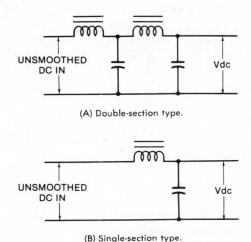

(A) Double-section type.

(B) Single-section type.

Fig. 9-20. Inductive input filters.

comes to full conduction and continues for a half-cycle, and then it cuts out; the action repeats again. The current through the inductor is therefore substantially constant. The level of current through each rectifier does not follow the applied voltage sine wave. It is, in fact, approximately a square wave.

The volt-ampere demand on the transformer is less than for the capacitive input case. If the inductance is less than the critical minimum, the circuit then acts as the capacitive input type. This kind of filter is likely to be found in vacuum-tube circuits, and because there are still a few of these around, a few more details might be of interest.

V_{dc} Versus V_{rms}: Inductive Case

The curves of Fig. 9-9 are intended only for capacitive input circuits. For inductive inputs, the approximate secondary voltage, E_{rms}, needed for a given dc output, V_{dc}, is obtained from

$$\text{Open-circuit } E_{rms} = 1.1 \ (V_{dc} + IR) \qquad (9\text{-}8)$$

where

I is the dc load current in amperes,

R is the total series resistance of inductors, resistors, transformer impedance, etc., as marked in Fig. 9-21.

The transformer and rectifier resistances are represented by R_T and r_{ac}, respectively. In the case of a center-tapped full-wave circuit, E_{rms} is the voltage on one side of the center tap, and the transformer resistance is that referred to one-half of the secondary.

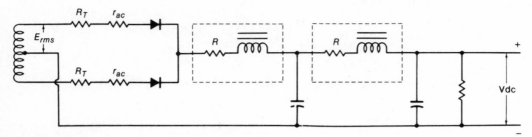

Fig. 9-21. Circuit quantities to use in Equation 9-8.

Critical Inductance

The minimum inductance for the inductance input filter for correct operation at 60 hertz in the full-wave circuit is given by

$$L_{min} = \frac{R_{total}}{1130}$$

where R_{total} is all the series resistance in the circuit.

Unlike the capacitive input case, the dc output voltage can never exceed the secondary rms voltage—so long as the inductance is above the critical value. This is easily seen if Equation 9-8 is rearranged:

$$V_{dc} = \frac{E_{rms}}{1.1} - IR \qquad (9\text{-}9)$$

If the load current is zero or very small, then IR is zero or very small and $V_{dc} = E_{rms}/1.1$, which must be less than E_{rms}.

In practice, however, a zero-current situation cannot coexist with an inductance above the critical value. This is because the critical value is related to the total resistance in the circuit in such a way that an infinitely high resistance (which would be the case if the current is zero) would require an infinitely large inductance for the circuit to act as an inductive input filter. At very low current, then, it will act as a capacitive input filter and the voltage will rise to a high value, which might be undesirable.

Bleeder Resistance

The solution to this problem is obtainable by referring to Equation 9-9. For a given value of inductance, all that need be done to ensure that the current will always be sufficient, regardless of changes in load, is to connect a resistance permanently across the output such that $R_{total} = 1130 \times L$.

Suppose, for example, that L is 20 henrys, a value determined for the maximum load current. Then, a resistance connected permanently across the output such that

$$R_{total} = 1130 \times 20 = 22,600 \text{ ohms}$$

will ensure that with smaller loads, even to the point of disconnecting the load entirely, the circuit will still work effectively as an inductive input filter.

This same result is obtainable from the equation

$$I_{minimum} = \frac{V_{dc}}{L \times 1130} \tag{9-10}$$

For example, if V_{dc} is 200 volts and L is 20 henrys,

$$I = \frac{200}{20 \times 1130} = 0.0088 \text{ ampere or } 8.8 \text{ mA}$$

In the first example if V_{dc} is 200 volts, then

$$I = \frac{200}{22,600} = 8.8 \text{ mA}$$

as above.

Swinging Inductor or Choke

From Equation 9-10, it can be seen that when the load current is large, the minimum inductance is small, and vice versa. By taking advantage of the fact that the inductance of an iron-core inductor (choke) decreases with increased dc current in the winding and increases as the dc current decreases, it is possible to design an inductor that will automatically swing in inductance value in the right direction to meet the requirement for critical inductance.

RECTIFIERS AND INSULATION

The way in which a transformer is connected to external circuits and the nature of the circuits affect the insulation requirements of the transformer. Good examples of this are found in rectifier circuits. Examples are shown in Fig. 9-22.

In the circuits of Fig. 9-22, both the rms and peak voltages are marked on the transformer secondary, while the dc voltage across the filter capacitor is the maximum possible on no load. The voltage obtainable on load would be less than this, and its value would depend on the amount of current being drawn.

The 1000-volt points are due to the addition of 500 volts peak ac across the secondary in series with the 500-watt dc voltage across the capacitor. It is a simple matter to formulate an insulation plan for any transformer in which the interwinding and winding-to-core voltages are analyzed in this fashion.

Note the difference in core-to-secondary voltage stress between Fig. 9-22A and Fig. 9-22B due simply to moving the rectifier. Similarly, note the shift from an interwinding stress of 1000 volts to a winding-to-core stress in Figs. 9-22C and 9-22D.

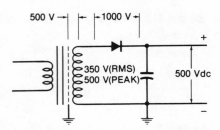

(A) With core-to-secondary voltage stress of 500 V.

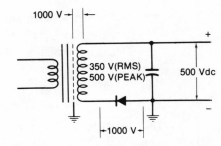

(B) With core-to-secondary voltage stress of 1000 V.

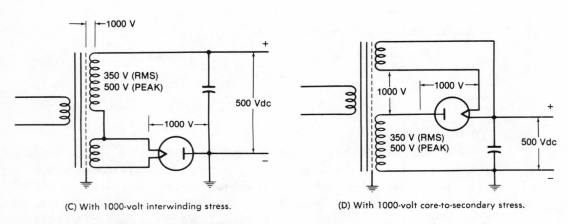

(C) With 1000-volt interwinding stress.

(D) With 1000-volt core-to-secondary stress.

Fig. 9-22. Transformer connections in rectifier circuits.

VOLT-AMPERES AND CURRENTS

As previously explained, the heating effects due to I^2R losses in the windings finally determine the volt-ampere capability of a transformer. But in a rectifier circuit with its highly distorted current waveforms it is a mathematically complex matter to predict accurately what the losses will be. This is especially true when a capacitive input filter—probably the most common type used today—is involved. The inductive input case is somewhat easier.

There is a practical and generally satisfactory solution to the problem (satisfactory at least to all but those with inveterate academic minds). It is certain that a rectifier transformer must be designed along more generous lines than for a resistive load; in other words, it must have a greater volt-ampere capability than seems necessary from consideration of its load watts. Therefore, all you need do is to design as though for a resistive load of higher power.

As always, the power capability of the rectifier transformer is governed by temperature rise and permissible voltage drop in the windings. Unfortunately, a useful prediction of these factors is not easy to obtain. For most purposes, though, a usually satisfactory guideline is simply to choose the wire gauges on the basis of a generous

1000 to 1200 CM/A. Another rule of thumb frequently used with the common full-wave bridge circuit with a capacitive input filter is to design for a rating of about 1.4 times the dc power output. For the center-tapped full-wave arrangement, this could be, say, 1.5 times the dc output. In the latter case, each half of the secondary winding carries the high charging current for a portion of the half cycle only, and for the purpose of choosing a wire gauge the current can be reckoned as 70 percent of that in the bridge case.

For safety, a bench test of temperature rise, such as that discussed in Chapter 16, is advisable, or failing that, conduct a careful "soak" test using frequent "hand" and "nose" checks for the feel and odor of excessive heat.

TEN

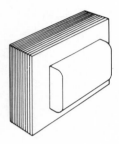

Transformers in Converters and Inverters

In a book on transformers, there is very little space to spare for discussion on solid-state circuitry and associated theory. It has to be assumed that anyone interested in designing transformers either has the required circuitry knowledge or intends to get it. Usually, then, the circuitry attached to the transformer can be represented by a simple resistance. As with rectifier circuits, however, there is more involved in discussing converter and inverter transformers. In these circuits, the transformer is the cornerstone of the entire function in that it not only transforms power, but in some types, it governs the operational frequency and overall efficiency. In short, to design the transformer, it is essential to know how it ties in with the entire circuit operation, and this in turn demands some knowledge of solid-state principles.

Certainly it would be inappropriate to cover transistor theory here. Right now it is assumed that the reader is acquainted with transistor circuit design. To avoid sending some reader chasing through libraries on impromptu refresher courses, however, I provide a descriptive resume of the functions and principles of these interesting circuits.

USES OF CONVERTERS AND INVERTERS_____

The purpose of the *converter* is to transfer power from a dc source, frequently a low-voltage battery, to a load requiring a different (usually higher) dc voltage. For this reason, it is sometimes called a *dc transformer*.

266

For example, a converter might work from a 12-volt battery and deliver one or more dc outputs at different dc voltage levels to power, say, radio transmitting and receiving equipment or other dc-powered devices, as in Fig. 10-1.

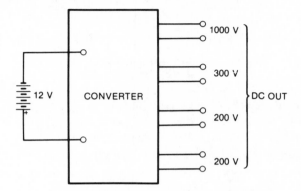

Fig. 10-1. Converter "converts" dc input to dc outputs.

Using the same kind of dc source, the *inverter* is required to transfer power to a load requiring ac voltage as in Fig. 10-2. For instance, it might deliver, say, 110 volts ac at 60 hertz to power standard domestic equipment such as TV sets, radios, kitchen appliances, and the like, in a cottage, boat, or trailer. In aircraft, it might supply power at relatively high frequencies of, say, 400 or 1000 hertz. This has the big advantage that the equipment it powers can be made smaller and lighter.

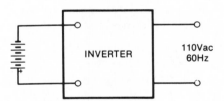

Fig. 10-2. Inverter "inverts" dc input to ac output.

In this context, the use of the word "converter" to describe the function of transferring power from one dc level to another seems reasonably justified. However, using the word "inverter" to describe the action of changing dc to ac may cause a sharp pain in the heart of the language purist. Never mind—it will pass with time.

A less frequent but completely practical use for inverters is as a frequency changer (Fig. 10-3). Here the frequency to be changed is rectified to dc in the conventional way and then the dc is used to power an inverter designed to put out the frequency required.

HISTORY AND BASIC PRINCIPLES

Most readers of this book are familiar with both converters and inverters although they might not recognize them by these names.

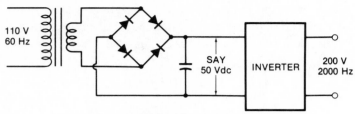

Fig. 10-3. Rectifier-inverter changes 100 V, 60 Hz, to 200 V, 2000 Hz.

The induction (Rumkorff) coil discussed in Chapter 2 is a form of inverter circuit as is its progeny, the automobile ignition system. These devices "invert" a dc battery voltage to a much higher level of pulsed voltage. The design emphasis in these cases is on achieving a very high output voltage pulse, but there are also power considerations involved. For instance, it takes power to create that fat, blue spark that jumps the gap and ignites the fuel mixture in a gas engine.

To accomplish this, the circuit employs an interrupter to chop up the dc applied to the primary winding of the coil, thus creating the varying current needed to induce voltage in the secondary, while the turns ratio of the primary to secondary transforms the voltage to a higher level.

A descendant of the induction coil known as the *ringing vibrator* is shown in Fig. 10-4. This circuit, which at one time found uses in small telephone exchanges, is a true power inverter. The interrupter, or trembler, works on the same principle as the electric bell. It is arranged to switch the battery voltage first to one half of the primary winding and then to the other, thus creating an alternating input to the transformer. A transformed, or inverted, alternating output then comes from the secondary winding.

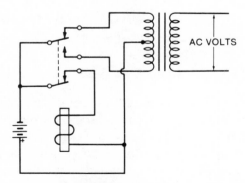

Fig. 10-4. Ringing vibrator inverter.

These features—the interruption of the primary winding circuit and voltage transformation—are characteristic of the types of converters and inverters to be discussed (there are such things as rotary converters and inverters, which will not be considered here).

In the area of converters, the vibrator power unit used years ago in automobile radios and other portable electronic equipment is probably the best known. This arrangement, shown in Fig. 10-5, is an extension of the ringing vibrator. Again, the circuit features the interrupter and the transformer but has the additional feature of a vibrating rectifying device at the output that converts the alternating voltage back to dc. The set of contacts marked B are mechanically part of the vibrator unit and are driven in synchronism with contacts A. Contacts B change over at the instant the voltage in the secondary winding is reversing. Therefore, the capacitor C is always charged in the same direction and the output is dc. This circuit is similar to the center-tapped rectifier arrangement of Chapter 9. It can also be used with a voltage doubling arrangement.

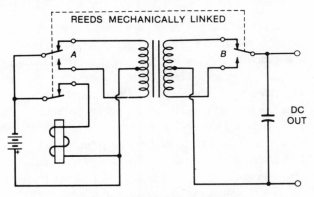

Fig. 10-5. Synchronous vibrator converter.

Frequently, the converter power unit consisted of a ringing vibrator similar to that in Fig. 10-4 but with vacuum-tube or copper-oxide rectifiers connected to the transformer secondary as in Fig. 10-6. (The heater circuit is not shown; it is supplied directly from the battery.) With this change, the circuit closely resembles that of the modern converter.

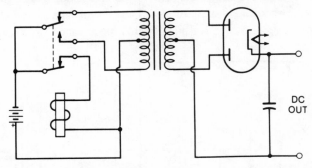

Fig. 10-6. Converter with vibrator input and vacuum-tube rectifier
full-wave output.

With the advent of solid-state electronics, it was a short step to replace the electromechanical vibrator with an electronic switching circuit and the vacuum tubes with solid-state rectifiers. In Fig. 10-7, an electronic switching arrangement resembling a multivibrator is shown. Many other types of solid-state converters are possible.

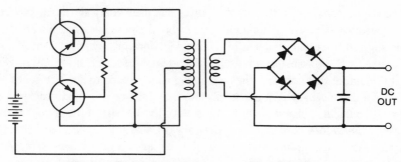

Fig. 10-7. Converter with transistor input vibrator and diode bridge rectifier output.

Remove the rectifiers and an inverter results—well, nearly. At first sight, one might think that the addition of rectifiers makes the converter circuit more complicated than the inverter. But in practice, it turns out that the reverse is more likely to be true.

The reason is not hard to find. The frequency at which a converter operates can theoretically be any value one wishes because the output is always dc. There are, of course, limitations in practice to the permissible frequency variations, but the frequency does not have to be controlled closely. In the case of the inverter, however, there is usually a very practical need to control the frequency quite closely to a predetermined figure because the equipment that it drives—the TV, electric razor, aircraft radio, or whatever—is designed to work from power supplies of a specific frequency. The circuitry used to achieve frequency stability is relatively complex.

There are also other considerations that can add to the complexity of an inverter as compared with a converter. For instance, highly reactive loads tend to upset the switching circuit, and a buffer amplifier might be needed in the circuit between the load and the oscillator as in Fig. 10-8. The output of an inverter is a square wave, which is satisfactory for most power applications. In some applications, though, this presents difficulties. Sometimes it is necessary to modify the waveshape by introducing a shaping circuit before the load. The circuit of Fig. 10-9A, for instance, is designed to produce

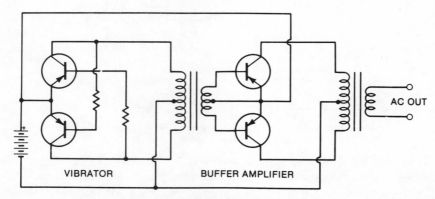

Fig. 10-8. Solid-state inverter with buffer amplifier.

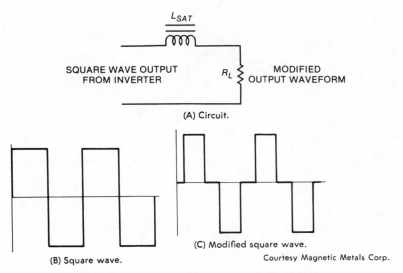

(A) Circuit.

(B) Square wave.

(C) Modified square wave.

Courtesy Magnetic Metals Corp.

Fig. 10-9. Simulation of peak-to-average ratio of a sine wave.

a *simulated* sine wave. From a square wave input of peak value equal to the sine wave to be synthesized (Fig. 10-9B), it produces a waveshape like that of Fig. 10-9C. If the inductor is designed to absorb about 30 percent of the applied square wave, the waveform in Fig. 10-9C will have the same peak/rms voltage ratio as a sine wave.

Inverters, however, can be nearly as simple as converters. It all depends on what the inverter is needed for.

SATURABLE-CORE SWITCHING

Converter and inverter circuits come in many varieties, but among the most effective and simplest are those designed around a saturable-core transformer controlling a regenerative transistor switching circuit. The converter circuit in Fig. 10-10A and the inverter of Fig. 10-10B are representative of circuits of this type.

The term "saturable core" is rather forbidding and the literature on the subject deepens the mystery with talk of "square" core materials although the core might in fact be round or some other shape. Then there is talk of flux change in the core with respect to time, in webers per second, and volt-second saturation capability, all somewhat alarming.

If, however, we cut through the esoteric abracadabra beloved by the professional and draw aside the veil of mystery you have the same old transformer sitting there fluttering its eyelashes at us—the same that we started with back in Chapter 1, more or less. True, it is treated a little differently from the transformers discussed so far. For one thing, it is supplied with a square wave input, and it has a square wave output, as was mentioned a few paragraphs previously. Incidentally, the square wave output has an advantage in the converter circuit in that it is easier to smooth the output of the rectifier. In fact, if a perfect square wave were perfectly rectified, it would theoretically

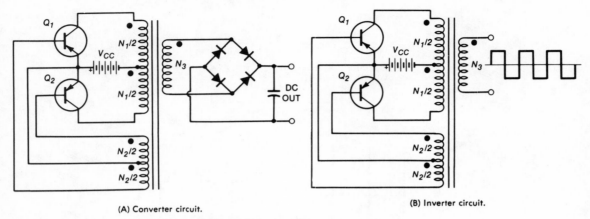

(A) Converter circuit.

(B) Inverter circuit.

Fig. 10-10. Saturable-core converter and inverter circuits.

need no smoothing at all as shown in Fig. 10-11. The two half-cycles would fit together like bricks to give a straight dc line.

But the chief difference is that in the saturable-core transformer, the core is deliberately run into saturation, as its name implies—a state of affairs I have tried to avoid up to now. Let's look at that situation.

Referring to Fig. 10-10 again, the circuit is seen to be a basic grounded-emitter configuration shorn of all components not immediately necessary to the explanation of how it works. The dots indicate the relative polarities of all the windings at any instant. In other words, when one dot is positive, they are all positive, and when one is negative they are all negative. First assume that the circuit is oscillating: The transistors are alternately turning off and on, and current is flowing alternately through the halves of

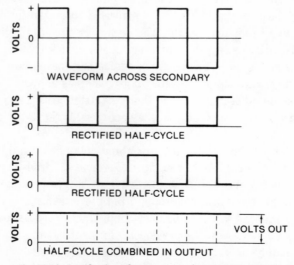

Fig. 10-11. Rectification of square wave in converter output.

the transformer primary winding N_1. The transistors are pnp and therefore conduct when the base is negative.

When Q_1 conducts, the collector end of N_1 approaches the positive potential and the battery voltage V_{CC} appears across $N_1/2$, assuming negligible resistance in the transistor. The flux in the core increases with the current, and voltages are induced across the other windings with relative polarities as shown. It can be seen that the base of Q_1, which is connected to winding N_2, is negative, the condition needed for conduction, and the base of Q_2, which is connected to winding N_2, is positive, so Q_2 is turned off. When the core saturates, there is no further change in flux and the induced voltages fall to zero. This in turn reduces the base drive to Q_1 and the current falls. The falling current brings the core out of saturation, and because the field is collapsing, the induced voltages across N_1 and N_2 reverse.

The base of Q_1 now goes positive, switching Q_1 off. The base of Q_2 goes negative, switching Q_2 on. Supply voltage V_{CC} now appears across the other half of N_1 and the dots now represent negative polarity. In effect, each half winding $N_1/2$ acts alternately as a primary winding in a 1:1 ratio autotransformer with the other half as the secondary. The induced voltage in N_1 is always such as to add to the supply voltage V_{CC} across the "off" transistor, which therefore has $2 \times V_{CC}$ across it. This latter point is important in selecting transistors.

From the practical point of view, this circuit as it stands might have difficulty in starting when it is first switched on. This is because the base voltage on each transistor is zero and therefore the collector current in each is zero. A practical solution to this problem is to put a small forward bias on each base so that both transistors tend to conduct when the circuit is first switched on, as shown in Fig. 10-12. Because there is never absolute symmetry in any circuit, one transistor will always conduct more than the other and take over the start of the cycle.

This circuit is the basis for many excellent converters and inverters, but a dual-transformer circuit like that in Fig. 10-13 has many advantages. Here transformer T_1

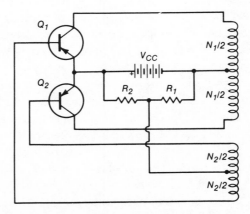

Fig. 10-12. A small bias on the bases of Q₁ and Q₂ assists starting.

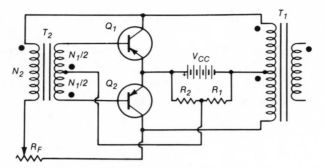

Fig. 10-13. Dual-transformer circuit.

does not saturate. It is of conventional design, exactly like those described in Chapter 8 and 9, but with square wave inputs and outputs instead of sine wave. Because it does not saturate, losses are small and the magnetizing current never rises to a high value as it does in the single-transformer circuit.

The saturable-core function is carried out by the relatively small transformer T_2 that needs to deliver only a few watts (only 3 or 4 watts in a 200- to 250-watt output circuit). The switching action is done by the small transformer and occurs when the core is saturated by the *base* current of the "on" transistor. In addition to greater efficiency, this scheme also reduces collector current spikes that threaten to destroy the transistor (but quite effective corrective circuits can be added to the single-transformer circuit to counter this problem). The detailed action of the dual arrangement is as follows.

Dual-Transformer Circuit

In Fig. 10-13, the small saturating transformer (T_2) is used to perform the switching action. The power transformer (T_1) handles only the output power and feedback. On switching the circuit on, one transistor conducts more heavily than the other. Supposing the former is Q_2; then negative voltages appear at the marked ends of the primary of T_1 and of the secondary of T_2. Transistor Q_2 is driven further towards saturation and Q_1 is switched off. Because Q_2 is fully conducting with negligible voltage dropped across it, its collector is essentially at the positive voltage V_{CC} while the collector of Q_1 is at the induced negative value of the top of T_1 primary winding which is equal to V_{CC}. Thus, the collector-to-collector voltage is twice V_{CC}.

This voltage is across R_F in series with the T_2 primary winding. Therefore, the voltage across the primary is $2 \times V_1$ less the voltage drop across R_F. The current through R_F is the base current of Q_2 referred to the primary winding. As the magnetizing current through T_2 increases, the voltage drop across R_F increases and the voltage drop across T_2 decreases. As T_2 reaches saturation, the magnetizing current increases very rapidly because there is now no change in the flux; the voltage drop across R_F increases rapidly and the voltage across the T_2 primary winding drops rapidly. This in turn reduces the base drive to Q_2, which comes out of saturation. The collector voltage of Q_2 rises as the current decreases. This in turn reverses the induced voltages in the T_1 primary winding and the regenerative action causes the transistors to rapidly change states to Q_1 conducting and Q_2 turned off. The action continues to repeat in this manner.

There are many variations and refinements of these circuits. Practical circuits are found in many publications, especially in manufacturers' transistor data manuals.

TRANSFORMER DESIGN CONSIDERATIONS_____

The Basic Equation

Referring back to Fig. 10-10, consider the basic transformer design equation that relates the voltage across a winding to the number of turns, flux density, etc., in a converter/inverter:

$$V = 4FfaNB \times 10^{-8}$$

and if this is compared with the general transformer equation (Equation 3-14) in Chapter 3, it will be found they they are identical. Only when numbers are assigned to the symbols is a departure from the previous equation observed. Recall that a value of 1.11 was assigned to the form factor F because for a sine wave, the ratio of rms to average value is 1.11 and multiplying the equation by it converted it from average voltage to rms voltage. But now, the transformer under consideration is to be designed for square waves, not sine waves. Therefore, F is made equal to 1.0 and the equation becomes

$$V = 4faNB \times 10^{-8} \qquad (10\text{-}1)$$

This is sometimes referred to as the *basic converter/inverter equation*. In this equation, V is the peak square wave voltage across the primary winding. It is in fact equal to the battery voltage less the voltage drop in the circuit. The symbol f is for the *desired* frequency of oscillation, a the cross-sectional area of the core, N the number of turns across half of the primary winding, and B the maximum flux density at which the core is to be run (in the present case, it is the flux density at which the core saturates).

It is clear from this that the transformer parameters will determine the frequency of oscillation, as well as the functioning of a power output transformer, as given by

$$f = \frac{V}{4aNB \times 10^{-8}} \qquad (10\text{-}2)$$

The equation also reveals a feature that is important in inverter design, namely that the frequency can be controlled by adjusting the voltage V.

The transformer design follows precisely the pattern used in previous "straight" transformer design examples. The values to be assigned to f and B, however, are arrived at in a slightly different way. Previously, there was no question about a value for f; it was fixed by the frequency of the supply line. Now it is whatever frequency desired for the oscillator. On the other hand there was a choice previously for flux density B. Now B *must* be the saturation value of the core material, so the choice of B is governed entirely by the choice of material.

When these have been established, the converter inverter counterpart of Equation 3-21, with the conversion factor of 6.45 included to permit the use of gauss and square inches, is

$$\frac{N}{V} = \frac{3.87 \times 10^6}{faB} \tag{10-3}$$

for the turns-per-volt figure of the transformer.

The Core

In all engineering problems, the choice of materials depends entirely on a previously defined set of limitations, which includes questions of cost, availability, dimensions, and performance. There are indeed core materials with characteristics designed especially for use in circuits like converters and inverters that permit the highest kind of performance to be designed in the component; that permit substantial reduction in size and weight for given results; that could prove less costly in the end because they permit savings in other components (although initially more expensive); and that are readily available to the professional.

But the experimenter often works by different rules. Procurement of certain materials might be difficult for him or her and, in any case, his delight may be to wrest an adequate performance from a collection of junk using his resources and knowledge. His principal motivation is technical interest, but he may also feel an underlying glee at having saved dollars and beat "the system."

So, with no apologies to the technical purists, it can be said that ordinary silicon iron, which is less than ideal, can be used with perfectly acceptable results. At the same time, the ideal characteristics cannot be ignored, because only by knowing them will we know the extent of the sacrifice being made, if any.

The *B-H* Curve

First, let us consider the question of the "square" core mentioned earlier. In Chapter 2, hysteresis was discussed in some detail and the general shape of a *B-H* curve as in Fig. 10-14A was covered. This curve is typical of silicon iron. When talking about a "square" core or "square" material, the term "square" describes a *B-H* curve like that in Fig. 10-14B. In Fig. 10-14B, the curve has steep vertical sides and an almost horizontal roof and floor, giving it a "square" appearance; this illustrates a "square" core. Certain materials are processed to emphasize this shape. The material is generally a 50-50 nickel-iron alloy and is frequently used in the toroidal format. These materials have names like Square 50, Square 80, and Supersquare 80 and are designed specifically for saturation operation. Other materials, however, are of the *square loop* type, such as grain-oriented silicon iron and even ordinary silicon iron in a modest way. Other kinds of material are known as *round loop* alloys, but they are not our concern at the present.

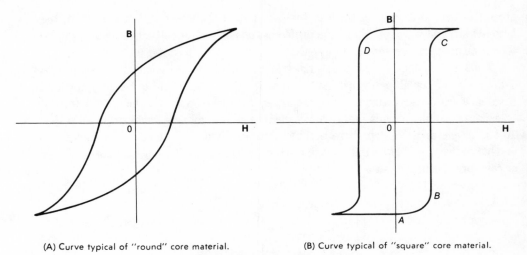

(A) Curve typical of "round" core material.　　　(B) Curve typical of "square" core material.

Fig. 10-14. Hysteresis (B-H) curves.

As discussed in Chapter 2, the permeability μ of a material is given by $\mu = B/H$, which means it is greatest when the slope of the B-H curves is steepest but is near zero at those points where the curve is nearly horizontal. The primary winding of a transformer using this material therefore has relatively high inductance (and thus low magnetizing current) at flux densities represented by the sides of the curve, and near-zero inductance (and thus high magnetizing current) at flux densities corresponding to the nearly horizontal top and bottom of the curve. Along the top and bottom of the curve, the material is in saturation. It is the abrupt foldover into saturation at the top right and bottom left corners (viewed as a counterclockwise movement from A to B to C to D to A) that triggers the fast switching action of the transistors.

Saturation Flux Density

It is essential, then, to know the value of flux density at which the core material saturates. The most satisfactory way of getting this figure is from the manufacturer. The home experimenter who wishes to use a core from an old transformer faces something of a problem here, because saturation density is not easily determined without special equipment.

Mind you, it is not impossible to arrive at a design using cut-and-try methods. It has been done. But when you consider that ordinary silicon iron is not an ideal material to start with and add to that fact the uncertainties of trying to establish by trial and error the saturation point of a particular specimen, you see that some difficulty could ·be encountered.

A possible approach for a core of this kind is to select a value of density that is almost certain to run the iron into saturation. For example, try a figure of around 18,000 gauss at 60 hertz. If the turns are worked out for this figure using Equation 10-3, then certainly a material with a saturation capability equal to or lower than this will saturate given the same applied voltage. But if it saturates at the lower end of the range, the large differential will probably create problems, possibly of a destructive nature, due to a high magnetizing current and losses.

Another problem is that a material that has an ill-defined point of transition into saturation—a "round loop" material in other words—is not likely to be very stable in frequency with load variation and, moreover, can develop rather dangerous (to the transistors) voltage spikes on the leading edges of the voltage waveforms, as in Fig. 10-15. Simple resistive-capacitive circuits, strategically placed, can eliminate this problem but perhaps not before the damage is done if one is experimenting with cores. These circuits are discussed later.

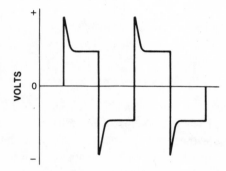

Fig. 10-15. Voltage spikes can "kill" transistors.

Of course, such gloomy thoughts do not deter home experimenters, who are traditionally an adventurous breed, but it does no harm to keep them in mind and approach the task with some caution.

Frequency

If an inverter rather than a converter is being considered, the choice of frequency is probably governed by the intended load. For example, if the load is to be standard domestic appliances, then the frequency must be around 60 hertz because the appliances are designed for this. On the other hand, if the inverter is to be used to drive equipment in aircraft where weight is a factor, the standard frequency in this situation might be 400 or 1000 hertz, and this governs the choice of inverter frequency.

The fact that it is an inverter might dictate the use of a core material with an emphatically square *B-H* characteristic in order to obtain frequency stability with load change—even at so low a frequency as 60 hertz if the point is felt to be critical. A nickel-iron alloy may be indicated for high frequencies in order to keep the losses low, in which case one that also possesses a square *B-H* characteristic will be chosen.

If it is to be a converter, the frequency is not so important nor is the frequency stability with load change. A high frequency, however, keeps the size of the transformer itself down. A typical frequency is on the order of 400 hertz even with nickel-iron toroids, which can operate if necessary into thousands of hertz. Silicon iron can usually be run up to 400 hertz although losses then become rather heavy, making 150 to 200 hertz, perhaps, a more practical figure, especially if the lamination thickness is around the standard one of 0.014 or 0.016 inch.

Then there are the ferrite molded cores that offer relatively low losses at frequencies to 10 or 20 kilohertz, are small in size, and are available at not too great a cost to boot.

How should you choose a core? There are many ways, and the design examples given later will help to show which one to use.

Spikes

One of the problems encountered with round loop materials and those that are not as square as they might be, such as ordinary silicon iron, is that the lack of squareness results in spikes on the output waveform. These transients heat the transistor junction and may prove to be terminal to the transistor operation in their effects if left uncorrected.

Strategically placed capacitors, as shown in Fig. 10-16, help to reduce these spikes, but understand that it might be too late to apply the cure if the problem is not anticipated.

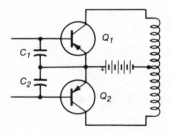

Fig. 10-16. Spikes are reduced by capacitors C_1 and C_2.

Transients of this kind can also be reduced by paying attention to the leakage reactances in the windings. For low leakage reactance, a tape-wound toroid is an excellent core with the windings interspersed bifilar fashion and balanced. For instance, the two halves of a center-tapped winding might be wound simultaneously (Chapter 4), or one half might be wound leaving spaces between the turns, and then the other half is wound in the spaces. The feedback windings should be coupled as closely as possible to the collector winding and balanced.

If laminations are used, great care should be exercised in assembling them to keep the effective gap as small as possible.

Winding Capacitance

If the secondary winding consists of many turns of wire (because of high voltage, perhaps), the capacitance in the windings could set up damped oscillations (called *ringing*) that appear in the output. Capacitance can be reduced somewhat by special winding techniques that are somewhat tedious to carry out. If possible, the numbers of turns should be reduced by using a larger core area. This, however, means reducing the number of turns on all the windings in order to keep the ratio the same as before. But the number of primary winding turns may be already quite low because a very low voltage dc source is being used. Further reduction could lead to switching problems—so be careful.

SPECIFIC DESIGN CONSIDERATIONS _____

The Current in the Primary

In Fig. 10-10, the primary winding is center tapped. The current in each half of the winding is the collector current of each transistor. The collector current must supply the input power to the primary winding; therefore these currents are given by input power divided by input volts, or

$$I_C = \frac{P_{in}}{V_{in}} \tag{10-4}$$

where

I_C is the collector current in amperes,
P_{in} is the input power in watts,
V_{in} is the battery voltage in volts.

This current is switched by the transistors to first one half of the primary and then the other. The effect is that of a half-cycle square wave of peak value I_C in each half of the winding, as given above.

In Chapter 1 (under "Different Kinds of Voltage") it was seen that the rms value of a half-cycle square wave is

$$rms = \frac{peak}{\sqrt{2}} = 0.707 \; peak$$

Thus, the effective current in the primary winding is

$$I_P = 0.707 \times I_C = 0.707 \; \frac{P_{in}}{V_{in}} \tag{10-5}$$

where I_P is given in amperes.

Base Drive

The required drive voltage for the bases of the transistors is obtained from the transistor data sheet. Keep in mind, however, that this figure is usually a minimum. Individual specimens might need more base drive to achieve the same results. Therefore, if the design is for mass production or a limited run, it is perhaps wise to increase the base drive figure to account for differences in individual components.

But an unnecessarily high drive means wasted power. Therefore, if the equipment is one circuit only, which is usually the case with experimenters, the thing to do would be to check the transistors and design accordingly.

The base drive requirements are related to the current gain (h_{fe}) of the transistor. Again, this figure is obtained from the data sheet, and again it must be kept in mind that this is usually a minimum figure (although a typical figure is sometimes given).

In practice, a transistor can have an h_{fe} several times greater than the stated figure. For base current calculation, the minimum h_{fe} number is commonly used. It is given by

$$I_B = \frac{I_C}{\text{minimum } h_{fe}} \qquad (10\text{-}6)$$

The power required to drive the base is then

$$P_B = V_B \times I_B \qquad (10\text{-}7)$$

where

P_B is the base power in watts,
V_B is the base drive voltage in volts, from data sheet,
I_B is the base current in amperes,
h_{fe} is the transistor current gain.

Collector-to-Emitter Breakdown Voltage

Recall from the description of the operation of the circuit that at least twice, the battery voltage appears across the quiescent transistor. An important transistor parameter is therefore the ability to withstand a voltage greater than $2 \times V_{CC}$.

Any spikes present on the voltage waveform contribute to breakdown, so it is essential to make due allowance in the choice of transistor.

Starting

The biasing arrangement shown in Fig. 10-12 is one of a number of possible methods of achieving reliable starting. The values chosen for the resistors depend on various factors, including the load current and the instantaneous surge imposed by the charging of the filter capacitor in the case of converters. This is one reason the filter capacitance should be kept low.

Because of the variables involved, it is not easy to evaluate the values required for the resistors. It is usually best to make adjustments on the bench under operational conditions.

The resistors form a potential divider across V_{CC}. Generally, the value of R_1 is high compared with R_2, and the total resistance has to be high enough to avoid an excessive current drain on the battery.

The object of the exercise is to place a voltage across the emitter-base junction that is just slightly greater than that needed for conduction. In germanium transistors, this voltage is 0.25 to 0.35 volt, while for silicon junctions it is about 0.5 to 0.6 volt. Taking account of battery drain and selecting a fairly high value for R_1, R_2 would be approximately

$$R_2 = \frac{R_1}{(V_{CC}/V_j) - 1}$$ (10-8)

where

V_j is the emitter-base junction voltage,
V_{CC} is the battery voltage.

Example—With a 12-volt battery, R_1 can be put at about 150 ohms. Then, if the transistor is a germanium type,

$$R_2 = \frac{150}{(12/0.3) - 1} = 3.8 \text{ ohms}$$

Remember that this is an approximation to be adjusted on the bench under operational conditions. A point to watch is the relative power dissipations in the resistors. The current through R_1 is given by $I = V_{CC}/(R_1 + R_2)$ or about 0.078 ampere. The power dissipation in R_1 is $I^2 R_1$ or $0.078^2 \times 150 = 0.9$ watt. A 1.0-watt resistor would be fine. However, the base current for the transistors flows through R_2. Depending on the power output of the circuit and the type of transistors, this could well be anything from a small fraction of an ampere to several amperes. Supposing it is 1 ampere, the dissipation in R_2 would be on the order of 4.0 watts!

Another consideration is the voltage drop across R_2 in the base circuit. The base drive winding must put out enough voltage to compensate for this.

DESIGN EXAMPLES

Single-Transformer Circuit

A basic inverter (Fig. 10-10B) must operate from an automobile battery of approximately 12.6 volts (a charged battery in good condition usually puts out this voltage within about 2 percent). The output required is 117 volts ac, 60 hertz, and 100 watts to supply power to small domestic appliances.

As always, a start is made by assuming an efficiency figure. This might be between 80 and 90 percent, say 85 percent. The input power would then be 100/0.85 = 117.6, say 118, watts.

The secondary current is the output power divided by the voltage, or 100/117 = 0.85 ampere. The collector current (Equation 10-4) is $I_C = 118/12.6 = 9.36$, say 9.5, amperes, and the effective primary winding current (Equation 10-5) is $I_P = 0.707 \times 9.5 = 6.7$ amperes.

If the chosen transistor requires a base drive voltage of 2.2 volts, increase this to, say, 3 volts or even 4 volts to compensate for the voltage drop in the base circuit and differences between specimen transistors. Assume the base drive voltage is 3 volts in this case. The minimum h_{fe} given in the data sheet is, say, 17.0, so from Equation 10-6, the base current must be I_B = 9.5/17 = 0.56 ampere and the power needed for the base from Equation 10-6 is P_B = 0.56 × 3 = 1.68 watts. The transistor selected must be able to withstand a collector-emitter voltage of twice the battery voltage, or 2 × 12.6 = 25.2 volts; for safety a 50-volt figure would be fine.

Wire gauges can now be selected for all windings. Assuming a conservative current density of 1000 CM/A,

Primary:	6.7 × 1000 = 6700 CM
Secondary No. 1:	0.85 × 1000 = 850 CM
Secondary No. 2:	0.56 × 1000 = 560 CM

From the wire tables (Table 6-1), the closest gauges are No. 12 (6529 CM) for the primary, No. 21 (812 CM) for secondary No. 1, and No. 22 (640 CM) for secondary No. 2.

The material for the core can be any one of the special "square" types, or, with suitable precautions, good-grade silicon iron, if you don't mind a little less efficiency (or even in an adventurous moment you could select an old power-transformer core of indeterminate characteristics, paying attention to the remarks made earlier on this subject).

At any rate, a figure for core saturation density is established. Suppose this is 18,000 gauss. The core cross-sectional area a can be worked out using the power capability equation or the simpler Equation 8-1. Thus, by Equation 8-1, $a = 0.16\sqrt{P} = 0.16\sqrt{118}$ = 1.74 square inches.

Then the number $N_1/2$ of turns in the primary winding (half of the total winding) is given by Equation 10-3, suitably rearranged as

$$N_1/2 = \frac{3.87 \times V \times 10^6}{faB} = \frac{60 \times 1.74 \times 18,000}{3.87 \times 12.6 \times 10^6}$$

$$= 25.95, \text{ say 26, turns}$$

making 52 turns for the total primary winding. Rearranging Equation 3-2 gives, for the secondary output winding,

$$N_3 = \frac{V_3}{V_1}N_1 = \frac{117}{12.6} 26 = 241 \text{ turns}$$

and for the base drive winding

$$N_2/2 = \frac{V_2}{V_1}N_1 = \frac{3}{12.6} 26 = 6.19, \text{ say 7, turns}$$

or this is 14 turns center-tapped for the total base winding.

From this point, the design follows the same lines as for other power transformers. Select a tentative lamination or core size and fit the windings to it. Make adjustments as necessary and check the losses. Make further adjustments to compensate for voltage drop and losses in the same manner as before.

Dual-Transformer Circuit

The power output transformer T_1 in this circuit (Fig. 10-13) is nonsaturable. Its design follows standard power transformer practice except that due consideration must be given to the center-tapped primary winding in exactly the same manner as was done for the single-transformer circuit.

Suppose the dc source is 12.6 volts and the power output 200 watts at 115 volts, 400 hertz, in the basic dual-transformer circuit of Fig. 10-13. Assume, initially, an efficiency of 85 percent for the transformer as before.

Then the input power is 200/0.85 = 235 watts. The current in the secondary winding is 200/115 = 1.74 amperes and the transistor collector current is I_C = 235/12.6 = 18.65, say 19, amperes. The effective primary winding current is I_P = 0.707 × 19 = 13.4, say 14, amperes.

A core material can be selected on the same basis as for any other transformer. For good efficiency at 400 hertz, a C core of grain-oriented silicon iron could be selected, but standard laminations of good-grade silicon iron could be used with somewhat less efficiency. Once more, the core size can be chosen on the basis of the Wa product. The numbers of turns of the primary and secondary windings are determined in the usual way from the basic equation and the voltage ratio.

The flux density will be the highest normal working value for the core material at 400 hertz. It is *not* the saturation value. The design is pursued to completion in the normal way.

NOTE: It might be advisable in some designs to include a small gap in the core to avoid dc saturation. This can be determined on the test bench. If laminations are used for the core, the inherent effective gap could be sufficient, or a somewhat larger gap can be created by simply not interleaving the laminations. If a larger gap is needed, then the latter course, with a thin paper or card inserted where the Is and Es butt, will do the job.

The Saturable Transformer

The method of calculating for the saturable transformer (T_2 in Fig. 10-13) is essentially the same as that used for the base winding in the single-transformer case except that the primary winding parameters have to be established.

The collector current I_C has been determined at 19 amperes. The h_{fe} (minimum) figure comes from the transistor data sheet. Suppose it is 15; then the base current is I_B = I_C/h_{fe} (minimum) = 19/15 = 1.27 amperes.

The base drive voltage is given in the data sheet as, say, 1.5 volts. Allow, say, 2.0 volts for voltage drop and differences between specimens, giving a required base voltage of, say, 4 volts. The output power to drive the bases is then P_o = I_B × V_B = 1.27

× 4 = 5.0 watts, where P_o is the output power, I_B the base current, and V_B the drive voltage for the base.

Assuming a good efficiency of, say 92 percent, the input watts to the primary winding N_1 is P_{in} = 5.0/0.92 = 5.43 watts. Because the base current like the collector currents, are squarish and alternate in the halves of the base winding, the effective current in the secondary winding is I_S = 0.707 × 1.27 = 0.898 (0.9) ampere.

Recall from the description of the circuit operation that twice the battery voltage, in this case 2 × 12.6 = 25.2 volts, is available across the input to the power transformer T_1. The base drive transformer primary is connected across the power transformer primary and so has this voltage available to it.

Allowing for the voltage drop across R_F and a little margin for adjustments, it would be reasonable to plan for 18.0 volts across the primary winding of T_2. Therefore the transformer ratio r needs to be

$$r = \frac{\text{Primary volts}}{\text{Secondary volts}} = \frac{18}{4} = 4.5$$

The primary winding input watts, P_{in}, was determined at 5.43. Therefore, the primary winding current is input watts divided by primary winding volts, or $I_P = P_{in}/N_P$ = 5.43/18 = 0.3 ampere.

Choosing a conservative current density of 1000 CM/A, the conductors work out to No. 25 (320 CM) for the primary and No. 20 (1024 CM) for the secondary.

In order to keep the transformer small and obtain high efficiency, use a toroidal core of a tape-wound high-performance "square" alloy, say a Magnetic Metals core in Square 50. This material is characterized by a high maximum flux density, excellent squareness of the hysteresis loop, and low loss. Manufacturers' data showing the loops and losses at various frequencies is shown in Fig. 10-17. The alloy enters saturation at a little under 15,500 gauss on the 400-hertz curve in Fig. 10-17A. In Fig. 10-17B, the loss is about 3 watts per pound at 400 hertz, 15,000 gauss (0.004-inch material).

Because the K factor is of considerable practical importance in terms of ease of winding in toroids, it is considered quite specifically here. For the wire gauges being considered, try a tentative 0.2 for the K factor on for size.

Now that numbers have been assigned to the K factor, wire sizes, frequency, saturation density, and input volts, all that is needed to determine the Wa product is the winding ratio r, which for the present purpose is determined to be the same as the voltage ratio V_P/V_S = 18/4 = 4.5. Hence,

$$Wa = \frac{V_P(A_1 + A_2/r)2.74}{KfB}$$

$$= \frac{18(320 + 1024/4.5)\,2.74}{0.2 \times 400 \times 15{,}000} = 0.0225 \text{ inch}^4$$

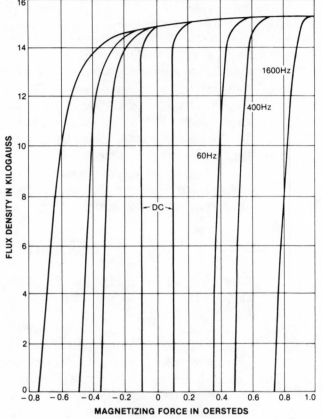

(A) Hysteresis loops in upper two quadrants at various frequencies.

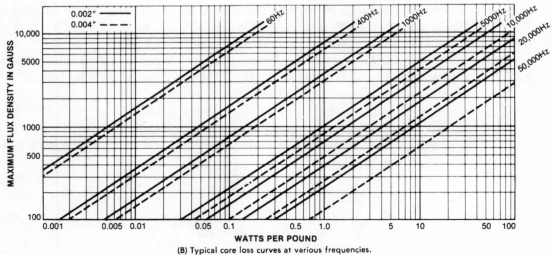

(B) Typical core loss curves at various frequencies.

Courtesy Magnetic Metals Corp.

Fig. 10-17. Manufacturer's data for Square 50 core material.

NOTE: This equation contains the conversion factors necessary to give the answer in inch^4 when A_1 and A_2 are in circular mils and B is in gauss.

A search of the vendor's core catalog that lists Wa as a specification reveals a number of possible candidates. One type, No. 382, comes very close with a Wa of 0.0230 (Table 10-1). The net core area is given as 0.363 sq cm, which is 0.0563 sq in.

The number of turns needed for the primary winding is then

$$N_1 = \frac{3.87 \times V \times 10^6}{faB} = \frac{3.87 \times 18 \times 10^6}{400 \times 0.0563 \times 15,000}$$

$$= 206 \text{ turns}$$

and $N_2 = N_1/r = 206/4.5 = 45.8$, say 46, turns, or, for the entire winding, $46 \times 2 = 92$ turns, center-tapped.

The actual K value works out as follows. The area of the primary winding copper on the center of the core is $N_1 \times A_1 = 206 \times 320 = 65,920$ CM, and the area of the secondary winding copper is $N_2 \times A_2 = 92 \times 1024 = 94,208$ CM. The total copper area is then $65,920 + 94,208 = 160,128$ CM.

From the catalog, the case window (in plastic) is 0.353 sq in (Table 10-1). Converting this to circular mils (see Conversions, Chapter 1) gives $1,273,240 \times 0.353 = 449,453$ CM and the K factor is

$$K = \frac{160,128}{449,453} = 0.356$$

This is large, compared with the estimate. It means that the hole left in the doughnut after winding will be much smaller than originally planned. If this factor is too large, adjustments must be made to the design. Possibly the current density can be reduced, leading to smaller-gauge wires, or a larger core can be used, or both.

Table 10-1. Data for Core 382 (Arrowed) for Saturable-Core Transformer Design

CORE SIZE	CORE DIMENSIONS IN			CASE DIMENSIONS IN						NET CORE AREA CM²			MEAN PATH LENGTH		CASE WINDOW AREA IN²		RATIO ID/OD	GR. CORE WEIGHT SQUARE 50		PRODUCT Wa X Ca IN⁴
	ID	OD	HT	ID Metal	ID Plastic	OD Metal	OD Plastic	HT Metal	HT Plastic	.001 SF=.75	.002 SF=.85	.004 .006 SF=.90	IN	CM	Metal	Plastic		Pounds	Grams	
433	.375	.438	.125	.310	.310	.513	.513	.197	.212	.01905	.0216	.0229	1.277	3.24	.0755	.076	.856	.00149	.680	.000297
343	.500	.563	.125	.435	.435	.637	.637	.197	.212	.01905	.0216	.0229	1.670	4.24	.1486	.149	.888	.00195	.889	.000585
356	.582	.625	.125	.498	.498	.700	.700	.197	.212	.01905	.0216	.0229	1.865	4.74	.1948	.194	.899	.00219	.992	.000767
482	.625	.688	.125	.560	.560	.762	.762	.197	.212	.01905	.0216	.0229	2.062	5.24	.246	.246	.908	.00242	1.098	.000970
431	.687	.750	.125	.622	.622	.825	.825	.197	.212	.01905	.0216	.0229	2.257	5.73	.304	.304	.916	.00265	1.202	.001197
38	.375	.500	.125	.312	.310	.563	.575	.197	.200	.0378	.0428	.0454	1.374	3.49	.0765	.076	.750	.00320	1.452	.000597
106	.438	.563	.125	.373	.375	.638	.630	.197	.205	.0378	.0428	.0454	1.572	3.99	.1093	.110	.778	.00366	1.661	.000854
47	.500	.625	.125	.473	.435	.688	.700	.197	.200	.0378	.0428	.0454	1.767	4.49	.1500	.149	.800	.00411	1.866	.001172
74	.625	.750	.125	.560	.570	.825	.820	.197	.195	.0378	.0428	.0454	2.160	5.49	.246	.255	.833	.00503	2.28	.001924
90	.750	.875	.125	.685	.665	.950	.945	.197	.195	.0378	.0428	.0454	2.553	6.48	.369	.347	.857	.00594	2.70	.00288
366	.875	1.000	.125	.810	.790	1.075	1.095	.197	.212	.0378	.0428	.0454	2.945	7.48	.515	.490	.875	.00686	3.11	.00403
153	1.000	1.125	.125	.935	.915	1.200	1.220	.197	.210	.0378	.0428	.0454	3.338	8.48	.687	.658	.889	.00777	3.53	.00536
2	.500	.750	.125	.435	.435	.825	.825	.197	.200	.0756	.0857	.0907	1.963	4.99	.1486	.149	.667	.00914	4.15	.00232
144	.625	.875	.125	.560	.560	.950	.950	.197	.200	.0756	.0857	.0907	2.356	5.98	.246	.246	.714	.01097	4.98	.00385
5	.650	.900	.125	.585	.575	.975	.985	.197	.200	.0756	.0857	.0907	2.435	6.18	.269	.260	.722	.01134	5.14	.00420
148	.750	1.000	.125	.685	.685	1.075	1.075	.197	.200	.0756	.0857	.0907	2.749	6.98	.369	.369	.750	.01280	5.81	.00576
483	.875	1.125	.125	.810	.790	1.200	1.220	.197	.212	.0756	.0857	.0907	3.142	7.98	.515	.490	.778	.01463	6.64	.00805
9	1.000	1.250	.125	.915	.920	1.340	1.342	.197	.200	.0756	.0857	.0907	3.534	8.98	.658	.665	.800	.01646	7.46	.01027
64	.500	.750	.250	.435	.420	.825	.840	.327	.330	.1512	.1714	.1815	1.963	4.99	.1486	.139	.667	.01829	8.29	.00464
67	.625	.875	.250	.570	.570	.945	.945	.327	.340	.1512	.1714	.1815	2.356	5.98	.255	.255	.714	.0219	9.95	.00797
79	.750	1.000	.250	.665	.670	1.085	1.093	.327	.330	.1512	.1714	.1815	2.749	6.98	.347	.353	.750	.0256	11.61	.01085
30	1.000	1.250	.250	.915	.920	1.340	1.343	.327	.330	.1512	.1714	.1815	3.534	8.98	.658	.665	.800	.0329	14.93	.0205
159	1.125	1.375	.250	1.060	1.040	1.450	1.470	.327	.342	.1512	.1714	.1815	3.927	9.97	.882	.849	.818	.0366	16.59	.0276
53	1.250	1.500	.250	1.170	1.170	1.570	1.570	.327	.320	.1512	.1714	.1815	4.320	10.97	1.075	1.075	.833	.0402	18.25	.0336
37	.625	1.000	.188	.570	.570	1.085	1.085	.265	.272	.1706	.1933	.205	2.553	6.48	.255	.255	.625	.0268	12.16	.00899
7	.750	1.125	.188	.665	.670	1.215	1.217	.265	.265	.1706	.1933	.205	2.945	7.48	.347	.353	.667	.0309	14.03	.01224
485	1.125	1.500	.188	1.060	1.040	1.575	1.595	.265	.280	.1706	.1933	.205	4.123	10.47	.882	.850	.750	.0433	19.65	.0311
489	1.625	2.000	.188	1.545	1.525	2.090	2.110	.265	.280	.1706	.1933	.205	5.694	14.46	1.875	1.827	.813	.0598	27.13	.0661
3	.625	1.000	.250	.570	.550	1.085	1.085	.327	.330	.227	.257	.272	2.553	6.48	.255	.238	.625	.0357	16.17	.01196
61	.750	1.125	.250	.665	.665	1.215	1.215	.327	.340	.227	.257	.272	2.945	7.48	.347	.347	.667	.0411	18.66	.01628
380	.875	1.250	.250	.810	.790	1.325	1.345	.327	.340	.227	.257	.272	3.338	8.48	.515	.490	.700	.0466	21.15	.0242
10	1.000	1.375	.250	.925	.920	1.455	1.468	.327	.330	.227	.257	.272	3.731	9.48	.672	.665	.727	.0521	23.64	.0315
16	1.625	2.000	.250	1.525	1.525	2.110	2.110	.327	.330	.227	.257	.272	5.694	14.46	1.827	1.827	.813	.0795	36.08	.0856
27	.750	1.000	.375	.665	.670	1.085	1.093	.452	.465	.227	.257	.272	2.749	6.98	.347	.353	.750	.0384	17.42	.01628
146	.625	1.125	.250	.560	.540	1.200	1.220	.327	.340	.302	.343	.363	2.749	6.98	.246	.229	.556	.0512	23.22	.01539
➤ 382	.750	1.250	.250	.685	.670	1.325	1.343	.327	.330	.302	.343	.363	3.142	7.98	.369	.353	.600	.0585	26.54	.0230
39	1.000	1.500	.250	.925	.920	1.570	1.592	.327	.330	.302	.343	.363	3.927	9.97	.672	.665	.667	.0731	33.18	.0420
13	1.250	1.750	.250	1.170	1.170	1.820	1.822	.327	.340	.302	.343	.363	4.712	11.97	1.075	1.075	.714	.0878	39.81	.0672
32	.625	1.000	.375	.570	.570	1.085	1.085	.452	.445	.340	.386	.408	2.553	6.48	.255	.255	.625	.0535	24.26	.01794
41	.750	1.125	.375	.665	.675	1.215	1.210	.452	.465	.340	.386	.408	2.945	7.48	.347	.358	.667	.0617	27.99	.0244

Courtesy Magnetic Metals Corp.

ELEVEN

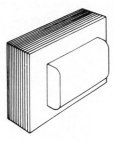

Inductors

From time to time in the previous chapters, the properties of magnetic induction, induced currents, and induced voltages were mentioned. These phenomena are all facets of the property called *inductance*. Inductance can be described as the property of a coil that causes it to oppose a change in the current through it. A coil is said to have an inductance of 1.0 henry if it induces an opposing or "back" emf of 1.0 volt when the current through it changes at the rate of 1.0 ampere per second.

In the present context, however, it is not necessary to be detailed with regard to what inductance is or is not. It is enough to recognize that it exists, that its unit is the henry, and that it occupies a position of considerable importance in electricity—particularly in transformer design.

The inductors dealt with in this chapter are the low-frequency type encountered in power supply filters, low-frequency resonant circuits, and those kinds camouflaged as the primary windings of audio and other types of low-frequency transformers. That is not to say the transformers already discussed do not possess inductance. They do—often large amounts of it. The difference is that in the components to be discussed now, an emphasis is placed on achieving specific values of inductance.

The reasons one value of inductance should be the target rather than another in various circuits is really not within the scope of this book (although in the next chapter this very point is dealt with in relation to certain low-frequency circuits). In this chapter, the concentration is on *how to obtain* the inductance after the required value has been determined. It is important to note, though, that the inductance of an iron-core component is not a constant. Its value in one circuit is not necessarily the same as it will be in another; so the circuit conditions are of much importance in the design.

THE INDUCTANCE EQUATION

Let us consider the following equation for inductance:

$$L = \frac{3.19 \times N^2 \times 10^{-8}}{(l/\mu a_1) + (g/a_2)} \tag{11-1}$$

where

- N is the number of turns of the inductor,
- μ is the permeability at the appropriate flux density,
- a_1 is the cross-sectional area of the iron in square inches,
- a_2 is the cross-sectional area of the gap in square inches,
- l is the mean magnetic path length in inches,
- g is the length of gap in inches.

This relation might seem like a lot to swallow, but taken in small bites, it is really quite simple. The discerning reader will recognize the term $l/\mu\, a_1$ as the reluctance of the iron (discussed in Chapter 2). The term g/a_2 is the reluctance of the air gap because the permeability of air is 1.0. At this point, it should be stressed that such equations are evolved by clever, cloistered chaps using carefully constructed mathematical models that rarely have direct application in the real world of copper and iron. *Their* theoretical model is a long, thin, perfect inductor, but the inductor of physical reality is short, fat, round, or square—all shapes, and riddled with imperfections.

So we must not expect to achieve high levels of accuracy using such equations. As a matter of fact, the design of inductors, like that of transformers, is a cut-and-try affair based to a great extent on empirical data and bench adjustments.

The equations do have their uses and also demonstrate the effects that gaps, permeability, reluctance, and so on have on inductance. If it is assumed that the iron and air gap areas are in fact the same (a reasonable practical assumption), the equation with the help of some algebraic manipulation becomes

$$L = \frac{3.19 \times a \times N^2 \times 10^{-8}}{(l/\mu) + g} \tag{11-2}$$

or

$$N = \sqrt{\frac{L[(l/\mu) + g]10^8}{3.19 \times a}} \tag{11-3}$$

and if there is no gap,

$$L = \frac{3.19 \times a \times N^2 \times \mu \times 10^{-8}}{l} \tag{11-4}$$

or

$$N = \sqrt{\frac{l \times L \times 10^8}{3.19 \times a \times \mu}}$$ (11-5)

Obviously, the greater the values of μ, a, and N, the higher is the inductance. Similarly, the greater the value of l and g, the smaller is the inductance. Note, however, that since μ is generally a high number, the reluctance of the iron is small, so small in fact that often the value of g is the overriding factor. If g is not very small, it will indeed reduce the inductance substantially.

Example 1—For example, certain nickel-iron alloys are treasured for their very high permeability at low magnetizing force. Suppose you want an inductance of 85 henrys. The magnetizing force is expected to be low, so select a nickel-iron alloy with a permeability of around 80,000 at the appropriate flux density (the figures are obtained from manufacturers' data). This should yield a high inductance with few turns of wire.

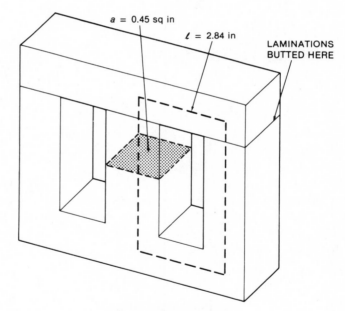

Fig. 11-1. Core for Example 1.

The core will consist of standard laminations, as shown in Fig. 11-1, with the ends butted hard together to eliminate the gap. Then, using Equation 11-5,

$$N = \sqrt{\frac{2.84 \times 85 \times 10^8}{3.19 \times 0.045 \times 80,000}} = 1449 \text{ turns}$$

But there is a fly in the unguent. It is not possible to eliminate the gap by simply butting the ends of the laminations together. There is always an effective gap, its size depending on how well the laminations are made and the care with which they are assembled.

No matter how well they are put together, it is difficult to reduce this gap to below about 0.002 inch (0.051 mm) at best and more than 0.003 inch (0.076 mm) is likely. If it were possible in the present case to achieve an incredibly small gap of 0.0003 inch (0.3 of one-thousandth of an inch), the result from Equation 11-2 would be

$$L = \frac{3.19 \times 0.045 \times 1449^2 \times 10^{-8}}{(2.84/80,000) + 0.0003} = 8.99 \text{ henrys}$$

This is a long way from the 85 henrys needed. To get the 85 henrys and taking the gap into consideration as in Equation 11-3 we have

$$N = \sqrt{\frac{85[(2.84/80,000) + 0.0003]10^8}{3.19 \times 0.045}}$$

$$= 4457 \text{ turns}$$

for the number of turns.

A similar result is obtained using the curves of Fig. 11-2, which are the same alloy. Here the changed permeability—the effective permeability, in other words—is shown for different gaps. The gap is expressed as a percentage of the mean magnetic path length of the core. Using the gap of 0.0003 inch and expressing it as a percentage gives

$$\% \text{ Gap} = \frac{g}{l} \times 100 = \frac{0.0003}{2.84} \times 100$$

$$= 0.01 \text{ percent}$$

for this example.

The permeability figure of 80,000 is located on the x axis. From this point, move vertically (dotted line) to meet the 0.01-percent curve, then across horizontally to read the effective permeability on the y axis. The latter figure is 9000. The turns required for 85 henrys are then calculated by Equation 11-5:

$$N = \sqrt{\frac{2.84 \times 85 \times 10^8}{3.19 \times 0.045 \times 9000}}$$

$$= 4322 \text{ turns}$$

somewhat less than our previous result.

It should now be obvious that the inductance obtained in practice is going to be widely different from the design figure unless the gap and permeability figure used can be guaranteed. This might not be easy for the experimenter. There are two ways to get more dependable results. One is to consider using tape-wound toroidal cores that have no gap and have guaranteed characteristics. The other is to put more turns on than seem necessary and then adjust the gap as discussed in Chapter 16.

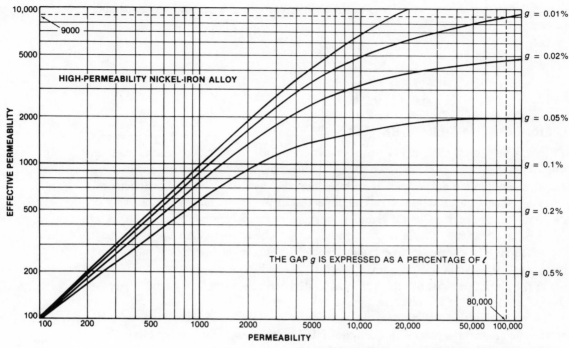

Fig. 11-2. Permeability versus effective permeability with gap.

For standard grades of silicon steel—the kind of material usually contained in salvaged power transformers—the permeability figures given in Table 11-1 can be used with a fair degree of accuracy in Equations 11-4 and 11-5. These figures assume a tightly interleaved core with absolutely minimal gapping and no dc in the winding.

Table 11-1. Very Approximate Permeabilities for Ordinary 4-% Silicon Steel With Minimal Gap and No DC

Gauss	Incremental Permeability
Very low	1000
100	1200
200	1300
400	1600
500	1700
1000	2000
2000	2700
4000	3000
7000	2000
10,000	1000

If the air gap is very large, the permeability approaches unity and the inductance is given by

$$L = \frac{3.19 \times a \times N^2}{g \times 10^8} \qquad (11\text{-}6)$$

with a in square inches and g in inches.

IRON-CORE INDUCTORS WITH DC

Many inductors, such as filter chokes and the primary windings of transformers in single-ended amplifiers, are required to carry both dc and ac in the windings, and in these cases, a gap is often deliberately introduced into the core to *improve* the inductance.

The inductance of an iron-core coil depends, as we know, on the permeability of the iron. And the permeability is given by the slope of the $B\text{-}H$ curve (Fig. 11-3A). The curve is not linear and obviously the permeability is highest where the slope is steepest and it falls rapidly wherever the curve inclines to the horizontal, which is at the top and bottom of the curve in this diagram. Therefore, the inductance is not constant. At the top of the curve, particularly, the iron is saturated and the permeability (and therefore the inductance) falls to a very low figure.

The effect of a dc current in the winding is to "bias" the iron. Suppose a dc magnetizing current is passed through the coil. This creates a magnetizing force, H_{dc}, as shown in Fig. 11-3B. If an ac magnetizing force, H_{ac}, is superimposed on the dc, the force will swing to $H_{dc} + H_{ac}$ and back to $H_{dc} - H_{ac}$. This in turn causes the flux density to swing from B_{dc} to $B_{dc} + B_{ac}$ back to $B_{dc} - B_{ac}$. Because the swing occurs on the straight (high-permeability) portion of the curve, the inductance remains high.

If the dc magnetizing force is increased to H_{dc}' and the same ac force is superimposed on it, the effect is as shown in the higher part of the curve in Fig. 11-3C. Because the iron is initially biased close to saturation, the flux swings are quite small (and also distorted), resulting in low permeability and therefore low inductance.

Now consider the effect of a gap in the core. The larger the gap, the more nearly the $B\text{-}H$ curve approaches that of an air-core coil. The dotted straight line in Fig. 11-4C is the $B\text{-}H$ curve of an air-core coil. An air-core coil cannot saturate, and whatever the value of the dc magnetization the flux density swing due to the ac remains more or less constant and the inductance remains constant although at a low value.

There is an optimum value for an air gap in a core for which the inductance is maximum. The effect of gaps on inductance is illustrated in Fig. 5-3, Chapter 5. Because the gap has a much higher reluctance than the iron, the effect of core permeability on the inductance becomes less as the gap is increased. In fact, there might be no advantage to be gained from the use of high-permeability alloys in the place of cheaper silicon steel. For the same reason, the length of the magnetic path l becomes relatively unimportant.

Also, it is necessary to be wary of the N^2 term in the inductance equation. When the High Priests of Theory solemnly intone that the inductance is proportional to the square of the turns, the congregation must respectfully chant in response, "Unless there

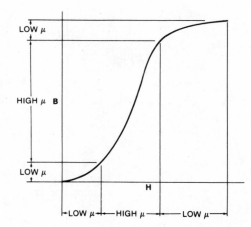

(A) Regions of high and low permeability.

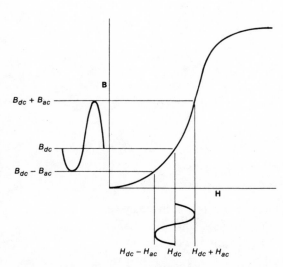

(B) Dc magnetizing force H_{dc} with superimposed ac magnetizing force H_{ac}.

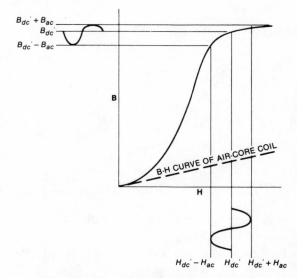

(C) High-level dc magnetizing force with superimposed ac.

Fig. 11-3. Effect of part of B-H curve used.

is dc in the windings." Why? Because with dc in the windings, the N^2 rule might not apply. With heavy dc currents and large gaps, the inductance is likely to be more proportional to N than N^2. Of course, the theorists know this, but the point is apt to be overlooked in practice.

HANNA CURVES

Designing inductors is made easier with the aid of the Hanna curve shown in Fig. 11-4. This graph is drawn for standard silicon iron at 60 hertz. It is used as follows. At the start of the design the required inductance L and the dc current I to be carried

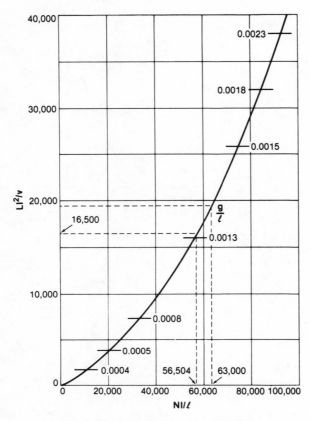

Fig. 11-4. Hanna curve for silicon iron at 60 Hz.

in the windings are known. A tentative choice of core dimensions is made; this may be based on past experience or simply on the fact that a core of these dimensions is on hand. The volume v of the core and the magnetic path length l are worked out as we have done in the past, and from these figures LI^2/v is worked out, where L is in henrys, I in milliamperes, and v in cubic inches. It is now possible to deduce the number of turns and the required gap size g directly from the curve.

Example 2—Suppose a choke is needed to have an inductance of 30 henrys and carry 50 milliamperes of dc current. A lamination is available as shown in Fig. 11-5 and a stack of 1.5 inches is tentatively chosen.

This lamination has a mean magnetic path length l of 3.797 inches. The core area is 1.125 square inches; multiply this by a stacking factor of 0.9 and the effective area a is 1.0125 square inches. The volume v is $a \times l = 3.844$ cubic inches.

Consider first the LI^2/v axis. The numbers are $L = 30$, $I = 50$, and $v = 3.844$, giving

$$\frac{L \times I^2}{v} = \frac{30 \times 50^2}{3.844} = 19,511$$

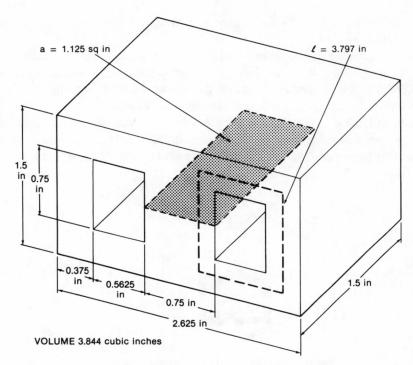

a = 1.125 sq in

l = 3.797 in

1.5 in

0.75 in

0.375 in

0.5625 in

0.75 in

2.625 in

1.5 in

VOLUME 3.844 cubic inches

Fig. 11-5. Magnetic path length, area, and volume for Example 2.

Following the dotted line from the 19,511 point in Fig. 11-4 the g/l line is intersected just above 0.0013. Therefore, $g = l \times 0.00135 = 3.797 \times 0.00135 = 0.005$ inch for the gap. From this point, the line is followed down to the base line for a reading of 63,000. Thus

$$\frac{N \times I}{I} = 63,000$$

Therefore

$$N = \frac{l \times 63,000}{I} = \frac{3.797 \times 63,000}{50} = 4784 \text{ turns}$$

for the coil.

The wire must now be selected to satisfy two conditions: the permissible voltage drop in the winding and the temperature rise. The latter factor is not usually a problem in inductors, but it should be kept in mind. Also, the designer must decide on the kind of wire to be used (single or heavy insulation) and what sort of winding technique (random or layer winding).

In the present example, a maximum permissible voltage drop of 25 volts will be assumed. This puts the maximum winding resistance at $V/I = 25/0.050 = 500$ ohms.

Because the voltage across the coil is quite low, random winding will be chosen using single-insulation wire.

There is an ohms figure to shoot for, so the best bet is to calculate the cubic content of the available winding space and use the random winding column of the wire table. If 1/16 (0.0625) inch is allowed all around for the bobbin and the cover, the winding area is 0.273 square inch (Fig. 11-6). The cubic space is the area times the length of the mean turn, and using the methods previously discussed, the mean turn works out to be 6.373 inches. The cubic space is then 6.373 × 0.273 = 1.74 cubic inches. The winding must be no more than 500 ohms, which works out at 500/1.74 = 287 ohms per cubic inch.

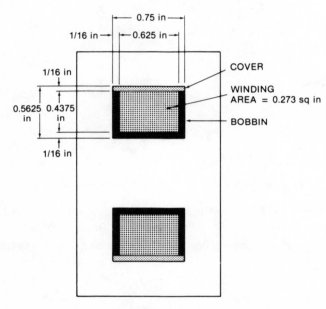

Fig. 11-6. Winding area.

Gauge No. 33 is given as 271.5 ohms per cubic inch. This is fine for resistance, giving a total of 1.74 × 271.5 = 472 ohms, but obviously it will be tight in the winding space, if indeed it will go in at all.

An additional check using the turns per square inch column can be made. Note that 14,500 turns per square inch is given for No. 33 gauge. The winding area previously calculated is 0.273 square inch. Therefore, the turns possible in this area is 0.273 × 14,500 = 3958 or 826 turns short (17 percent). The next thinner gauge is No. 34. For No. 34, the ohms per cubic inch is given as 425, making the total resistance if the space is filled 1.74 × 425 = 739 ohms. In terms of turns per square inch, No. 34 is given as 18,800 or 5132 turns if the space is filled. For the 4784 turns needed to give 30 henrys the ohms are approximately 660.

With this core, the precise specification cannot be met. Gauge No. 33 can be used at lower inductance. This inductance can be found by working backwards starting with NI/l = 3958 × 50/3.797 = 52,120. Tracing up to the curve then along to the perpendicular axis gives about 16,500. Thus

$$\frac{LI^2}{v} = 15,000$$

or

$$L = \frac{15,000 \times 3.844}{50^2} = 23 \text{ henrys}$$

Or, use the next thinner gauge of wire and live with the increased voltage drop, which will be approximately 34 volts. Or, the insulation dimensions can perhaps be pruned to permit more turns on to the bobbin, and perhaps you can squeeze the windings a bit to get them on to the laminations. Or the core size can be changed.

There are two ways of changing the core size. Alter the stack thickness or try a different size of lamination. If more of the same laminations are on hand, try increasing the volume v by adding laminations. One way of arriving at a value for v is to work backwards again. With No. 33 gauge wire, 3958 turns can be wound into the available space as has already been established, and this is satisfactory for resistance. With gauge No. 33,

$$\frac{LI^2}{v} = 15,000$$

Therefore, for 30 henrys

$$v = \frac{LI^2}{15,000} = \frac{30 \times 50^2}{15,000} = 5.0 \text{ cu in}$$

Volume v is stack thickness times tongue width times magnetic path length; therefore the stack thickness is

$$\text{Stack} = \frac{v}{\text{Tongue} \times \text{Path}} = \frac{5.00}{0.75 \times 3.797}$$

$$= 1.76 \text{ inches}$$

If the laminations are the usual 0.015 to 0.016 inch, then sixteen more will do the trick. The length of mean turn is now increased by two times the increase in stack thickness, 2 × 0.26 = 0.52, for a total of 6.373 + 0.52 = 6.893 inches. The cubic winding space is then 6.893 × 0273 = 1.88 cubic inches. As before, No. 33 gauge has a resistance of 271.5 ohms per cubic inch, giving a total resistance of 271.5 × 1.88 = 510 ohms. This almost exactly meets the original specification of 500 ohms. There are all sorts of permutations of core dimensions and turns possible to achieve a given objective.

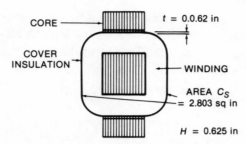

CORE → t = 0.0.62 in

COVER INSULATION → WINDING

AREA C_S = 2.803 sq in

H = 0.625 in

Fig. 11-7. Factors for calculating approximate temperature rise.

To wind up the present example, a check on temperature rise is advisable. The winding resistance is 510 ohms and therefore the dc watts dissipated in the winding are $I^2R = 0.050^2 \times 510 = 1.275$ watts. This very small power loss is unlikely to cause any problems; however, it can be readily checked in exactly the same manner as in previous transformer design, as follows.

There is only one winding. The only insulation involved is the cover. The enamel can be ignored. Referring to Fig. 11-7, all the watts must pass through the cover thickness, which is therefore t. The term H has the same meaning as in the case of the transformer (Chapter 6), as does C_S. For only one winding, the calculation is shortened to

$$T_T = 20 \frac{W}{C_S} (12.5t + H)$$

Make t equal to, say, 0.062 inch. Then using the values in Fig. 11-7 and $W = 1.275$ watts,

$$T_T = 20 \frac{1.275}{2.803} [(12.5 \times 0.062) + 0.625] = 12.7°C$$

is the total temperature rise.

As always, the insulation requirements must be checked. In this case, there is 0.062 inch on the cover, which is probably much more than is needed for most purposes, but check it against the criteria given in Chapter 6. Because it is a random-wound coil, the bobbin has cheeks—in this case $\frac{1}{16}$ inch thick; again this is ample for most uses.

To complete the design, locate the gap ratio on the curve where it was intersected. It is approximately $g/l = 0.0014$, then $g = l \times 0.0014 = 3.797 \times 0.0014 = 0.005$ inch. But if the final adjustment method given in Chapter 16 is used, the actual number obtained for the gap is unimportant.

It is important to note that the gap as calculated from the curve is the *total* gap in the magnetic circuit. In practice, the magnetic path is broken at two places (Chapter 5). The physical gap used must therefore be half of the calculated value at each break. In the example, the actual spacing would be $0.005/2 = 0.0025$ inch.

Comments

The curve used in this manner does not give precise results, but it is usually good enough for most practical purposes. At any rate, it puts the design into the ball park. An important point is that the curve was drawn for low incremental induction, which is usually but not necessarily the case with this type of component. (The word "incremental" in this context means the ac induction in the presence of dc.)

Strictly speaking, due account should be taken of the incremental induction level. For example, the two families of curves from Magnetic Metals for 0.004- and 0.012-inch grain-oriented silicon iron (Fig. 11-8) are drawn for various induction levels ΔB from 100 to 5000 gauss. There will be a considerable difference between the answers given by the curves at the extremes of the induction range.

Thus, by working with the curve closest to the expected figure (or by judging the position of a curve at an intermediate value), greater accuracy is obtained—at least on paper. In practice, the variables involved tend to make the exercise rather approximate in any case.

At low values of dc current, the required gap size is correspondingly small. In fact, it might be too small to achieve. Recall from Chapter 5 that even under the best conditions—that of two C cores with specially finished butting faces—the gap cannot be smaller than about 0.0012 inch. With standard laminations, much larger minimum gaps can be expected, especially if the greatest care is not exercised in assembly of the transformer.

Quite often, in order to achieve the optimum gap, it is necessary to interleave the laminations with the care usually reserved when attempting to eliminate the gap. Other times, the laminations will not be interleaved but simply inserted with the Es in the same direction and the Is forced close to them.

For relatively large gaps, the latter method of assembly is used with appropriate thicknesses of nonmagnetic material inserted between the lamination edges. This technique is discussed in Chapter 15, and a method of setting gaps under operational conditions is given in Chapter 16.

THE TRANSFORMER AS AN INDUCTOR

It is frequently necessary to design some kinds of transformers with specific values of inductance in the primary winding. When doing this, remember that only part of the winding space is available for the primary, so the inductance must be designed into this limited space.

Note also that some kinds of transformers, notably push-pull types in low-frequency amplifiers, can carry dc current in the primary winding without reducing the inductance. The method just described for finding inductance in the presence of dc therefore does not apply in these cases.

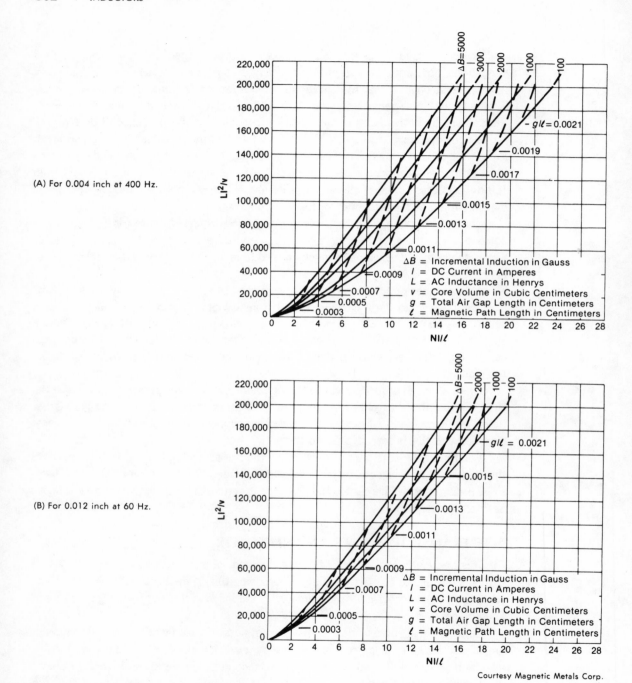

(A) For 0.004 inch at 400 Hz.

(B) For 0.012 inch at 60 Hz.

Courtesy Magnetic Metals Corp.

Fig. 11-8. Hanna design curves for grain-oriented silicon.

TWELVE

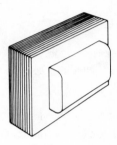

The Impedance Transformer

It is often necessary in circuit design to make the resistance, or impedance, of a load appear as some other value to the source that supplies it. A common example of this is the use of an output transformer to link a speaker, which by its nature is usually a low-impedance device, to an amplifier that may have a relatively high impedance (although low-impedance amplifiers that can be connected directly to speakers are common enough today). There are many other applications for the impedance transformer, but the audio types are among the most refined of the genre and for this reason receive much of the attention in this discussion. The principles remain valid for any type of impedance transformer.

IMPEDANCE MATCHING

These transformers have various aliases, depending on what they are used for. An *output transformer* has just been described, but an impedance transformer can also be referred to as a *matching transformer* if its purpose is to match one value of resistance to another to satisfy the fundamental law that to transfer the maximum power from a source to its load, the load resistance must equal that of the power source. For example, if the load resistance is 10 ohms and that of the power source is 100 ohms, there cannot be a maximum power transfer (Fig. 12-1).

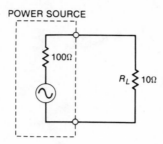

POWER SOURCE

Fig. 12-1. Since the load resistance is not equal to the resistance of
the source, there cannot be maximum power transfer.

Example 1—If a transformer with a ratio of 3.16:1 is inserted between the power source and the load (Fig. 12-2), the source, looking into the primary winding, sees a resistance R_L' equal to

$$R_L' = R_L \, r^2 = 10 \times 3.16^2 = 100 \text{ ohms} \qquad (12\text{-}1)$$

On the other hand, R_L, looking into the secondary, sees a source resistance of

$$R_L = \frac{R_L'}{r^2} = \frac{100}{r^2} = 10 \text{ ohms} \qquad (12\text{-}2)$$

In other words, both the source looking into the primary and the load looking into the secondary see the ideal condition for power transfer. Note that these equations are simply rearrangements of Equation 3-11 in which N_1/N_2 has been written as r.

Then there are *coupling transformers*, which link amplifier stages together, and *input transformers*, which provide ideal impedance conditions between, say, a microphone or other device and an input stage. Then there are *microphone transformers* . . . !

These various transformers do not necessarily equalize the resistance of load and source, nor are they necessarily concerned with transferring power. Sometimes they are more concerned with voltage than with power. But they all have the common property of changing one value of resistance to another, as seen looking into the primary or back into the secondary.

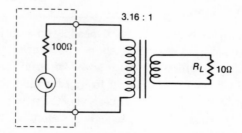

3.16 : 1

Fig. 12-2. Transformer "matches" the load resistance to that of the
power source for maximum power transfer.

Here is Equation 3-11 again:

$$R_L' = R_L \left(\frac{N_1}{N_2}\right)^2$$

where
R_L is the load resistance,
R_L' is the resistance the source sees looking into the primary,
N_1/N_2 is the primary-to-secondary turns ratio.

Because $N_1/N_2 = V_1/V_2$, as we saw back in Chapter 3, the impedance equation can also be stated as

$$R_L' = R_L \left(\frac{V_1}{V_1}\right)^2$$

where V_1 and V_2 are the voltages across the primary and secondary, respectively.

Example 2—If R_L is 100 ohms, what voltage ratio is needed to match the load to 1000 ohms? Note that R_L' is thus 1000 ohms. Rearranging the preceding equation slightly gives

$$\frac{V_1}{V_2} = \sqrt{\frac{1000}{100}} = 3.16$$

Example 3—Suppose the primary and secondary voltages of a transformer are V_1 = 50 volts and V_2 = 5 volts and R_L = 50 ohms. What apparent resistance will be in the primary winding? Now,

$$R_L' = R_L \left(\frac{V_1}{V_2}\right)^2 = 50\left(\frac{50}{5}\right)^2 = 5000 \text{ ohms}$$

is the apparent resistance of the primary winding.

TAPPED WINDINGS

The Tapped Primary

Example 4—Consider Fig. 12-3. This is a common configuration in which the primary winding is tapped in the center. Here N_1 is the total number of turns in the primary winding, say 5000. If N_2 is 500 turns and R_L is 100 ohms, then the reflected resistance R_L' across the entire primary winding (dotted line) is

$$R_L' = 100\left(\frac{5000}{500}\right)^2 = 10,000 \text{ ohms}$$

But what is R_L' across *half* the primary winding?

$$R_L' = 100\left(\frac{2500}{500}\right)^2 = 2500 \text{ ohms}$$

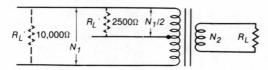

Fig. 12-3. The reflected load R_L' across half the primary is only one-fourth of that across the entire winding.

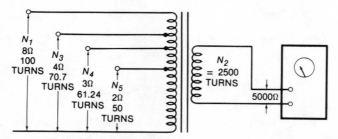

Fig. 12-4. Turns and "reflected" resistances in multitapped primary winding.

Across half of the primary winding, R_L' is just one quarter of what it is across the entire primary.

Of course, the primary winding can be tapped anywhere and frequently this is done. Figure 12-4 shows a multitapped primary winding used on a transformer designed as an adapter to convert an ac voltmeter to an output power meter. The meter has a resistance of 5000 ohms on its 5.0-volt range. It is required to match this to a choice of lower resistance inputs, such as across the output transformer secondary windings on various amplifiers. The values 2, 3, 4, and 8 ohms have been chosen. The secondary-to-primary turns ratios r are then

$$\text{For 2 ohms:} \quad \frac{N_2}{N_5} = \sqrt{\frac{5000}{2}} = 50.00 = r_5$$

$$\text{For 3 ohms:} \quad \frac{N_2}{N_4} = \sqrt{\frac{5000}{3}} = 40.82 = r_4$$

$$\text{For 4 ohms:} \quad \frac{N_2}{N_3} = \sqrt{\frac{5000}{4}} = 35.36 = r_3$$

$$\text{For 8 ohms:} \quad \frac{N_2}{N_1} = \sqrt{\frac{5000}{8}} = 25.00 = r_1$$

If the total primary winding, representing the 8 ohms position, is 100 turns, then the required number of secondary turns is $N_2 = 100 \times 25 = 2500$ turns. And the primary winding must be tapped as follows:

$$\text{For 2 ohms: } N_5 = \frac{N_2}{r_5} = \frac{2500}{50} = 50 \text{ turns}$$

$$\text{For 3 ohms: } N_4 = \frac{N_2}{r_4} = \frac{2500}{40.82} = 61 \text{ turns}$$

$$\text{For 4 ohms: } N_3 = \frac{N_2}{r_3} = \frac{2500}{35.6} = 71 \text{ turns}$$

The Tapped Secondary

Example 5—Suppose the secondary winding has multiple taps as in Fig. 12-5. The purpose of the taps can be to permit a choice of different loads on a power source having one fixed impedance. To illustrate that the relationships that exist between the primary and the secondary windings also exist between secondary and secondary windings, a different approach is used.

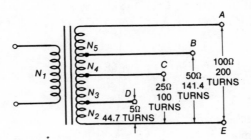

Fig. 12-5. If the turns and load resistance are known for one section of a multitap secondary, the turns needed for the other resistances can be found.

If R_{LT} is the load resistance of the total winding A-E and R_o the load resistance for any intermediate tapping point, and N_T is the total number of turns of the winding, and N_o the number of turns of the tap point for R_o, then

$$\frac{R_{LT}}{R_o} = \left(\frac{N_T}{N_o}\right)^2$$

or

$$N_o = \frac{N_T}{\sqrt{R_{LT}/R_o}}$$

Suppose in this case the total number of turns on the secondary N_T is known to be 200 for the 100-ohm load. What are the tapping points for the other loads? For the 50-ohm load:

$$N_{BE} = \frac{200}{\sqrt{100/50}} = \frac{200}{1.414} = 141.4 \text{ turns}$$

For the 25-ohm load:

$$N_{CE} = \frac{200}{\sqrt{100/25}} = \frac{200}{2} = 100 \text{ turns}$$

For the 5-ohm load:

$$N_{DE} = \frac{200}{\sqrt{100/25}} = \frac{200}{44.7} = 44.7 \text{ turns}$$

Exactly the same approach could have been used on a tapped primary winding.

In transformers of this type, it is often necessary to know what other sections, or combinations of winding sections, will provide. For example, what value of R_{Lo} can be used across the section of N_5 (A-B) of Fig. 12-5? The number of turns in N_5 is 200 − 141.4 = 58.6. Therefore

$$R_{Lo} = \left(\frac{N_5}{N_T}\right)^2 \cdot 100 = \left(\frac{58.6}{200}\right)^2 100 = 8.58 \text{ ohms}$$

is the resistance from A to B.

Thus, if the number of turns is known, it is possible to calculate R_{Lo} for any combination of tappings. It can also be done very easily from the impedances (or resistances), because the square root of the total impedance is equal to the sum of the square roots of the impedances for each section in series. Thus, in the last example,

$$\sqrt{100} = \sqrt{50} + \sqrt{R_{AB}}$$

$$\sqrt{R_{AB}} = \sqrt{100} - \sqrt{50} = 10 - 7.07 = 2.93$$

$$R_{AB} = 2.93^2 = 8.58 \text{ ohms}$$

TWO OR MORE LOADS

When two or more loads are driven simultaneously from separate secondaries as in Fig. 12-6, the problem is a little more complex, especially if the loads have different impedances and power requirements. Suppose that two power outputs, P_2 and P_3, are

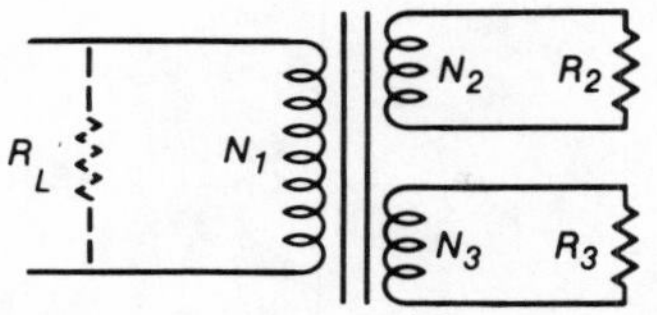

Fig. 12-6. Separate secondary windings driving different load
impedances at different power levels.

to be delivered to two loads, R_2 and R_3, respectively, from a source impedance $R_L{}'$. The required ratios are given by

$$\left(\frac{N_2}{N_1}\right)^2 = \frac{R_2}{R_L{}'} \left(\frac{P_2}{P_2 + P_3}\right)$$

and

$$\left(\frac{N_3}{N_1}\right)^2 = \frac{R_3}{R_L{}'} \left(\frac{P_3}{P_2 + P_3}\right)$$

We know from previous discussions that the input power (ignoring losses for the moment) is equal to the sum of the output powers. Thus, $P_1 = P_2 + P_3$. The equations therefore become

$$\left(\frac{N_2}{N_1}\right)^2 = \frac{R_2}{R_L} \times \frac{P_2}{P_1}$$

or

$$N_2 = N_1 \sqrt{\frac{R_2 P_2}{R_L{}' P_1}}$$

and

$$\left(\frac{N_3}{N_1}\right)^2 = \frac{R_3}{R_L{}'} \times \frac{P_3}{P_1}$$

or

$$N_3 = N_1 \sqrt{\frac{R_3 P_3}{R_L{}' P_1}}$$

Example 6—If $R_L{}' = 8000$ ohms, $R_2 = 400$ ohms, and $R_3 = 600$ ohms, and if $P_2 = 5$ watts, $P_3 = 4$ watts, then $P_1 = 5 + 4 = 9$ watts, and if $N_1 = 2500$ turns, then

$$N_2 = 2500 \sqrt{\frac{400}{8000} \times \frac{5}{9}} = 417 \text{ turns}$$

and

$$N_3 = 2500 \sqrt{\frac{600}{8000} \times \frac{4}{9}} = 456 \text{ turns}$$

In general form, the equation is

$$\frac{N_o}{N_1} = \sqrt{\frac{R_o P_o}{R_L' P_1}}$$

where

N_o is the number of turns on considered secondary winding,
N_1 is the number of turns on primary winding,
R_o is the load on secondary winding N_o,
R_L' is the reflected load to primary winding,
P_o is the watts in load R_o,
P_1 is the total watts input to primary winding.

This equation can be used for transformers having any number of secondaries. Note, however, that the correct value of R_L' is obtained only if *all* the secondaries are correctly loaded. If a secondary load (or several) is removed for some reason—for example, in switching speakers in and out of circuit—a resistance equal to the load resistance must be switched across the winding, as in Fig. 12-7. Resistors R_{S2} and R_{S3} are resistances that substitute for the loads. (In practice, it might be important to ensure that there is no open circuit while switching.)

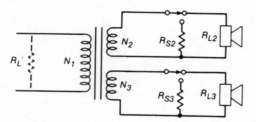

Fig. 12-7. If load is switched out, substitute load is switched in.

OTHER CONSIDERATIONS

As discussed earlier, the function of an impedance transformer is never confined simply to impedance transformation. The transformer is required to transform voltage, and usually current, and frequently deliver power as well.

Because it is governed by the same laws that govern other types of transformers, the selection of the turns ratio to provide a given impedance transformation automatically fixes the voltage and current ratios. (Although consideration of voltage and current requirements might cause the designer to reconsider his or her choice of turns ratio and therefore of the voltage and current transformation ratios.)

Similarly, the choice of current density in the windings, core size, core materials, and so on, reflect the need to handle given amounts of power, minimize copper and iron loss, control the size and weight, and meet costs limits in exactly the same way as for other kinds of transformers. The same methods and formulas are used here as in power transformer design.

In some applications, notably in amplifiers operating in the audio range of frequencies, however, other parameters have an important place in the design deliberations. For example, an output transformer might be called on to provide a substantially straight-line response extending from a few hertz up to 30 kilohertz or more. Generally in this case, a much higher primary inductance is needed than is required for, say, a run-of-the-mill power transformer in order to sustain the low-frequency response. This will then govern the choice of core, the kind of core material, and the number of turns required on the primary winding.

If dc current is flowing unidirectionally in the primary winding, as in a single-ended amplifier, the circuit designer's enthusiasm for lots of henrys might be curbed by the restraining influence of the transformer designer (assuming that two different persons are involved) who wants to avoid ending up with a transformer so big that a special hole has to be cut in the basement wall to get it into the building.

Similarly, a need to avoid deterioration of the response at the high-frequency end of the spectrum dictates the winding configuration, among other things, to be used to minimize leakage inductance.

In this type of application, the flux density in the iron must be based on the *lowest* frequency in the range to be covered, as defined by our old friend, Equation 3-14:

$$B = \frac{V \times 10^8}{4FfaN}$$

This says that at the lowest frequency the flux density will be greatest. Therefore to avoid running the core into saturation, it should be designed for the low frequencies; then it will be fine at the high frequencies. If there is dc current in the windings as well as ac, provision must be made for this in the choice of core area. That is to say, the iron must be run well below ac saturation at low frequency or distortion of the output waveform will result.

Impedance transformers used on the inputs of circuits and for coupling stages are generally run at very low inductance because the voltages are low in the preliminary stages. But the same consideration is given to inductance and leakage inductance as for output transformers and for the same reasons. It is usually possible to consider the use of high-permeability, low-loss core material because of the low inductions, and because power requirements are generally small, very small components result.

Obviously, then, the largest problem in designing transformers of these types is not the transformer design itself. The problem lies in determining from the circuit and its performance requirements what sort of characteristics the transformer should have. In other words, the circuit designer must join forces with the transformer designer and decide what kind of component is needed.

To repeat a precaution given in Chapter 11 in the section "The Transformer as an Inductor," although inductance is designed into the primary winding in exactly the same way as for other inductors, only part of the core window space is available for the primary inductor and space must be reserved for the secondary windings.

THE EQUIVALENT CIRCUIT

In principle, the audio output transformer can be considered to be representative of most types of impedance transformers. However, the requirements imposed on it the amplifier are among the most stringent. For this reason, the following simplified analysis is offered as a guide to the design of impedance transformers in general. The equivalent circuit is developed in a manner similar to that in Chapter 4.

Figures 12-8A and 12-8B depict a transformer in the collector circuit of a transistor and its equivalent circuit. It is possible, however, to simplify this circuit without great loss of accuracy in the final result as shown in Fig. 12-9. First, the core loss can reasonably be neglected, and in the first analysis, the capacitance can be omitted. This gives Fig. 12-9A.

The next step is to reduce the transformer to unity turns ratio and modify the load, R_L, and loss elements L_S' and R_S accordingly. The reasoning for this is as follows. In Fig. 12-9A, the impedance R_L' as seen by the transistor looking into the primary winding is $R_L' = R_L (N_1/N_2)^2$. This applies also to the loss elements on the secondary side. Thus, the transistor will also see the winding resistance as $R_S (N_1/N_2)^2$ and the leakage inductances as $L_S' (N_1/N_2)^2$. It will be obvious, therefore, that if the load resistance is increased to a value $R_L (N_1/N_2)^2$, and if the loss elements are increased by the same factor, the transformer can be replaced by one having unity turns ratio, as in Fig. 12-9B. The transistor looking into the primary will not notice the change, for the load resistance reflected into the primary is still $R_L \times (N_1/N_2)^2$.

Because the ratio is now 1:1 and contributes nothing to the impedance transformation, we can also dispense with the secondary winding and represent the transformer by a single inductance as shown in Fig. 12-9C. For all practical purposes, this is the equivalent of the circuit of Fig. 12-8A.

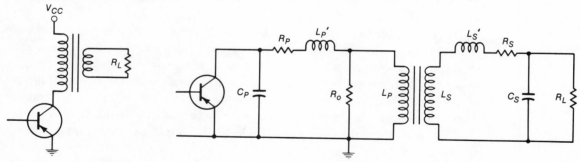

(A) Transformer in single-ended amplifier.   (B) Equivalent circuit of (A).

Fig. 12-8. Use of audio output transformer.

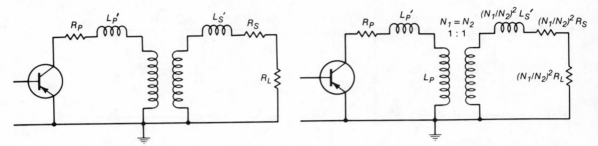

(A) Circuit neglecting core loss and capacitances.   (B) With unity turns ratio and modifying load and loss elements.

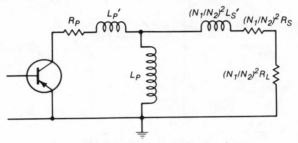

(C) Circuit disregarding secondary winding.

Fig. 12-9. Development of equivalent audio output transformer circuit.

FREQUENCY RESPONSE

It is now possible to check the performance of the transformer at spot frequencies, as it were, identified simply as low, medium, and high frequency. According to the same reasoning as in Chapter 4, the circuits representing each frequency condition emerge as shown in Fig. 12-10. From consideration of these circuits, a general picture of the transformer, and of the most desirable properties it should have, can be deduced as follows:

1. It should have high primary winding inductance to maintain the response at low frequencies.
2. It should have low leakage inductance to maintain the response at high frequencies.
3. The winding resistance should be low.

It can also be shown that at the low frequency at which the transistor in parallel with the load resistance is equal to the primary winding reactance, the response is down 3 dB from that at midfrequency. Or, restated, the output drops to 70.7 percent (3 dB) of the midfrequency output at a frequency f_1 when

$$R_{\text{parallel}} = 2\pi f_1 L$$

where
  $2\pi f_1 L$ is of course the reactance of the primary inductance,
  R_{parallel} is the parallel combination of the transistor resistance r_T and the reflected load resistance R_L'.

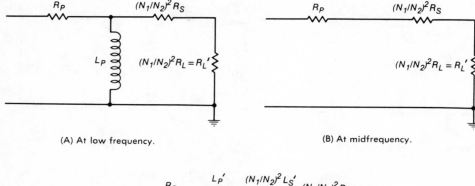

(A) At low frequency.

(B) At midfrequency.

(C) At high frequency.

Fig. 12-10. Transistor output circuit at three different frequencies.

Thus

$$R_{\text{parallel}} = \frac{r_T R_L}{r_T + R_L'}$$

An equation for the inductance of the primary winding can be written:

$$L = \frac{R_{\text{parallel}}}{2\pi f_1}$$

Thus, given the high frequency at which the output should be not more than 3 dB down, the required primary winding inductance can be found. This result stems from a further simplification of the equivalent circuit shown in Fig. 12-11A.

Similarly, it can be demonstrated that at the high frequency f_2 such that the leakage reactance L' is equal to the sum $r_T + R_L' = R_{\text{series}}$, the output is 3 dB down from its value at midfrequency, as in Fig. 12-11B. Or as an equation

$$R_{\text{series}} = 2\pi f_2 L'$$

This can be restated as

$$L' = \frac{R_{\text{series}}}{2\pi f_2}$$

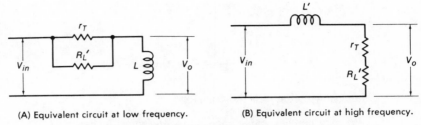

(A) Equivalent circuit at low frequency. (B) Equivalent circuit at high frequency.

Fig. 12-11. Further simplification of transistor output circuit.

Thus, given the high frequency at which the output should be not more than 3 dB down, the maximum leakage inductance that can be tolerated can be calculated.

At a low frequency $3f_1$, the output can be considered very nearly equal to that at midfrequency, and at a high frequency $f_2/3$, the output is nearly equal to that at midfrequency. These results form the basis for output transformer design at audio frequency, and the methods used to arrive at them are valid for any kind of low-frequency transformer design.

Example 7—Suppose that a transformer is required to work in the circuit of Fig. 12-8A. The following circuit parameters have been established. The power in the 8-ohm speaker is required to be 1.0 watt. Assuming a transformer efficiency of 80 percent, the input power will be 1.25 watts. The voltage swing across the primary winding is calculated to be 11.2 volts rms. The collector quiescent dc current I_C has been established at 0.2 ampere. Because $P = V^2/R$ and $R = V^2/P$,

$$R = \frac{11.2^2}{1.25} = 100 \text{ ohms}$$

is the resistance, R_L'. So the transformer with its 8-ohm load, R_L, must present to the transistor. Therefore the required turns ratio is

$$\frac{N_1}{N_2} = \sqrt{\frac{R_L'}{R_L}} = \sqrt{\frac{100}{8}} = 3.5$$

You want to achieve the modest response of no more than 3 dB down from 33 Hz to 8 kHz. If $r_T = 2000$ ohms, then r_T in parallel with R_L' gives R_{parallel} thus:

$$R_{\text{parallel}} = \frac{2000 \times 100}{2000 + 100} = 95 \text{ ohms}$$

This must be the reactance of the primary winding at the frequency $f_1 = 33$ Hz, or

$$95 = 2 \times 3.14 \times 33 \times L$$

Therefore

$$L = \frac{95}{6.28 \times 33} = 0.46 \text{ henry}$$

Similarly, at the high-frequency end, at a frequency f_2 of 8 kHz the leakage inductance must be

$$L' = \frac{2000 + 100}{6.28 \times 8000} = 0.042 \text{ henry}$$

We now have the following data on the transformer.

Primary winding inductance: 0.46, say 0.5, henry
Maximum leakage inductance: 0.042 henry
Dc current, I_C, in primary: 0.2 ampere
Primary winding ac voltage: 11.2 volts
Turns ratio N_1/N_2:   3.5

The output from the secondary winding is 1.0 watt into 8 ohms. Because $P = I^2R$, therefore $I = \sqrt{PR}$ and

$$I_S = \sqrt{\frac{P}{R}} = \sqrt{\frac{1.0}{8}} = 0.353 \text{ ampere}$$

is the load current in the secondary winding.

Good-grade silicon-iron laminations are usually adequate for this kind of design. Also, the core should be run at a low incremental flux density because of the dc in the primary winding. In addition, there is the need to achieve a specific inductance that probably will require many more turns than would a conventional power transformer with the same primary winding voltage. In fact, for the relatively small power involved, the need for a fairly large transformer is likely.

Consider this trial design with a transformer of the dimensions shown in Fig. 12-12. Turning to the Hanna curve in Fig. 11-4, Chapter 11, the known transformer data can be used to find the ordinate:

$$\frac{LI^2}{v} = \frac{0.5 \times 200^2}{3.844} = 5203$$

Locating this point on the vertical axis, the curve is intersected at approximately $g/l = 0.0006$, or $g = 3.797 \times 0.0006 = 0.0023$ inch. Dropping a perpendicular line from this point to the horizontal axis shows

$$\frac{N_1I}{l} = 25,000$$

or

$$N_1 = \frac{3.797 \times 25,000}{200} = 475 \text{ turns}$$

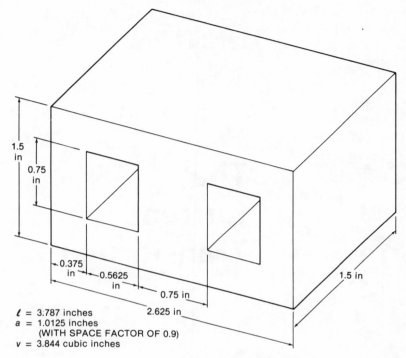

1.5 in

0.75 in

0.375 in

0.5625 in

0.75 in

1.5 in

2.625 in

l = 3.787 inches
a = 1.0125 inches
(WITH SPACE FACTOR OF 0.9)
v = 3.844 cubic inches

Fig. 12-12. Core for Example 7.

for the number of turns in the primary winding. Because the turns ratio is 3.5, the number of turns in the secondary winding is

$$N_2 = \frac{N_1}{3.5} = \frac{475}{3.5} = 136 \text{ turns}$$

Assume $\frac{1}{16}$ inch all around the window for bobbin and cheeks, if random wound. Then the available space works out at 0.273 square inch. Assigning half of this to the primary winding, there is 0.136 square inch available.

The wire gauges can now be tentatively selected on the basis of, say, 1000 CM/A. The primary winding conductor then requires an area of 0.2 × 1000 = 200 CM, and the secondary needs 0.353 × 1000 = 350 CM. For the primary winding, the closest gauge is No. 27 (201 CM) and for the secondary, No. 24 (404 CM) or No. 25 (320 CM) are acceptable.

THIRTEEN

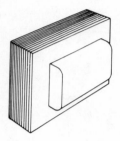

The
Current
Transformer

As stated in Chapter 3, the current transformer finds uses in adapting low-current protective devices and instrumentation to work in high-current circuits. One aspect of the latter application is likely to be of particular interest to the experimenter, namely, the adaptation of low-reading dc moving-coil (D'Arsonval) meters to read relatively high *alternating* currents.

First, let's look at how these meters are used to read high currents in dc circuits. As shown in Fig. 13-1, the meter is shunted by a resistance R_S so that a known proportion of the total circuit current flows through the meter. For example, if the ratio of R_S to the meter resistance R_M is such that 0.1 of the total current I_T flows through the meter and 0.9 through the shunt, a 100-milliampere meter can read currents up to 1.0 ampere. The meter could be calibrated in terms of 0 to 1.0 ampere instead of 0

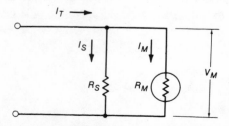

Fig. 13-1. Direct-current ammeter with shunt.

to 100 milliamperes. This current ratio will remain fixed for all values of I_T. For instance, if I_T is 10 amperes, the meter current will be 0.1 × 10 = 1.0 ampere, and the meter will be damaged because it is designed for only 100 mA full scale.

AC CURRENT MEASUREMENT

In order to use a dc meter in an ac circuit, however, it is necessary to rectify the current as shown in Fig. 13-2. Recall the full-wave bridge circuit described in Chapter 9. Here the meter is the load, and the rectifier ratings must be chosen to suit the meter full-scale current. If the shunt principles were to be applied here in order to extend the range of the meter on ac, the resistance R_S would have to be placed across the rectifier as indicated in dotted line—and there is the rub.

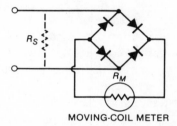

Fig. 13-2. A shunt across the rectifiers will result in nonlinear
meter readings.

Unlike the meter resistance, the resistance of a rectifier is not constant; it varies with the current passing through it. Therefore, a shunt resistance calculated to achieve given proportions of current in the shunt and rectifier legs of the circuit at a specific value of I_T will not result in the same proportions at some other value. The solution to the problem is to use a current transformer instead of a shunt.

One might ask why a dc meter should be used for ac measurements when there are all sorts of specifically ac-type meters around that will not only read ac directly but can read high values of current without the need of shunts or rectifiers or other attachments; for instance, moving-iron and thermal types can do this. It's a good question, but there is a good answer.

In the field of analog indicators, there is nothing quite so good as the ubiquitous moving-coil meter. It is surprisingly rugged, relatively inexpensive, and accurate. The additional cost of rectifiers and a current transformer is usually justified by the flexibility of the instrument, which can be switched effortlessly from dc to ac, and vice versa. Comparable all-round performance is hard to come by in other kinds of meters.

BASIC PRINCIPLES

Unlike in other transformers, the primary winding current in a current transformer is not dependent on the secondary winding load. It is a "forced" value dependent only on the circuit into which the primary winding is connected. The method of connection

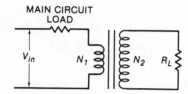

Fig. 13-3. Current transformer is used in series with main circuit.

is shown in Fig. 13-3. Here the primary winding is in series with the main circuit; it is designed to have a low impedance and has negligible effect on the current. Therefore, regardless of what changes may be made to the secondary winding load, the primary winding current is always the same as that of the main circuit.

But, certainly, if the load on the secondary winding is changed, something in the primary winding has to change; that "something" is a change in the relationships between the several components of the primary winding current. The primary winding current is made up essentially of three components; the loss current that supplies iron and copper loss, the magnetizing current that establishes the flux in the core, and, of course, the load current. However, these currents are not in phase with each other; there are differences in the phase angles, and these angles change when the load changes. The primary winding current, as always, is the *vector sum* of these currents.

If the load is removed from the secondary winding while the main circuit current is flowing, most of the primary winding current becomes magnetizing current, but the vector angles change in such a way as to keep the total current in the primary the same as before. Because the main circuit is now mostly magnetizing current, the flux in the core shoots up to a high level—there is now no opposing current in the secondary winding to prevent this—and a very high voltage appears across the secondary.

This can be seen from the basic transformer equation, $V = 4FfaNB \times 10^{-8}$. Obviously, if the flux B increases, so will the voltage (assuming that the core is not already saturated). In fact, due to the high turns ratio usually found in these transformers, the voltage in this condition can reach a dangerously high level.

A current transformer, like any other transformer, must satisfy equation $I_1/I_2 = N_2/N_1$ (Chapter 3), and this gives the key to the design procedure. The value of I_1 is determined by the circuit into which the primary winding is connected and I_2 by the load that is to be supplied; in the present context, this is the rectifier-meter circuit. For example, if a current of 2.5 amperes is to be transformed down to supply a secondary winding load of, say, 0.001 ampere, then the turns ratio must be basically

$$\frac{I_1}{I_2} = \frac{N_2}{N_1} = \frac{2.5}{0.001} = 2500$$

If high accuracy is not needed, it remains only to select a core and specific numbers of turns for the primary and secondary windings, put them together, and then we are in business.

However, the current-turns relationship refers only to load currents. It does not take account of magnetizing and loss currents, and, if these are not negligible, ratio errors become evident. If accuracy is required, as it usually is for instrumentation, the loss and magnetizing currents must be kept small.

One way of doing this is to use a low-loss, high-permeability nickel-iron alloy. A familiar name in this area is SuperPerm 80 by Magnetic Metals Corp.; others in the same general category are Mumetal (Alleghany Steel Corp.) and Hy Mu (Carpenter Steel Co.), to mention only a few. With such core materials, iron losses are so low and the permeability so high that the loss and magnetizing components of the primary winding current can usually be ignored in the design.

RECURIFIERS AND VOLTAGES

The voltage across the secondary winding is obviously the same as that across the rectifier (Fig. 13-4). And if the circuit is drawn for each half-cycle as in Fig. 13-5, it is seen that the voltage across the secondary winding is in fact that across two rectifiers and the meter in series, or $V_S = V_1 + V_2 + V_3$.

The values of V_1 and V_3 depend only on the characteristics of the rectifiers and remain substantially constant with changes in current. (Because the voltage is constant

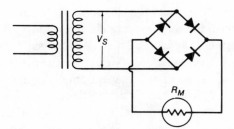

Fig. 13-4. Secondary voltage equals that across the rectifiers.

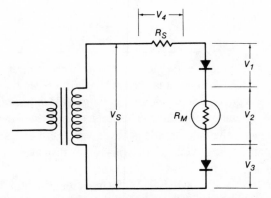

Fig. 13-5. On half-cycle the voltage is distributed as shown.

with a change in current, the resistance of the rectifier changes in accordance with $R = E/I$, as stated earlier.) This constancy of voltage can be seen in the typical diode curve in Fig. 13-6. When the diode conducts at voltage x, any attempt to increase the voltage results only in an increase in current.

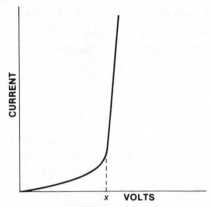

Fig. 13-6. Typical diode curve.

Voltage V_2 across the meter depends on the current and the resistance of the meter. The resistance of the meter depends on the design of the instrument and varies from one to another, even of the same model and current range. A good-quality 1.0-mA meter might have a voltage drop at full scale of 0.1 volt or less; its resistance is therefore $R = E/I$ or $0.1/0.001 = 100$ ohms or less. Sometimes the resistance is marked on the scale. Obtain the values of the rectifier voltages from manufacturers' data; they might be on the order of, say, 0.4 volt each. Typically, then, the voltage across the meter-rectifier circuit with a 0.001-mA meter might be about 0.9 volt at full-scale reading. There is a further drop in the secondary winding resistance itself, represented by R_S in Fig. 13-5, and this should be made small; at 0.001 mA the voltage drop in R_S might be on the order of 0.1 volt. The total secondary winding voltage in this example is therefore 1.0 volt.

AVERAGE VERSUS RMS

Note in Fig. 13-4 that there is no electrical filter in the output of the rectifiers as was discussed in Chapter 9. The output is supplied to the meter in the form of a pulsating direct current. At first sight this does not look like a good idea because it might disrupt the steady reading on the meter. But the mechanical inertia of the moving-pointer system prevents it from following the fast variations of current and it takes up an average, or mean, position and reads average voltage. But rms rather than average values are usually desired so we cheat a little.

The ratio of rms to average value is 1.11 (form factor, Chapter 1) for a sine wave. In the design of the transformer, it is arranged that the secondary winding current is actually 11 percent higher than the nominal full-scale reading of the meter. For instance, the secondary current would be made 1.1 mA if the meter full-scale reading is 1.0 mA. The meter is then calibrated in rms.

With devilish cunning, this is arranged quite simply by modifying the number of turns in the secondary. If the true turns ratio indicates N turns, the actual winding would be $N/1.11$ turns.

A MOVING-COIL AC AMMETER

Example 1—Using a basic 0–1-milliampere meter movement, let's design an instrument that reads from zero to 3.0 amperes rms. The circuit is shown in Fig. 13-7. In order to keep the iron loss and magnetizing current small for reasons of accuracy, the core is of nickel-iron alloy, such as SuperPerm or Mumetal. The use of these high-permeability, low-loss materials also permits the windings to have the smallest number of turns for ease of construction and it minimizes copper loss. In fact, the primary winding can conveniently be a *single turn*. How about that for keeping the number of turns small?

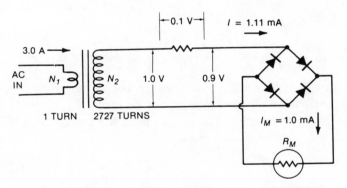

Fig. 13-7. Circuit for design example showing voltages and currents.

Because there is only one turn on the primary that must carry 3.0 amperes (3000 milliamperes), the milliampere-turns figure for the primary winding is 3000 × 1.0 = 3000 milliampere-turns.

The meter is 1.0-milliampere, full scale. But the meter and rectifier together give a deflection equal to the average current, which bears a ratio of 1.11 to the rms value. The ac load current in the secondary winding is thus 1.11 mA rms when the meter is reading full scale.

Because the secondary winding milliampere-turns must equal the primary winding milliampere-turns, it follows that the secondary winding must have 3000/1.11 = 2727 turns.

The voltage across the secondary winding is that across the rectifiers and meter. The rectifier data is obtained from the manufacturer of the specific rectifier being used; this rectifier will be an instrument-type bridge rated in the present case for 1.0 milliampere.

Suppose that the voltage across the rectifier-meter circuit is 0.9 volt at the full-rated current. Additionally there will be a voltage drop in the secondary winding. Assuming tentatively that this will be on the order of 0.1 volt, the total voltage across the secondary will be 1.0 at full scale. Since the voltage drop should be kept small, this figure will be checked later in the design to ensure that it is not being exceeded.

Using Equation 3-32, Chapter 3, the total flux Φ can now be determined. Assigning 1.11 for form factor F, then

$$\Phi = \frac{V_2 \times 10^8}{4.44 \times f \times N_2}$$

where

  V_2 is the voltage across the secondary in volts,
  f is the supply frequency in hertz,
  N_2 is the number of turns on the secondary.

Note that in previous examples, the primary winding voltage and number of turns in the primary were used to calculate flux or flux density, while here the secondary winding terms are used. In practical terms, one method is as good as another. Assuming the design is for 60 hertz, we get

$$\Phi = \frac{1.0 \times 10^8}{4.44 \times 60 \times 2727} = 139 \text{ maxwells or lines}$$

This is a low total flux. By using a relatively large core area, the flux density can also be kept small and take full advantage of the exceedingly high initial permeability characteristics of the core material being used. Suppose a low flux density figure of 100 gauss is tentatively chosen. Then from Equation 3-34

$$a = \frac{\Phi}{6.45B} = \frac{139}{6.45 \times 100} = 0.215 \text{ square inch}$$

is the core area.

If a square format is chosen for the core area, the copper loss will be minimized. Ideally, the core would be a tape-wound toroid for minimum effective gap; but to wind the large number of secondary turns needed onto a toroid by hand is a rather formidable task, so laminations or perhaps a C core can be considered for ease of winding. A check of Magnetic Metals' catalog turns up lamination No. 21 EI (Fig. 13-8), which has a square cross section of 0.25 square inch. This area is somewhat greater than was calculated but that is not such a bad thing, so we'll stick with it.

Again for ease of winding, the random-wound style will be adopted; in any case the secondary voltage is very low, so this method should be safe.

In order to keep the resistance low, the conductors chosen are heavier than would be the case for the same current in, say, a power transformer. For the single-turn primary windings, No. 10 AWG is selected. From the wire table (Table 6-1), this has a diameter of 0.106 inch.

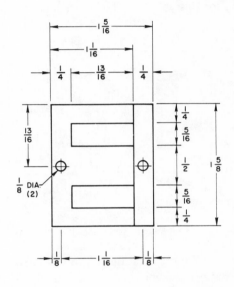

Mᴀɢɴᴇᴛɪᴄ Mᴇᴛᴀʟs

LAMINATION TYPE 21 EI

Part Shown Actual Size

CHARACTERISTICS OF A CORE STACK HAVING A SQUARE CROSS SECTION

VOLUME AND WEIGHT

VOLUME	– .815 in.³	– 13.3 cm.³
WINDOW AREA	– .253 in.²	– 1.63 cm.²
WT. SUPER Q 80	– .256 lb.	– 117 g.
WT. SUPERPERM "49"	– .240 lb.	– 109 g.
WT. SUPERFLUX	– .237 lb.	– 108 g.
WT. SILICON	– .219 lb.	– 99.3 g.

MAGNETIC DESIGN FORMULAE

Properties of Core Stack with Winding of "N" Turns

$$B_{max} = \frac{233 \times 10^3}{K_1 N} \quad \text{Gausses Per Volt at 60 Hertz}$$

$$H_o = (.151 \times 10^{-3})N \text{ Oersteds}$$
(Gilberts per centimeter) per milliampere of direct current

$$L_o = (.2431 \times 10^{-8})K_1 N^2 \mu_{ac} \text{ Henries}$$

MAGNETIC PATH DIMENSIONS

l = 3.27 in.	8.30 cm.
A = .250 in.²	1.61 cm.²

K₁ (STACKING FACTOR)

Thickness	Butt Jointed	Interleaved one per layer
.004″	.90	.80
.006″	.90	.85
.014″	.95	.90
.0185″	.95	.90

PERFORMANCE DESIGNATION	MATERIAL TYPE	THICKNESS (Inches)	CATALOG NUMBER	WEIGHT AND COUNT	
				LBS./M PCS.	PCS./LB.
SUPERPERM 80	HyMu 80	.004	21EI8404	2.02	494
SUPERPERM 80	HyMu 80	.006	21EI8406	3.00	334
SUPERPERM 80	HyMu 80	.014	21EI8414	6.99	143
SUPER Q 80	HyMu 80	.004	21EI8004	2.02	494
SUPER Q 80	HyMu 80	.006	21EI8006	3.00	334
SUPER Q 80	HyMu 80	.014	21EI8014	6.99	143
SUPERTHERM 80	HyMu 80	.006	21EI7406	3.00	334
SUPERTHERM 80	HyMu 80	.014	21EI7414	6.99	143
SUPERPERM 49	49	.004	21EI4904	1.88	531
SUPERPERM 49	49	.006	21EI4906	2.85	351
SUPERPERM 49	49	.014	21EI4914	6.66	150
SUPERFLUX	PERMENDUR	.006	21EIVP06	2.81	356
SUPERFLUX	PERMENDUR	.010	21EIVP10	4.69	213
MICROSIL	Gr. Or. Silicon	.004	21EI3304	1.75	571
MICROSIL	Gr. Or. Silicon	.006	21EI3306	2.63	381
SILICON	Non Or. Silicon*	.014	21EI**14	6.13	163
SILICON	Non Or. Silicon*	.018	21EI**18	7.89	127
SILICON	Non Or. Silicon*	.025	21EI**25	10.95	91.3
HYPERTRAN	Low Carbon	.025	21EI2125	11.18	89.0

* Customer to designate AISI grade of material desired.
** See "How To Order Section" for Code Number.

Courtesy Magnetic Metals Corp.

Fig. 13-8. Lamination data for current transformer.

This single turn is most easily wound outside the secondary winding with the secondary wound next to the core (Chapter 16). If 1/16 inch is allowed all-around for the bobbin, plus, say, 0.015 inch for the secondary cover and a similar amount to separate the primary winding from the core, the area left in the window for the secondary winding will be 0.079 square inch, as shown in Fig. 13-9. The numbers are in Fig. 13-10. Incidentally this design sheet blank, which is really intended for power transformers, does fine for this and other types, with appropriate penciled modifications as shown.

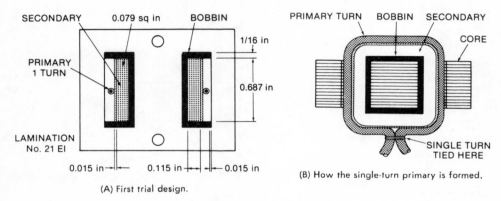

(A) First trial design.

(B) How the single-turn primary is formed.

Fig. 13-9. Winding the transformer.

Now, 2727 turns must be fit into this area. From Table 6-1 it is seen that No. 38 gauge in heavy insulation random winds at 39,300 turns per square inch of winding area, and 0.079 square inch accommodates 0.079 × 39,300 = 3105 turns. Gauge No. 37 gives 30,900 turns per square inch, or 0.079 × 30,900 = 2441 turns in the space available. The No. 38 gauge looks about right.

The secondary winding resistance in this case works out to approximately 425 ohms to give a secondary winding voltage drop of $V = R \times I = 425 \times 0.0011 = 0.467$ volt. This is considerably higher than originally specified.

To reduce the secondary winding resistance and voltage drop, a larger lamination is considered in a second trial design. This is type No. 68 EI (Fig. 13-11). Staying with the 0.25 square inch area, a stack 0.36 inch thick is required, and dividing this by a stacking factor of 0.9 inch gives a stack 0.4 inch thick. This does not result in an ideal proportion, but it is acceptable.

Retaining No. 10 gauge for the primary winding, the total space occupied by the bobbin, insulation, and the primary winding is 0.198 inch (Fig. 13-12). Deducting this from the lamination window width of 0.343 inch leaves 0.145 inch for the secondary winding. This works out to a winding area of 0.145 × 0.906 = 0.131 square inch. Gauge No. 35 with a random-wind figure of 20,600 per square inch comes in quite nicely with 2698 turns, almost 2727 as required. But what about the resistance and voltage drop? The resistance now works out at 230 ohms for a voltage drop of 230 × 0.011 = 0.25 volt—still more than double the desired value. These numbers are shown in Fig. 13-13.

TRANSFORMER DESIGN SHEET

Input volts		Hertz	60			Est. Efficiency			Turn/volt		
Lamination or core No.		21E1	SUPER	PERM	LAMINATIONS				B		
Window dimensions		G 0.8125 (13/16) INCH		F 0.3125 (5/16)		W			M_P		
C.S.A. dimensions		E 0.5		D 0.5		a 0.25			v		
		Wt			Watts/lb			Iron loss			

Windings			W_1	W_2							
Coil			SEC	PRIM							
Volts			1.0								
Amperes			.0011	3.0							
Turns			2727	1							
Gauge			38	10							
Turns/inch ^{2}			39300								
Margins											
Winding ~~length (L)~~ AREA in^2			.079								
Turns/layer											
Number of layers											

Build											Total
Copper			.115	.106							
Paper											
Cover			.015	.015							
Total			.130	.121							
Bulge Percent											
Total (R)			.130	.121							.251

Bobbin PLUS clearance W   Total bobbin (B) .062   Total depth .313

Losses											
Length mean turn (M)			2.895								
Total wire length (inches)			7816								
Ohms/ft			.648								
Resistance ~~(hot)~~			425								
Voltage drop			0.42								
Copper loss (I^2R)											

Iron loss / Total loss / Efficiency

Temperature Rise

Coil	t	C_S	W	W/C_S	°C T_o
1					
2					
3					
4					

$M = 2(E+D+4B)+6.28d$

WHEN $d = W/2$

THEN THE SECONDARY M IS

$2(.5+.5+4\times.062)+(6.28\times.065) = \underline{2.895}$
$(\underline{1}\quad .248) + .4$
2.495

Temp. Rise = T_o(Total) + 20HW/C_S =

Fig. 13-10. Design sheet for current transformer No. 1.

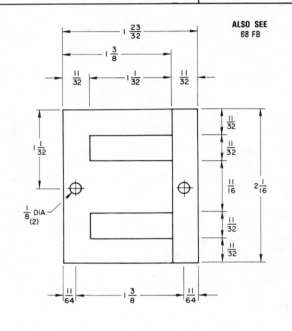

				WEIGHT AND COUNT	
PERFORMANCE DESIGNATION	MATERIAL TYPE	THICKNESS (Inches)	CATALOG NUMBER	LBS./M PCS.	PCS./LB.
SUPERPERM 80	HyMu 80	.006	68EI8406	5.33	188
SUPERPERM 80	HyMu 80	.014	68EI8414	12.44	80.4
SUPER Q 80	HyMu 80	.006	68EI8006	5.33	188
SUPER Q 80	HyMu 80	.014	68EI8014	12.44	80.4
SUPERTHERM 80	HyMu 80	.006	68EI7406	5.33	188
SUPERTHERM 80	HyMu 80	.014	68EI7414	12.44	80.4
SUPERPERM 49	49	.006	68EI4906	4.96	202
SUPERPERM 49	49	.014	68EI4914	11.70	85.5
SUPERFLUX	PERMENDUR	.006	68EIVP06	4.90	204
SUPERFLUX	PERMENDUR	.010	68EIVP10	8.17	122
MICROSIL	Gr. Or. Silicon	.006	68EI3306	4.62	246
MICROSIL	Gr. Or. Silicon	.014	68EI3314	10.77	92.8
SILICON	Non Or. Silicon*	.014	68EI**14	10.77	92.8

* Customer to designate AISI grade of material desired.
** See "How To Order Section" for Code Number.

Courtesy Magnetic Metals Corp.

Fig. 13-11. Lamination data for second trial design.

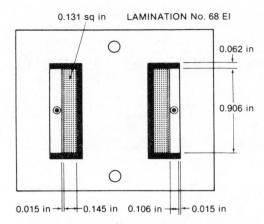

Fig. 13-12. Second trial design of current transformer.

What to do now? Try pruning bobbin thickness and reducing cover insulation to make room for heavier wire, or look for a core with a larger window, or simply live with the larger voltage drop. Let's opt to look for a more appropriate core.

This time, however, the cut cores are considered and No. MCL-121F (Table 13-1) is chosen for a trial. This core has an area of 0.111 square inch. Therefore, two of them, combined to yield a shell type configuration (Fig. 13-14), result in an area of 0.222 square inch.

Once more deducting the 0.198 required for the primary winding, bobbin, and insulation, 0.302 inch is left in the window width for the secondary winding. The area available to the secondary is then 0.302 × 1.25 = 0.377 square inch.

The wire table (Table 6-1) reveals that No. 31 gauge at 8090 turns per square inch will permit 8090 × 0.377 = 3050 turns to be fit into this area; therefore the 2727 turns will fit comfortably.

The resistance works out at 102 ohms for a voltage drop of 102 × 0.0011 = 0.112 volt; these numbers are shown in the design sheet of Fig. 13-15. This voltage drop is very close to the original specification of 0.1 volt.

Note that in working out the resistance the "cold" resistance figure of 12 inches to the foot is used rather than the "hot" resistance figure used in former examples, because it is expected that this transformer with its heavy conductors will run quite cool.

Intermediate Ranges

It is quite simple to incorporate intermediate ranges that can be switched in as required, as shown in Fig. 13-16. For example, two additional ranges of, say, zero to 0.3 ampere and zero to 0.03 ampere, with switching, could easily have been designed into the transformer as follows.

For the 0.3-ampere range, the milliampere-turns of the primary winding must be made equal to 3000 as before. Therefore, the number of turns needed for the primary

TRANSFORMER DESIGN SHEET

Input volts		Hertz			Est. Efficiency			Turn/volt			
Lamination or core No.	68 E1 SUPER PERM LAMINATIONS								B		
Window dimensions	G 1.031 (1 1/32)		F 0.343 (11/32)		W				M_P		
C.S.A. dimensions	E 0.6875 (11/16)		D 0.4			a 0.25			v		
	Wt			Watts/lb			Iron loss				

Windings												
Coil		SEC	PRIM									
Volts		1.0										
Amperes		.001	3.00									
Turns		2727	1									
Gauge		35	10									
Turns/inch		20,600										
Margins												
Winding length (L)												
Turns/layer												
Number of layers												

Build												Total
Copper			.106									
Paper												
Cover			.015									
Total			.121									
Bulge Percent												
Total (R)												
									Total bobbin (B)			
Bobbin ____ clearance ____									Total depth			

Losses												
Length mean turn (M)												
Total wire length (inches)												
Ohms/ft												
Resistance ~~(hot)~~		230										
Voltage drop		.23										
Copper loss (I^2R)												
									Iron loss			
									Total loss			
									Efficiency			

Temperature Rise						
Coil	t	C_S	W	W/C_S	°C T_o	
1						
2						
3						
4						

Temp. Rise = T_o(Total) + $20HW/C_S$ =

Fig. 13-13. Design sheet for current transformer No. 2.

Table 13-1. Data for Core MCL-121F (Arrowed) for Third Trial Design

SINGLE-PHASE CUT CORES

MATERIAL- SUPERPERM 80 2 MIL, STACKING FACTOR .89

- NOTE, THIS LISTING SEQUENCED ON CATALOG NUMBER -

CORE LIMITS BASED ON .70 VA/LB & .20 WATTS/LB @ 400 HZ, 4.0 KILOGAUSS
AMPERE TURNS ASSUMES .001 IN. AIRGAP, EXCEPT .002 IN. WHERE STARRED*

MAGNETIC METALS CATALOG NUMBER	DIMENSIONS, INCHES				DEFG PRODUCT	...IRON... WEIGHT LBS.	LEG AREA SQ.IN.	CORE DESIGN ...LIMITS... WATTS LOSS	AMPERE TURNS	TURNS PER VOLT	MAGNETIC METALS CATALOG NUMBER
	D	E	F	G							
MCL-1	.250	.125	.250	.500	.004	.016	.028	.003	6.00	79.0	MCL-1
MCL-1A	.250	.250	.250	.500	.008	.038	.056	.008	6.15	38.9	MCL-1A
MCL-3B	.437	.250	.187	.625	.013	.071	.097	.014	6.20	22.5	MCL-3B
MCL-3J	.188	.312	.187	.937	.010	.051	.053	.010	6.55	41.5	MCL-3J
MCL-4B	.250	.375	.250	.781	.018	.083	.084	.017	6.60	26.1	MCL-4B
MCL-5	.375	.250	.250	.875	.021	.077	.084	.015	6.50	26.1	MCL-5
MCL-9	.500	.375	.375	1.187	.083	.221	.167	.044	7.10	13.0	MCL-9
MCL-9A	.812	.375	.421	1.625	.208	.442	.271	.088	7.60	8.03	MCL-9A
MCL-9K	.250	.250	.375	1.312	.031	.071	.056	.014	7.00	38.9	MCL-9K
MCL-12	.500	.437	.500	1.125	.123	.277	.195	.055	7.25	11.2	MCL-12
MCL-15A	.375	.500	.500	1.562	.146	.295	.167	.059	7.80	13.0	MCL-15A
MCL-17	1.000	.500	.500	1.562	.391	.786	.445	.157	7.80	4.90	MCL-17
MCL-21D	.687	.375	.625	1.937	.312	.449	.230	.090	8.10	9.50	MCL-21D
MCL-35	1.250	1.250	1.000	3.000	4.69	5.15	1.39	1.03	15.9*	1.57	MCL-35
MCL-38D	1.000	.375	.750	2.312	.650	.747	.334	.149	8.50	6.53	MCL-38D
MCL-38E	1.000	.500	.750	2.312	.867	1.05	.445	.210	8.70	4.90	MCL-38E
MCL-78D	1.000	.500	.312	2.000	.312	.856	.445	.171	8.05	4.90	MCL-78D
MCL-84B	.750	.625	.937	2.250	.988	1.07	.417	.214	9.00	5.23	MCL-84B
MCL-85A	1.500	.750	1.375	2.250	3.48	2.97	1.00	.594	9.60	2.18	MCL-85A
MCL-85B	.750	.750	1.375	1.750	1.35	1.34	.501	.269	9.20	4.35	MCL-85B
MCL-85C	.750	.500	1.375	1.500	.773	.761	.334	.152	8.55	6.53	MCL-85C
MCL-85D	.375	.500	1.375	1.500	.387	.380	.167	.076	8.55	13.0	MCL-85D
MCL-85F	.375	.750	1.375	1.750	.677	.672	.250	.134	9.20	8.72	MCL-85F
MCL-85G	1.000	1.000	1.250	3.500	4.38	3.49	.890	.699	16.2*	2.45	MCL-85G
MCL-85J	.750	.500	1.500	2.250	1.27	.934	.334	.187	9.35	6.53	MCL-85J
MCL-97	.750	.750	1.125	1.750	1.11	1.27	.501	.253	8.95	4.35	MCL-97
MCL-97A	.750	.500	1.125	1.500	.633	.708	.334	.142	8.35	6.53	MCL-97A
MCL-106	.500	.500	.750	2.375	.445	.535	.223	.107	8.75	9.80	MCL-106
MCL-106A	.500	.250	.750	2.500	.234	.249	.111	.050	8.50	19.6	MCL-106A
MCL-107B	.250	.250	.625	1.500	.059	.087	.056	.017	7.45	38.9	MCL-107B
MCL-115B	1.000	1.000	1.000	3.500	3.50	3.35	.890	.671	16.0*	2.45	MCL-115B
MCL-121D	.250	.250	.500	1.312	.041	.076	.056	.015	7.15	38.9	MCL-121D
MCL-121F	.500	.250	.375	1.375	.086	.156	.111	.031	7.25	19.6	MCL-121F
MCL-121G	.250	.250	.500	1.375	.043	.078	.056	.016	7.20	38.9	MCL-121G
MCL-121H	.250	.500	.500	1.375	.086	.183	.111	.037	7.60	19.6	MCL-121H
MCL-123F	.687	.312	.375	.875	.070	.203	.190	.041	6.70	11.5	MCL-123F
MCL-123P	.375	.250	.375	1.000	.035	.090	.084	.018	6.75	26.1	MCL-123P
MCL-131C	.625	.500	.500	1.500	.234	.480	.279	.096	7.70	7.83	MCL-131C
MCL-137	.750	.750	2.500	5.000	7.03	2.71	.501	.542	13.4	4.35	MCL-137
MCL-137A	.625	.625	2.000	4.000	3.13	1.51	.348	.302	11.7	6.27	MCL-137A
MCL-137B	.750	.750	1.625	2.375	2.17	1.60	.501	.321	10.0	4.35	MCL-137B
MCL-138A	.500	.250	.500	1.000	.063	.129	.111	.026	6.85	19.6	MCL-138A
MCL-138B	.375	.375	.500	1.000	.070	.161	.125	.032	7.05	17.4	MCL-138B
MCL-147	.250	.125	.312	1.000	.010	.026	.028	.005	6.55	79.0	MCL-147
MCL-157C	1.000	1.000	1.000	1.750	1.75	2.40	.890	.480	14.3*	2.45	MCL-157C
MCL-157D	.495	.495	1.000	1.000	.368	.444	.218	.089	8.20	10.0	MCL-157D
MCL-165F	1.500	1.500	.625	2.500	3.52	6.80	2.00	1.36	15.4*	1.09	MCL-165F
MCL-166F	1.000	.375	.625	1.125	.264	.482	.334	.096	7.30	6.53	MCL-166F
MCL-173B	.625	.625	1.000	2.750	1.07	1.02	.348	.203	9.55	6.27	MCL-173B
MCL-190M	1.000	1.000	.906	3.500	3.17	3.30	.890	.660	15.9*	2.45	MCL-190M

MATERIAL- SUPERPERM 80 2 MIL

Courtesy Magnetic Metals Corp.

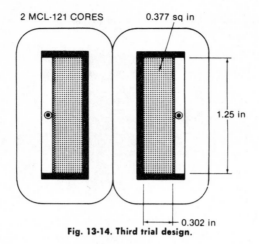

2 MCL-121 CORES   0.377 sq in

1.25 in

0.302 in

Fig. 13-14. Third trial design.

winding on this range is N_3 = 3000/300 = 10 turns. For the 0.03-ampere range, the number of primary winding turns must be N_4 = 3000/30 = 100 turns.

As a practical point, note that the turns are added when switched. Thus, on the 0.3-ampere position, the single turn of the 3.0-ampere range is added to nine more to give the 10 needed for the second range and these are added to a further 90 turns when switched to the third range to make the 100 turns needed for that range.

Example 2—In the last example, great care was taken to keep losses very small in the interests of achieving a good level of accuracy, which is usually a requirement in metering applications. But this is not always the case. In many applications, a fair degree of iron and copper loss can be tolerated, and this makes a large difference to the design and construction of the transformer. Often the number of turns required on both the primary and secondary winding is quite small, thus making a toroidal core a feasible choice.

Consider, for example, a current transformer to be placed in series with a 10-ampere load to drive a device requiring 100 mA (0.100 ampere) at 60 hertz. In this case, no rectification is needed. The load resistance is, say, 10 ohms. Therefore, the voltage across the load, and across the secondary winding, is $V_S = I_S \times R_S = 0.1 \times 10 = 1.0$ volt.

In this design, top accuracy is not a requirement. Therefore, the core material does not need to have the highest permeability or the lowest loss. A grain-oriented silicon steel, such as Microsil, Silectron Z, or Oriented T-S, can be used. A toroidal format will permit the material to be used to the best advantage.

Using a single-turn primary winding as before, the ampere-turns of the primary are $1.0 \times 10 = 10$ ampere-turns. The secondary winding must be the same; therefore it needs

$$N_S = \frac{\text{Ampere-turns}}{I_S} = \frac{10}{0.1} = 100 \text{ turns}$$

TRANSFORMER DESIGN SHEET

Input volts		Hertz		Est. Efficiency		Turn/volt	
Lamination or core No.	MCL - 121F x 2.		SUPER PERM			B	
Window dimensions	G 1.375		F 0.5		W	M_P	
C.S.A. dimensions	E 0.5 (.25 × 2)		D 0.5		a 0.222 (.111 × 2)	v	
	Wt		Watts/lb		Iron loss		

Windings		W₁	W₂								
Coil		SEC	PRIM								
Volts		1.0									
Amperes		.0011	3.0								
Turns		2700	1								
Gauge											
Turns/inch											
Margins											
Winding length (L)											
Turns/layer											
Number of layers											

Build											Total
Copper		.302	.106								
Paper											
Cover		.015	.015								
Total		.317	.121								
Bulge Percent											
Total (R)		.317	.121								.438

Total bobbin (B) .062
Total depth .500

Bobbin _____ clearance _____

Losses											
Length mean turn (M)		3.45									
Total wire length (inches)		9315									
Ohms/ft		.131									
Resistance ~~(hot)~~		102									
Voltage drop		.112									
Copper loss (I^2R)											

Iron loss
Total loss
Efficiency

Temperature Rise

Coil	t	C_S	W	W/C_S	°C T_o
1					
2					
3					
4					

$$M = 2(E + D + 4B) + 6.28\, d$$
$$= 2(.5 + .5 + .248) + 6.28 \times .151$$
$$2 \times 1.248 \qquad + \quad .95$$
$$2.496 + .95 = 3.446$$

Temp. Rise = T_o(Total) + 20 HW/C_S =

Fig. 13-15. Design sheet for current transformer No. 3.

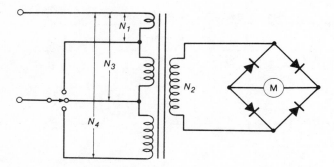

Fig. 13-16. Primary winding with switch to give choice of ranges.

This presents no winding difficulties, even on a toroid. The total flux is

$$\Phi = \frac{V_S \times 10^8}{4.44fN_S} = \frac{1.0 \times 10^8}{4.44 \times 60 \times 100}$$

$$= 3753 \text{ maxwells}$$

Although this core material can be run at 15,000 or 16,000 gauss in power applications, it would be advisable to keep the losses at a reasonably low level and run it at, say, 10,000 gauss in this case. The larger core will also permit relatively large gauge wire and thus minimize copper loss. The required cross-sectional core area is then

$$a = \frac{\Phi}{6.45B} = \frac{3753}{6.45 \times 10,000} = 0.058 \text{ square inch}$$

From this point, the design proceeds in the usual way by selecting a core, conductors, and so on, and checking the fits.

OLD TV TRANSFORMER LAMINATIONS

Note that the current transformers in the examples tend to be rather large in relation to the power output. In the first example, the power output is $P_o = I^2R = 0.0011^2 \times 100 = 0.00011$ watt or 110 microwatts! This low output is the penalty for obtaining low loss. If losses (that is to say, accuracy) are not so important, then one can be a little more liberal with the specification.

Certainly these old silicon-steel laminations can be used if the design is not too critical. After all, they were used in the old days before these fancy alloys were invented. And what was good enough for the old-timers is good enough for us—sometimes, that is.

METER RESISTANCE MEASUREMENT

If the resistance of a meter is not known, it can be measured using the method given in Chapter 16. It is important not to attempt a direct measurement of the coil resistance using an ohmmeter or similar current-operated device. If this is attempted, the chances are very good that the meter movement will be wrecked in short order.

FOURTEEN

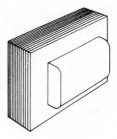

Salvage, Construction, and Sources

This chapter is written primarily for the experimenter who doesn't have access to the material sources and equipment usually available to someone who is in the electricity/electronics business. I discuss the science of scavenging and the con man or woman's gentle art of separating valuable material from people who have come to love and trust him or her. Then I explain about taking components apart and the technicalities of putting them together. And, finally, methods of pinpointing sources of new material are dealt with.

JUNK

As in the recipe for rabbit stew, that says, "First, catch your rabbit," you must first get your core before making a transformer. Wire is not usually a problem; it is easy to purchase in small quantities. So also is the case with paper and tape and other sundries. But core material might not be easy for some people to come by—new, that is. Fortunately, there is an almost limitless supply of used core material available at next to no cost in abandoned TV sets, defunct radios, and the like, obtainable at dumps, garage sales, church rummage sales, and surplus dealers.

The attics and basements of friends are rich sources of material, but here is a tip. Along with old TVs and radios that contain transformers, people tend to hang onto broken toasters, worn vacuum cleaners, burned-out washing machines, cracked hair dryers, and rusted electric curling tongs—stuff which is of no earthly use to you but you will

335

get anyway. In discussing your needs with your friends, be sure to impress on them that only equipment that contains transformers is truly garbage to be disposed of. The rest should be left in the attic to mature, like casks of old wine, until it reaches the antique stage, when it will rapidly appreciate in value.

To some people, "junk" is just another four-letter word, but to you it can be the open sesame to a transformer maker's Aladdin's cave. For example, an old TV set will yield at least one large power transformer, a filter choke, and a speaker output transformer, all in good-grade silicon steel, plus a flyback transformer that might be in one of the nickel-iron alloys or even a ferrite. In the deflection yoke, the same material might be found, although the shape is a little odd for transformers; but it is a great demagnetizer of tools and watches. Then there is a nice supply of wire contained in the scanning coils and on the transformers themselves. Also available are fiber board and other items that are not always easy to buy.

The transformers and chokes have the great advantage that they come equipped with mounting brackets, shrouds, clamping bars, and other factory-made hardware that fits the particular core exactly. Within a day or two, it is possible for you to accumulate enough transformer material to keep you supplied for several years.

The chances are that most everything you get will be good-grade silicon-steel laminations; certainly anything from old tube-type equipment is likely to be in this category. The more exotic alloys and ferrites are much less common in junk but not unheard of—C type cores and toroids are possible if one keeps one's eyes open. And, of course, minitransformers are commonplace in old transistor radios.

SALVAGING CORES

If the transformer is from something unfamiliar, first try to figure out its purpose in the circuit. This gives a clue to the kind of material in the core. Power transformers, filter chokes, and most output transformers have silicon-steel cores. If the core is a C core configuration, it is likely to be a grain-oriented material and most likely silicon steel.

Having identified the material, or at least being prepared to take a chance on it, your next step is to prepare it for use. The first rule is: "Don't take it apart until you want to use it." Why? Because you might not have to take it apart at all. It could be quite suitable for your purpose as it is; or, if not entirely suitable, it might be 50 percent suitable. For instance, the primary winding could be satisfactory and only the secondaries need be stripped off and replaced. The same applies to plain inductors (chokes). The inductor winding frequently has enough turns on it to allow it to be reduced and used as a primary. Add the secondary, or secondaries, and you are in business. In the case of chokes, however, beware of gaps in the core; if there is one, be sure to eliminate it by interleaving the laminations, or readjust it if it is to be used with dc.

The first step in disassembly is to strip off the nuts, bolts, clamps, mounting brackets, etc., and put them aside to be used again—if not on this transformer, then on another. Next, look for wedges between the bobbin and the laminations in the center limb of the core. If the laminations are not interleaved, as will be the case in a gapped transformer or inductor, half of them can usually be separated immediately from the rest. If they are Ts and Us, the Us can be pulled away, or, if Es and Is, the Is can be pulled away (see Fig. 14-1). If they are interleaved, every second U or I can be separated at either end of the core. If the core has been impregnated in varnish or wax, use a thin-blade knife to split the laminations apart.

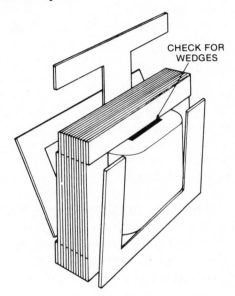

CHECK FOR
WEDGES

Fig. 14-1. Disassembling the core.

In a tightly cored transformer, be prepared to sacrifice a lamination or two by bending it in order to get a grip with pliers. An ideal tool for pulling out laminations is a pair of end cutters. These get a good grip and their shape contributes to leverage. The action is precisely the same as that used in pulling teeth, but if the transformer screams at you, leave the job until morning. Once one or two laminations have been removed, the rest are usually easy.

If the windings are not to be used as is, consider salvaging the wire if it is clean and appears good. However, heavy wire can be a nuisance to reuse, because it is kinked where it bent around the corners of the bobbin. Fine wire is easier, but be sure there is enough of it to do the job. All things considered, though, salvaging wire is not highly recommended unless you are really keen on saving money or the wire is hard to come by. But do save the bobbin to be used again.

MAKING A NEW BOBBIN

Start by making a wooden block to be used first as a form for making the bobbin, and later as a mandrel for winding. The length should be a mite shorter than the length of the center limb of the core, and dimensions A and B (Fig. 14-2A) just a shade larger. Two wooden cheeks are also made as shown in Fig. 14-2B. These act as guides for

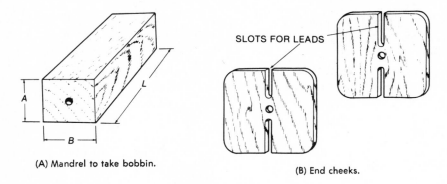

(A) Mandrel to take bobbin.

(B) End cheeks.

Fig. 14-2. Making a new bobbin.

putting on the interlayer paper in layer-wound coils and as supports for the bobbin cheeks in random winding. Slots or a series of holes must be cut in these pieces for the leads coming out of the coil.

A hole is drilled down the center of the block so that it can be mounted on the winder spindle. A little care in making the block will pay off later in ease of winding. The faces should be square and the hole should be truly down the center of the block so that the block doesn't wobble.

The bobbin itself can be made from heavy paper, card, fiber board, or plastic sheet of whatever thickness seems reasonable for the job on hand. For small-gauge wire, light materials would be used, and for heavy-gauge wire, stouter material is needed. Generally the consideration is mechanical rather than electrical, except at high voltages. If heavy paper is used, make several turns of it around the mandrel with a spot of adhesive on each layer. If card is used, cut a strip as shown in Fig. 14-3A. Score it lightly and bend it around the block. Then put adhesive in the overlap (Fig. 14-3B). Arrange the overlap so that it will lie on the face of the lamination and not inside the core window (Fig. 14-3C). Otherwise, its thickness will occupy valuable winding space inside the window.

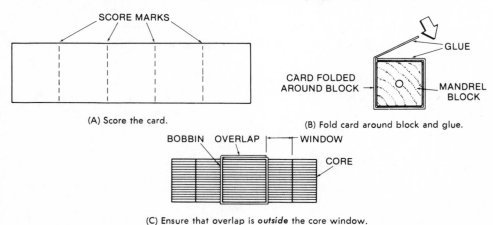

(A) Score the card.

(B) Fold card around block and glue.

(C) Ensure that overlap is *outside* the core window.

Fig. 14-3. Using a card or heavy paper on the bobbin.

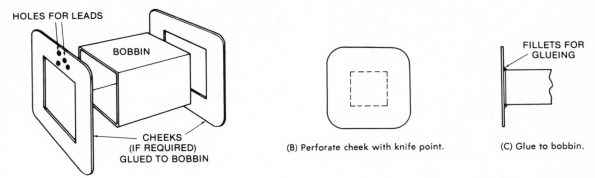

(A) Bobbin assembly with cheeks.

(B) Perforate cheek with knife point.

(C) Glue to bobbin.

Fig. 14-4. Putting card cheeks on bobbin.

If cheeks are required, as will be the case for random winding, use the shape shown in Fig. 14-4A. Holes or slots are required in the cheeks to bring the leads out of the windings; they should coincide with the slots or holes in the wooden cheeks.

If heavy card is used for the cheeks, the square hole needed in the center (Fig. 14-4A) is best made by marking the square on the card, then perforating the card along the sides of the square with the point of a knife (Fig. 14-4B). Don't slice the card, but rather stab it. This results in a rough edge on the reverse side of the card that forms an excellent fillet for gluing and adds strength to the assembly (Fig. 14-4C).

THE COIL WINDER

For many years, the hand-drill has been pressed into service by experimenters as a coil winder. But the writer's preference is the style shown in Fig. 14-5A in which a small hand-grinder is used. Here the stone grinder acts as a flywheel and gives a nice, smooth feel to the winding operation. A revolution counter attached to the end of the shaft is a luxury worth investing in if you intend to wind more than a few coils. The coupling between the grinder and the winder shaft is simply a 1-inch strip of tinplate cut from a can, folded over, and then soldered. A hole is then drilled through it to take the threaded shaft, after which it is spread apart to form the coupling as shown in Fig. 14-5B. A slot

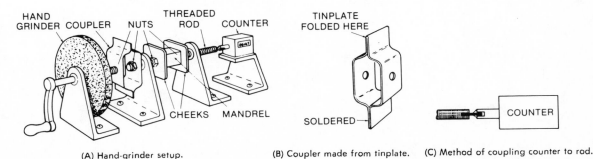

(A) Hand-grinder setup.

(B) Coupler made from tinplate.

(C) Method of coupling counter to rod.

Fig. 14-5. Hand-grinder as coil winder.

in the other end of the threaded shaft fits into a fin soldered to the revolution counter (Fig. 14-5C). No doubt readers will figure out more sophisticated devices for themselves, including, perhaps, power drive with foot control, but I used the grinder for many years with no problems at all.

WINDING THE COIL

The start of the winding must be firmly anchored. This is done by folding a piece of tape around the wire as in Fig. 14-6A and winding over the tape. The tape I prefer is the fabric type used by electricians, but the plastic type will do. When several turns have been wound on, cut the tape. Be sure to start the winding on the correct face of the transformer—that is to say, the face that lies outside the core window. The leads should come out on this face.

The turns are then laid on neatly side by side (assuming that it is to be a layer-wound coil). With a little practice, you can do it quite rapidly, especially when using the finer-gauge wires. Set the revolution counter to zero, or the number noted, before winding. When you reach the end of a layer, a precut strip of interleaving paper, the same width as the mandrel, is wound on with the first turn of the next layer, as shown in Fig. 14-6B. At the end of the next layer, the same procedure is followed with the next strip of paper and so on until the end of the final layer.

Several turns before the end of the final layer a loop of tape is laid down as shown in Fig. 14-6C. The last several turns are wound over the tape and then the wire is cut and the end taken through the loop. The loop is then pulled tight, thus anchoring the last turn.

If fine wire is being used, it is sometimes advisable to solder stranded flexible leads to it instead of bringing the wire itself out. In this case, a piece of tubular insulation is slipped over the joint as shown in Fig. 14-7A.

It is sometimes necessary to tap the winding and bring a lead out. There are several possible ways of doing this, as shown in Figs. 14-7B, 14-7C, and 14-7D. The method of Fig. 14-7D is particularly good when heavy wire is involved. Here the wire is wound on up to the point where the tap is required. A strip of thin copper foil is then soldered to the wire at the point where the tap is required. The foil and wire are laid on the coil so that the foil protrudes from the windings. A thin sheet of paper is laid on the foil and

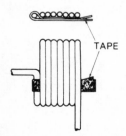

(A) Tape anchors first turn.

(B) Paper wound on with first turn of next layer.

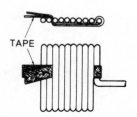

(C) Tape anchors last turn of the winding.

Fig. 14-6. Putting the coil on the transformer.

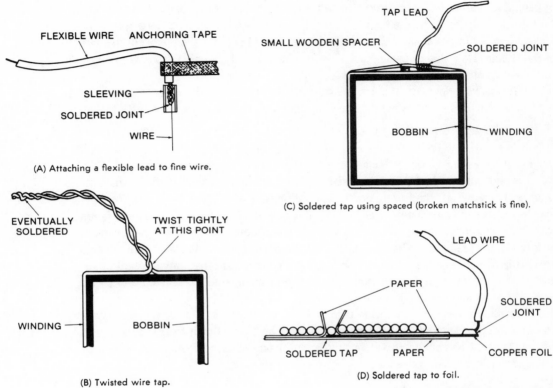

FLEXIBLE WIRE   ANCHORING TAPE

SLEEVING

SOLDERED JOINT

WIRE

(A) Attaching a flexible lead to fine wire.

EVENTUALLY
SOLDERED

TWIST TIGHTLY
AT THIS POINT

WINDING   BOBBIN

(B) Twisted wire tap.

TAP LEAD

SMALL WOODEN SPACER

SOLDERED JOINT

BOBBIN   WINDING

(C) Soldered tap using spaced (broken matchstick is fine).

LEAD WIRE

PAPER

SOLDERED
JOINT

SOLDERED TAP   PAPER   COPPER FOIL

(D) Soldered tap to foil.

Fig. 14-7. Bringing leads off the coil.

the winding continued on top of it, thereby anchoring it. When the winding is finished, a lead is soldered to the projecting foil. Of course, the foil must be heavy enough to carry the current.

As discussed elsewhere in the book, a screen is sometimes required between the primary and secondary windings. This can be simply a piece of copper foil taken for one turn around the winding (with appropriate insulation between the winding and the foil). Great care must be taken to prevent the overlap of the foil from touching and thus creating a shorted turn. One way to prevent this is to wind on a strip of paper with the foil (as shown in Fig. 6-14). The foil usually has to be grounded, so don't forget to solder a lead to it and bring the lead out of the windings.

Another effective and very simple method of making a screen is to wind on a layer of wire but bring only one end out of the coil. The other end is left unconnected inside. The wire used for this purpose can be quite thin because it carries no current, and thicker wire will not be more effective (Fig. 6-15).

Appropriate insulation is put on top of the primary, and again on top of the secondary (or secondaries). A layer of tape overall finishes the winding.

In random winding, which is faster, it is essential that the reel from which the wire is being taken should spin quite freely, especially if it is fine-gauge wire. Arrange this by mounting the wire reel on a rod clamped in a vice with a washer under the reel. The reel should be quite central on the rod so that it doesn't wobble. Before starting to wind, make a number of preliminary turns of the winder before anchoring the start of the wire to the bobbin. Hold a piece of sandpaper against the inside of the cheeks to remove burrs that could snag the wire.

When slowing down or stopping the winding, be especially careful not to permit the wire reel to overrun, or the wire is likely to end up on the floor in a very nasty tangle.

ASSEMBLY

Finally, the laminations are inserted into the coil. This is best done by laying the coil on its side on the bench and using both hands to control the insertion of the laminations. If no gap is required, then each kind of lamination, a T and U or E and I as the case may be, is inserted alternately at each end, taking care not to damage the coil insulation with the sharp edges of the metal. The final few laminations will be fairly tight and must be forced into the bobbin. When the bobbin is packed, the edges of the laminations should gently tap together—the operative word here is ''gently''—so that they are butting together with minimum gap. Then mount the clamping arrangement on the core together with brackets, and so on.

A terminal board with soldering tags on it, as in Fig. 14-8, is a convenient way of finishing the project. Failing this, the wires should be long enough to reach the points where they will be connected without having to make joints, and the leads should be protected with tubular insulation.

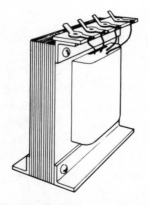

Fig. 14-8. Leads taken to tag board.

C Type Cores

C type cores are generally held together by means of a metal band around the two halves (Fig. 14-9). The banding is done with a special tool in much the same way that bands are put around crates in a packing department. However, the force with which the band is applied is important. The tool can be set to apply a predetermined amount of force.

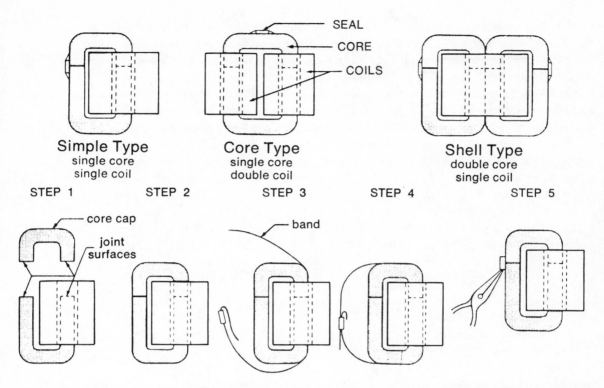

Core Cross-Section (D × E) —in²	Core Strip Width (in.)	Band Size (in.)	No. Bands Required	Seal Dimension (in.)	Banding Force (lb.)
.188 or less	Any	3/16 x .006	1	3/16 x ¼	37.5
.188 to .375	⅜ or larger	⅜ x .006	1	⅜ x ⅜	75
.375 to .75	⅜ to 1½	⅜ x .012	1	⅜ x ⅜	150
	1⅝ or larger	⅜ x .006	2	⅜ x ⅜	75
.75 to 1.5	½ to 1⅛	⅜ x .012	1	⅜ x ⅜	150
	1¼ or larger	⅜ x .012	2	⅜ x ⅜	150
1.5 to 3.0	¾ or larger	¾ x .023	1	⅞ x 1⅞	600
3.0 to 4.25	¾ or larger	¾ x .035	1	⅞ x 1⅞	900
4.25 to 6.0	2 or larger	¾ x .023	2	⅞ x 1⅞	600
6.0 to 9.0	3¼ or larger	¾ x .023	3	⅞ x 1⅞	600
9.0 to 13.5	3¼ or larger	¾ x .035	3	⅞ x 1⅞	900

NOTE: Cut Cores should be banded with the proper banding force to insure good seating of the mating surfaces.

Courtesy Magnetic Metals Corp.

Fig. 14-9. Banding procedure for C cores.

The force required is related to the dimensions of the core, and the manufacturer's recommendation, if available, should be followed. The reason for this is that although the mating surfaces of the core are precisely finished to obtain the best possible fit (minimum equivalent gap), exactly the right amount of force is needed to guarantee it. The faces might tilt slightly with respect to each other if too little or too much force is used, thus making the effective gap larger than it should be. And we now know how devastating the effect is of even a small gap on the performance of a core.

Should the proper banding tool not be available, however, it is possible to use other methods. For example, I have used jubilee clips (those metal bands with the tightening screw for attaching hoses in automobile engines). These devices are available in many different sizes, and if one is not enough in a given case, use two. But how to measure the force? Don't bother. Measure instead the *effect* of the force on the electrical performance using the method given for adjusting gaps in Chapter 16. When the core has been adjusted for the best performance, solder the band with a heavy soldering iron, as indicated in Fig. 14-10. Apply the minimum of heat and avoid heating up the entire core, which is not good for maintaining the characteristics of the material.

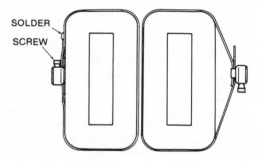

SOLDER

SCREW

Fig. 14-10. Banding with hose clips.

WINDING TOROIDS BY HAND

As stated earlier, anyone who proposes to wind large numbers of turns onto a toroid is subjecting himself to cruel and unusual punishment. But, few or many turns, there is only one way to wind them.

First calculate the amount of wire needed for the winding and cut this amount from the spool. Rewind it onto a short, small-diameter rod or similar object (even a small piece of wood or a piece of paper folded a number of times into a short spill). Then carefully thread it through the center hole and around the core. As the winding goes on, the center hole grows smaller, but the coil or hank of wire grows smaller too. If only a short piece of wire is involved, it can be threaded through the hole point first and around the core.

If the core has no plastic covering, be sure to wind insulating tape around it to protect the wire.

SOLDERING

Believe it or not, there are people in this world who cannot solder—people who try and try but never succeed. Why this is so is difficult to say, because soldering is indeed one of the simplest of operations, especially when the metal to be soldered is copper. If you are one of these unfortunates, won't you try just once again?

The rules are simple. The surfaces to be soldered must be scraped absolutely clean, free from enamel, dirt, and oxidation—metallic bright. The soldering iron must have enough heating power and mass to prevent it from cooling too much when it contacts the work, but not so hot as to burn the solder or overheat the job. About 100 watts is usually sufficient for most wire-to-wire soldering unless very heavy gauges are involved, when 150 might be necessary. But even 100 watts is much too much for printed-circuit boards and the wires of solid-state devices; these components are easily destroyed by heat, and 25 watts is usually enough.

The soldering-iron bit or tip must also be clean, free from dirt and oxidation, and metallic bright. How to clean the soldering iron depends on the type. If it has a plain copper bit that has pitted badly, the bit must be filed smooth and clean (Fig. 14-11) and *tinned immediately*—not in 30 seconds, or in 10 seconds, but now, this very instant, even as the bit is being cleaned it must be tinned. Tinning is done by applying electrical-grade rosin-core solder to the clean, hot bit. Only electrical-grade solder, please. The solder should flow over the bit like water and leave it silver bright all over. The bit should be hot enough that the solder flows instantly without passing through a discernible plastic, putty-like state; if the solder does become plastic-like, the bit is not hot enough.

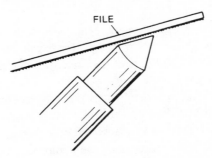

FILE

Fig. 14-11. File the soldering-iron tip clean and bright (this does not apply to the corrosionproof type).

If the bit is one of the stainless types (which are well worth the money), filing is not necessary and should be avoided. All you need to do is wipe the hot bit over a damp sponge. Again, the bit should be tinned immediately.

Suppose the job on hand is to solder two wires together. The iron is hot and freshly tinned. The wires have been cleaned. Now tin each wire separately. To do this, support the part to be tinned lightly on the iron bit and flood solder over the wire. Note the use of the word "flood"; this is the key—the solder should run like water. If you have to plaster the solder on, then there is not enough heat. The wire should now be silver bright.

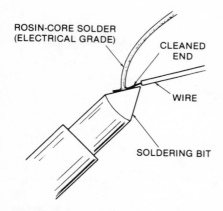

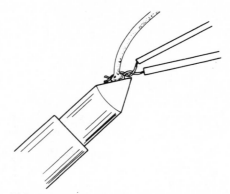

(A) Tinning the wire: Apply clean, bright end of wire and rosin-core solder to hot bit.

(B) Twist bright tinned ends together. Apply hot bit and flood rosin-core solder over the bit and wire.

(C) Remove the hot bit and solder. Hold joint steady until it cools.

Fig. 14-12. Correct soldering procedure.

Twist the wires together lightly, support the joint on the hot bit, and flood solder over it. Take the iron away and don't move the joint until you see a sudden but sometimes faint change in the appearance of the joint that indicates that the solder has solidified. That's it.

For soldering large masses of metal, large irons are needed or even a blowtorch, but the principles are the same. The tip should be clean, bright, and hot; solder should flow like water, flooding over the work piece; and when the heat is withdrawn and you see the haze of solidification the joint should be immobile.

Drain any excess solder from the joint by simply holding the hot iron against the workpiece from below so that the solder flows back onto the bit. The process is illustrated in Fig. 14-12.

SOURCES OF NEW MATERIAL

If you live near a large town, it is likely that you live close to sources of at least some kinds of material that you need—if only you knew where to look. This is not such a big problem. Suppose for example that you want a core material—one of the hard-to-find kinds, which means almost any kind. Laminations, for instance, of common silicon steel are rarely stocked by run-of-the-mill electronic or electrical parts retailers. So how do you go about it?

First, the problem is a lot easier if you know what to call the stuff you are looking for. You and I know that they are called transformer laminations or cores, but in fact the words "transformer" and "core" and "lamination" do not appear very often in advertisements in trade directories. So what are those hunks of metal called? What about Electrical Steel, Magnetic Steel, Electrical Alloys, Magnetic Alloys, Silicon Steel, Cobalt Alloys, Ferrites, Ferromagnetic Metals, Ferromagnetic Alloys? All of these names apply to or are related to the material you are looking for, and these are the names that appear most often in guides and directories.

After a few minutes of chasing through the yellow pages of the phone directory, the chances are that you will spot one, two, or all of these names. If you do and the place is within easy reach, go there. Don't phone. As you stand outside the 10 billion dollar complex sprawling over 50 acres you might feel that your search for $1.50 worth of material is not likely to meet with much success inside. And you may be right. Nevertheless, you are on the track.

Don't be intimidated by the rows of blank impersonal windows glinting in the afternoon sun. Behind every one of them is a human being just like yourself, anxious to help—just as you would help a stranger who asks for directions. Walk right in and ask to speak to the sales or marketing manager. You may be surprised at the cordial reception you receive.

This person is in the business. If he or she can't give you what you want, he should direct you to someone who can. And if he can't do that, ask for useful literature, catalogs, or technical writeups. Perhaps he will give you names and telephone numbers to follow up on. If there are local agents for his products, he will certainly give you their names. Hence, at best he will have what you want, at second best he will tell you where to get it, and at worst you spent money on gas in exchange for the valuable experience of trying.

Suppose you had no luck! Then sit by the telephone and call electronics manufacturing concerns around the area. The larger ones have their own transformer departments; ask them where they get their materials. Speak to purchasing agents. Keep asking and within a very short time you will know pretty well all there is to know about the availability of the material in your area and elsewhere too.

Where else can you get leads to pursue? Try the local reference library and ask to see the *Thomas Register of American Manufacturers*, which includes the *Thomas Register Catalog File*. This multivolume publication lists manufactured products first by product category, then by state and city, and then by manufacturer. For example, under "Laminations: Electric Steel" we find under, say, New York State the cities in New York State in which laminations are manufactured and for each city a list of the manufacturers operating there, together with addresses and telephone numbers. Also included in the catalog are all the other states and cities and manufacturers from north to south and from coast to coast. Further, in the catalog file section are pictures and descriptions of the manufacturers' products. For Canada, *Frazers Canadian Trade Directory* performs a similar service.

Any product made in the U.S. or Canada can be tracked down through these volumes, including, for instance, magnet wire, insulating materials, electronic, and nonelectronic materials of all kinds. And if there is something you cannot find, you are invited by the publishers to write them for information at the following addresses:

> Thomas Register Information Service
> Mr. T. J. Veidt
> Thomas Register
> One Penn Plaza
> New York, NY 10001

and

> Frazers Canadian Trade Directory
> 481 University Avenue
> Toronto, Ontario
> Canada

Note, however, that these manufacturers are mainly trade suppliers and not likely to fill casual small orders from private individuals. But they can direct you to agents and supply clues to the locations of small-quantity suppliers in your area, if there are any. They will also often supply valuable information of a technical sort. Magnetic Metals Corporation, for instance, has indicated that it will furnish catalogs requested by private individuals. It has excellent publications available on laminations, cut cores, and tape-wound cores (toroids).

Be sure to check out all the alternative names for magnetic materials listed earlier. Whomever you speak to, always put the question: Do they know of any place that might have the material?

The same procedure is used for wire. But this is usually much easier. The big component retailers who cater to the hobbyist carry limited numbers of gauges in stock as a rule. The methods outlined for magnetic steels will soon turn up leads to local vendors or manufacturers in your area, if there are any. In addition, check with electrical rewind shops; they will probably help you with other materials as well—insulating materials, varnish, etc. Often they will help you out with small quantities.

FIFTEEN

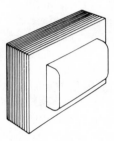

Transformers
Off the Shelf

It often makes more sense to buy a transformer or to use one that happens to be on hand, even if it does not exactly fill the bill, than to design and make one from scratch. A mass-produced or surplus component might cost you less than you can make it for, especially if you put a dollar value on your time, and the saving might be worth some sacrifice in size or performance.

Again, in the hookup stage of a project, it is frequently better to improvise with readily available off-the-shelf units until the circuit design gels. When the load requirements and so on have been firmly established, the transformer can then be precisely designed or purchased for the job. Again, it is often possible with little effort to modify a ready-made transformer to exactly meet a particular need, or to interconnect its windings in such a way as to achieve results which were not in the original specification. Sometimes it is possible to combine two or even three inexpensive units to do the job of one expensive component.

This chapter, then, deals with other people's transformers and how to use them, sometimes in less orthodox fashions than they were designed for.

POWER TRANSFORMERS

Vendor's Specifications

More often than not, power transformers are retailed on the basis of current and voltage ratings (volt-ampere or wattage ratings) only. No other information is available.

This is particularly true for units purchased in those chain stores that cater to the casual experimenter. Such niceties as efficiency, temperature rise, and regulation figures are not guaranteed and, indeed, are not even known by the retailer.

This, in fact, is one reason why the store can sell at the low price it does. Another reason is that in relation to the stated volt-ampere rating, the transformer is likely to be rather skimpily designed and could run rather hot at full load. And, to be perfectly fair, another reason is that factors like regulation, efficiency, and temperature rise depend so much on the kind of circuitry to which the transformer is connected that guarantees of performance are rather complicated. Nevertheless, these transformers are good values, and, if due attention is paid to their probable performance limitations, they can be used to good effect.

In contrast, the more "pricey" lines of transformers are often supplied with data that gives conservative guaranteed performance limits for specific conditions. Or, if the data is not supplied with the transformer, it is available on request or in catalogs. The finish of the units is usually very good and can include such refinements as protective shrouds, vacuum impregnation against moisture, and lug or screw terminations instead of simply the tail ends of the windings as is usual in the inexpensive types. Whether all this is worth paying a good deal of extra money for is something the experimenter must decide for himself or herself.

In short, what you get in the way of data, guaranteed performance, and finish depends on what you pay. If you know exactly what you want and if it is available, then the right price can give you a guaranteed performance. If your budget is tight or if your requirements are nonstandard, it is still possible to get what you want at comparatively small cost in the popular stores or from surplus dealers or even scrapheaps, by applying some of the design information already discussed.

ESTIMATING PERFORMANCE

First, a word on the staffing of stores. While some stores might have at least one technically oriented clerk or a knowledgeable manager, this is not always, or even usually, the case. Most often what the staff knows about the product can be written rather easily on the head, if not the point, of a small pin, so their statements should be taken with a sizable pinch of salt.

Fortunately, it is not difficult to make a very rough estimate of some important performance criteria, and usually the clerk can be persuaded to help out. What you need is a reading of the no-load secondary voltage. Ask him or her to connect the primary winding to a supply source (or allow you to do so) and make the measurement. At the same time, check the main supply voltage across the primary so that you get a true voltage ratio.

A comparison of the no-load voltage V_N with the rated full-load voltage, V_L leads to the following approximations:

$$V_d = V_N - V_L$$

$$R_T = V_d/I_o$$

$$W_L = I_o^2 \times R_T$$

where
  V_d is the transformer voltage drop in volts,
  R_T is the transformer resistance in ohms,
  W_L is the copper loss in watts,
  V_L is the rated output volts at full load,
  I_o is the rated full-load current in amperes.

It is even possible to get a rough idea of the order of temperature rise T by measuring the area C_S of the cooling surface of the coil and then calculating

$$T = \frac{20HW_L}{C_S}$$

where H is the "height" of the transformer winding as given in Chapter 6. Add, say, 20 percent to the calculated figure to get the hot-spot temperature rise. This method, of course, is very rough but it gives a "feel" for the temperature rise. An easy way to estimate the cooling surface C_S is to mark the length of the coil perimeter on a strip of paper as shown in Fig. 15-1. Measure between the pencil marks and then multiply this by 2 and again by the width B of the coil to obtain C_S.

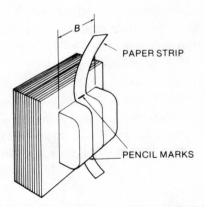

PAPER STRIP

PENCIL MARKS

Fig. 15-1. Paper strip used to measure periphery of the cooling surface.

If the no-load voltage cannot be measured in the store, then discuss the possibility of purchasing the transformer on a return basis if it should prove unsuitable. This is not an unreasonable request and is not usually a problem so long as the unit is returned within a few days, as agreed with the store. In this case, you have the opportunity of applying some of the tests discussed in Chapter 16, including the very important one of temperature rise at the load you intend to use with it.

As mentioned previously, these inexpensive transformers tend to run hot at their rated loads. It is not a bad rule, therefore, to purchase a transformer with a rating somewhat higher than is actually needed. The resulting cooler running will contribute to longer life. However, it probably will also give a higher output voltage than specified, because the voltage drop is less at the lower load.

All of this hassle could cause you to be branded an eccentric, possibly with some justification. But it will also earn you a reputation for knowing what you want and insisting on getting it, and this means better service from the store the next time you buy something.

ADDING AND SUBTRACTING TURNS

When a transformer with poor regulation is run at a much lower volt-ampere than its maximum rating, there is a tendency for the output voltage to increase substantially. If this happens, it is not usually a great problem to remove a few turns to compensate. This can sometimes be done without removing the core.

An alternative is to *add* a few turns and connect them as a "choke" winding as in Fig. 15-2A (assuming there is enough space to spare in the window). Here the extra turns are connected in phase opposing so that the voltage developed across the extra turns subtracts from the output voltage. This also can sometimes be done without removing the core.

By the same token, if the voltage is low, the extra turns connected in phase adding will increase the voltage. This question of phase or polarity of windings is dealt with in more detail shortly. The additive mode is shown in Fig. 15-2B. The number of turns to be added or subtracted is, of course, determined from the turns-per-volt figure for the transformer, obtained as in Chapter 16.

THE OLD AND THE NEW

Vacuum-tube heater and filament transformers putting out 6.3 volts and multiples thereof (12.6 and 25.2 volts) are still being made and are sold by many outlets, including popular chain retail stores. Also, they are frequently found in old equipment, as would

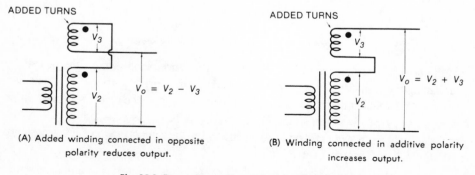

(A) Added winding connected in opposite polarity reduces output.

(B) Winding connected in additive polarity increases output.

Fig. 15-2. Decreasing and increasing transformer output voltage.

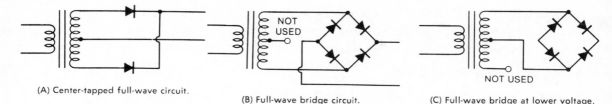

(A) Center-tapped full-wave circuit.

(B) Full-wave bridge circuit.

(C) Full-wave bridge at lower voltage.

Fig. 15-3. Use of heater and filament transformers.

be expected. It just so happens that these transformers, legacies from a vanishing era, are very useful for dc power units in many solid-state projects. They are available in current ratings from about 0.3 ampere to 25.0 amperes or so. The usefulness of many models is further extended by the fact that they are center tapped. When these transformers are used in rectifying circuits, this feature permits a choice of arrangements as illustrated in Fig. 15-3. (Refer also to Chapter 9.)

Another useful transformer for solid-state dc power units, especially of the 5- and 12-volt regulated variety, is the 18-volt center-tapped type, also commonly available in various current ratings from about 2 amperes up.

Of course, all these can be used for their original purpose of supplying loads with raw ac. A word of warning, though: Don't try to increase the current capacity of the secondary winding by paralleling the two halves as in Fig. 15-4. This connection clearly gives a direct short across the entire winding.

Fig. 15-4. This connection results in a short across the total winding.

RATINGS WITH RECTIFIERS

Keep in mind that the volt-ampere rating for a transformer used with rectifiers must be greater than the dc output. For the ordinary full-wave circuit, an ac rating of at least 1.3 times the dc load is usually safe. For example, a dc load of 50 watts probably requires a transformer rated for $1.3 \times 50 = 65$ VA.

Observe that a center-tapped winding used with a full-wave circuit using the center tap provides only half of the unsmoothed dc voltage obtained from the same winding when it is used with a bridge rectifier (center tap unused). On the other hand, the center-tapped full-wave arrangement, as a rule, provides approximately twice the current of the bridge circuit for the same primary winding volt-ampere rating.

INTERCONNECTIONS

Transformers can be connected in groups of two or more to provide a variety of outputs, just as several windings on any one transformer can be interconnected to give different outputs. However, care must be taken to connect in the correct polarities (or phase). Other precautions are also necessary. Chapter 16 gives methods for determining winding polarities. Once these are known and the leads coded in some way you can proceed.

Series-Connected Secondaries

The secondary windings of a transformer can be connected in series to provide sum or difference output voltages, depending on the winding polarities. The dots in the diagrams identify the instantaneous relative polarities of the windings; in other words, at the instant a dot-end is positive, all other dot-ends are positive, and when a dot-end is negative, all other dot-ends are negative. For example, the arrangement of Fig. 15-5A gives an output of 25 volts by series connecting a 20-volt winding and a 5-volt winding in the proper polarity as shown. In Fig. 15-5B, a difference voltage is obtained because the windings are connected in opposing polarity.

A variety of voltages can be achieved by various polarity combinations, such as the one shown in Fig. 15-5C. Here the windings are connected to give 20 + 5 − 7 = 18 volts.

Parallel-Connected Secondaries

Secondary windings can be parallel connected in order to get increased current capacity *but only if the windings have identical voltage outputs* and are connected in the correct phase. In Fig. 15-6A, the result of the parallel connection will be disastrous because the higher-voltage winding would pump a massive current through the low-impedance low-voltage winding. This is perhaps more easily seen if the secondary windings are redrawn as in Fig. 15-6B, leaving out the primary and the load for clarity.

The circulating current is the effective voltage, which is 25 − 5 = 20 volts, divided by the sum of the winding resistance. If the resistances are, say 0.05 and 0.01 ohm, as they may well be in low-voltage windings, the current is $I = V/R = 20/0.06 = 333$ amperes. Even if the voltages are not that far apart, as in Fig. 15-6C, the current is still substantial. Here we have 0.2/0.02 = 10 amperes. But if the voltages are identical, the current is zero. This current is a circulating current that is over and above the load current and, of course, flows whether or not the load is connected.

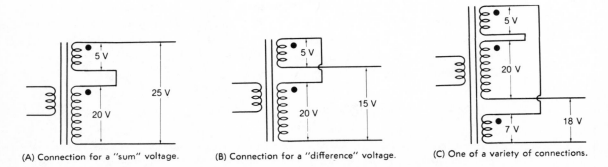

(A) Connection for a "sum" voltage.   (B) Connection for a "difference" voltage.   (C) One of a variety of connections.

Fig. 15-5. Secondary windings connected in series.

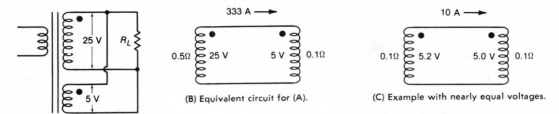

(A) Paralleled windings.

(B) Equivalent circuit for (A).

(C) Example with nearly equal voltages.

Fig. 15-6. Secondary windings connected in parallel.

The effect of connecting the windings in opposite polarity was discussed earlier (Fig. 15-4).

The secondary windings of separate transformers can also be connected in parallel using the same rules. Assuming that the secondaries have identical voltages but do not belong to identical transformers, the greater share of the load will be borne by the transformer that has the best regulation. Therefore, this practice should be used with caution.

Primaries in Series—Two or More Transformers

Primary windings in parallel are, of course, the normal method of connection for more than one transformer, but they can also be connected in series. For example, two 110-volt transformers may be used on a 220-volt supply as in Fig. 15-7.

If the primaries are not identical, however, the voltage will be higher across the one with the higher impedance. This results in the secondary voltages that are different than expected. If the impedances are widely different, one transformer core might be driven into saturation with a consequent increase in iron loss.

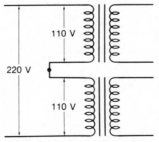

Fig. 15-7. Voltage will not divide equally unless the transformers are identical, even though they are both designed for 110 V.

Booster and Reducer Connections

A brief study of the so-called booster circuit in Fig. 15-8 reveals that it is identical with the autotransformer arrangement discussed in Chapter 3. The only difference, in fact, is that here a standard two-winding transformer has been converted to the

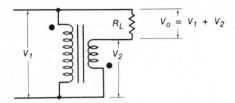

Fig. 15-8. Standard transformer connected as an autotransformer
or booster.

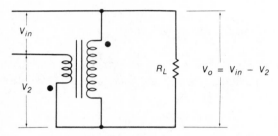

Fig. 15-9. Reducer circuit.

autotransformer principle by the method of connection rather than having been designed specifically for the purpose.

The current-voltage-turns relationship is the same as for the autotransformer and need not be repeated. The required transformer power capacity is arrived at in the same fashion. Briefly, it is determined by $V \times I$. If it is required to boost a 110-volt input to 120 volts to supply a load that draws 5.0 amperes at 120 volts, the transformer must have a secondary winding output of 10 volts at 5 amperes, or 50 VA.

By the same token, a reduced voltage can be obtained in the reducer circuit of Fig. 15-9.

Choke Circuit

Reduced output can also be obtained by using the arrangement of Fig. 15-10. The difference between this circuit and the booster circuit of Fig. 15-10, which at first sight seems the same, is that the windings are connected in opposing polarities.

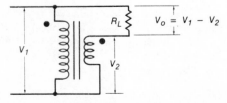

Fig. 15-10. Choke circuit.

IMPEDANCE TRANSFORMERS

Impedance transformers are most often encountered in low-frequency amplifiers and are usually broadband components designed to provide a level response over a relatively wide range of frequencies. Some types, such as the audio output transformer, must in addition handle power. In most cases, the important parameters, as discussed in Chapter 12, are primary inductance, primary and secondary leakage inductances, and the turns ratio.

Run-of-the-mill impedance transformers, however, are generally retailed in terms of the secondary and primary impedances and the power capability (if this is a factor). A small output transformer might be rated typically at 8.0 ohms secondary, 500 ohms center-tapped primary, and 0.2 watts output. Here there is no mention of inductance or leakage inductance, and there is no mention of performance in terms of response or distortion.

In short, if excellent response characteristics are important to your project, it is essential that you obtain much more information about the transformer. Of course, for the more expensive types, this information is likely to be available, but if the store can't give it to you or it is not inside the box or in the company's catalog, you might have to go directly to the company. If a specific response range is not the prime factor, however, then it can usually be assumed that the product on sale is adequate. If in the end the transformer doesn't seem to work, *then* you can complain bitterly.

In lines of standard transformers, it is not possible to provide models with impedance and output specifications that exactly suit every possibility. It is generally most convenient, therefore, to select the nearest power capability, the exact secondary impedance (if available, which it usually is in standard practice), and the closest primary impedance. The primary winding circuit can then be adjusted to the transformer.

Any Transformer Will Work as Anything

Before leaving this subject let us recall a statement made in an earlier chapter. Any kind of transformer will work as an impedance transformer, provided the ratio is right for the job on hand. This can be useful in an emergency—on Sunday when the shops are closed, or when the output transformer on the public-address system gives out in the middle of an important rally, or whatever. But you must be careful. A serious mismatch can damage the amplifier, and not enough power capability in the transformer can cause roasting of the transformer.

By the same token, an impedance transformer will transform voltage and power, care being taken not to overload it, of course. At this very moment, I have an output transformer from an old TV set that provides 36 volts from 110 volts on a permanent basis in a piece of equipment.

Interconnected Windings

See Chapter 12 for examples.

OFFBEAT SOURCES OF TRANSFORMERS

At the beginning of this chapter, I referred to the popular electronics retail store as a source of power transformers. These stores usually also carry a limited (very) stock of impedance transformers, usually of the audio output variety. But when the possibilities of this type of store have been researched and found wanting, where does one look?

Electrical Wholesalers

Take instrument transformers, for instance, such as current and voltage types. They don't grow on trees, for sure, but certain standard lines are available through wholesale electrical and some electronic parts distributors. The ratings might not be too useful for other than their intended purpose, which is for service in power systems, but then an experimenter could have just such a requirement.

The standard secondary winding rating for current transformers is 5 amperes, and the primary winding ratings range into hundreds of amperes. Voltage transformers, on the other hand, have standard secondary winding ratings of 120 volts with primary winding ratings ranging up to thousands of volts.

The electrical wholesaler also stocks other types that might have occasional interest for the experimenter. Small power transformers with ratings ranging from 75 to 300 VA and secondary voltages of 6, 32, 64, and 115 volts are available. Standard bell transformers with a single 10-volt secondary or three secondaries of 6, 12, and 18 volts are obtainable. These are low-power units, but they can often prove useful as is or modified.

Then if you want to be really adventurous and possibly find an early grave in a careless moment, transformers for gas-discharge lamps can be obtained with secondary winding ratings up to 15,000 volts at 120 mA.

Manufacturers' Repair and Parts Depot

Very useful sources of small transformers of all kinds are the repair and parts depots of large electronic equipment manufacturers. Choose a name, look in the telephone directory for the address, and there you are with a mother lode of electronic riches, including transformers. Replacement components for TV sets, radios, hi-fi equipment, and a host of domestic appliances are available here. Although these depots usually supply the trade market, your money is as good as anyone else's—more welcome, in fact, because you will pay the retail price, whereas their regular customers pay a smaller, wholesale price.

COLOR CODING

A system of color-coded leads exists for transformers, although not all manufacturers use it or have used it in the past. Nevertheless, it can be useful to know, especially when dealing with salvaged transformers. These color-coded transformers are generally

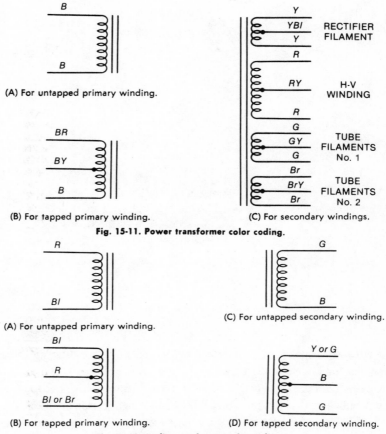

(A) For untapped primary winding.

(B) For tapped primary winding.

(C) For secondary windings.

Fig. 15-11. Power transformer color coding.

(A) For untapped primary winding.

(C) For untapped secondary winding.

(B) For tapped primary winding.

(D) For tapped secondary winding.

Fig. 15-12. Audio transformer color coding.

power units designed to feed vacuum-tube rectifier circuits or audio transformers—mostly output types for speakers but sometimes interstage-coupling or input transformers.

Figures 15-11A and 15-11B show two color-coded arrangements for the *primary* winding of power transformers. Figure 15-11A is for an untapped winding and Fig. 15-11B is for a tapped one. Figure 15-11C shows a set of typical secondary windings.

Letters and combinations of letters denote colors. If two capital letters are combined, the first denotes the background color and the second the tracer, or stripe. The color symbols are as follows:

B	black	*R*	red
Bl	blue	*Y*	yellow
Br	brown	*G*	green

Figure 15-12 shows arrangements for audio transformers. In the diagrams for both power and audio transformers, the secondary winding is omitted when illustrating the primary, and the primary is omitted when illustrating the secondary. This is done to avoid repetition and for clarity.

SIXTEEN

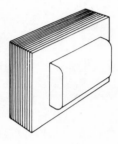

Tests and Measurements

If not handled with respect and care, transformers can deliver shattering shocks—even those designed for low-voltage outputs.

Many experimenters, including those who should know better, carry out their tests using what is euphemistically called a bench hookup—a nightmarish tangle of wires loosely knitted into the form of a rat's nest on top of the bench. The input leads, emerging at random points on the periphery of this mess, trail to a distant power outlet where the bare ends of the wire are jammed with broken matchsticks into the socket holes. The output leads don't emerge at all. They and the load to which they are attached with bare twist joints are buried under it all. The object of this hookup seems to be to instil a little danger into what would otherwise be a moderately safe occupation.

WORDS OF CAUTION

There are all sorts of hidden hazards waiting to trap the unwary. Consider the current transformer. This component might have 1.0 volt or so across its secondary winding on load. But disconnect the load with the primary winding still energized and the voltage can increase suddenly to—would you believe it?—hundreds of volts. This is not good for the transformer, and it is even worse for you. The theory of this phenomenon is given elsewhere in the book, but for the moment take my word. The secondary winding of a low-voltage power transformer might not normally give you a shock, but if it is joined to the primary winding in autotransformer style (or due to a fault), it can give a severe shock.

Remember, too, that it is not always the shock that does the most damage but rather the violent, involuntary spasm that sends your arm or body slamming into something else.

In short, be careful. Keep the test layout clean and uncluttered. You should be able to follow every wire at a glance. Connections should be properly made and insulated. Screw-type connectors, such as those electricians use, are very handy for this. And, above all, keep children away from work in progress.

Excessive Current

Most people tend to think of voltage as the danger, but current is also a real danger. A short circuit across the output of a transformer can present many amperes in the secondary, possibly 100 or more in a power transformer. But this current will not necessarily, perhaps even rarely, blow the fuse protecting the wall socket circuit. Consider a short in the secondary winding of a 117- to 6-volt transformer. If as much as 250 amperes flow in the secondary winding, the current in the primary winding is still only 12.5 amperes, which is not enough to blow a standard 15-ampere fuse.

The primary winding should therefore have its own fuse, based on the normal current in the primary. This is easily determined from the output volt-amperes. For example, if the transformer secondary winding is to deliver 28 volts at 4 amperes from a 117-volt supply, the primary winding current will be around 1.2 ampere (allowing 80 percent efficiency). A 1.5- or 2.0-ampere fuse should be fine in the primary winding—but no higher. If this is not done, the transformer will BURN, possibly taking the house and all you possess with it.

Tests

Outside of commercial practice, the usual test applied by an experimenter is simply to connect the transformer to the circuit for which it is designed and check to see that everything performs according to plan. If smoke doesn't curl out of the transformer after a while, if there is no "hot" smell, or if the transformer feels fairly cool to touch, the temperature rise is usually considered to be satisfactory. And this is not a bad test if the design is initially sound and care is taken in the construction.

Many experimenters, however, might want to apply more searching tests, if only to check to their own satisfaction how closely they have succeeded in meeting the original specifications. Also, in the event that the circuit doesn't perform satisfactorily, suspicion might fall on the transformer; then it becomes necessary to carry out tests in troubleshooting the problem.

OUTPUT VOLTAGE AND REGULATION

The most important characteristic of regulation is expressed as follows:

$$\% \text{ Regulation} = \frac{\text{No-load voltage} - \text{Full-load voltage}}{\text{Full-load voltage}} \, 100$$

or, simply as a decimal,

$$\text{Regulation} = \frac{\text{No-load voltage} - \text{Full-load voltage}}{\text{Full-load voltage}}$$

While doing this test, check the voltages.

The circuit is shown in Fig. 16-1. Resistance R_L is chosen to pass the rated current at the rated voltage. However, a problem might be encountered here. If the rated current is, say, 5 amperes and the voltage is 8 volts, the resistance would have to be $R_L = V/I = 8/5 = 1.6$ ohms; the power dissipated in the resistor will be $8 \times 5 = 40$ watts, and 1.6-ohm resistors to dissipate 40 watts are neither cheap nor plentiful. The best bet is therefore to connect the transformer into the circuit for which it is designed. The circuit is then R_L in the diagram. A switch S, temporarily connected as shown, permits the on-load and off-load voltages to be read. This in fact has a large advantage over the artificial load in that the figure obtained on test is what you get in practice.

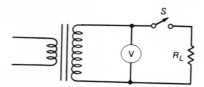

Fig. 16-1. Measuring regulation.

In making this test, allow the transformer to reach its operating temperature on full load, because the temperature changes the resistance of the windings and therefore the voltage reading. The "cold" no-load voltage versus the "hot" full-load voltage is in itself a form of regulation that can be useful for some applications.

If the readings are not to your liking, check the resistance of the windings. Possibly a wrong-gauge of wire was used, causing a larger drop than was anticipated on full load. Note that the wire resistance will not appreciably affect the no-load voltage.

RESISTANCE READINGS AND TEMPERATURE RISE _____

Carrying out this test provides the opportunity to measure the temperature rise of the primary winding. As explained in the design process, increased temperature rise causes increased resistance. Thus, a measure of the increased resistance is also a measure of the temperature rise.

Using the circuit shown in Fig. 16-2, the bridge is switched on to the primary winding before power is applied, and the resistance is measured. Record this "cold" resistance along with the ambient temperature. (If the primary winding resistance is small, a direct connection instead of the switch might be advisable to avoid error due to the switch contacts.) The bridge is then switched out and the power switched to the primary.

The transformer is allowed to heat up to its stable temperature. This point is determined by checking the resistance from time to time. The stable point is when there is

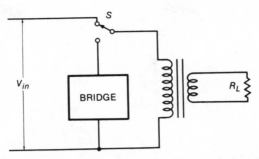

Fig. 16-2. Measuring "cold" and "hot" resistance of primary winding.

no perceptible change in resistance over a period of, say, 2 hours. When the component is stable, note the "hot" resistance. Calculate the percentage increase in resistance from

$$\%\text{Increase} = 100 \; (R_R - R_C)/R_C$$

where

R_R is the final resistance reading,
R_C is the starting resistance (ambient)

Turning now to the graph in Fig. 16-3, find the percentage on the horizontal axis. Move vertically from this point to intersect the curve that corresponds to the ambient temperature, T_C. From this intersection, move over to the vertical axis and read the temperature.

If scale A is selected on the horizontal axis, then read the answer on scale A of the vertical axis. The B scales are used together in the same way. Never combine an A scale and a B scale.

Example—Suppose R_C = 2 ohms and R_R = 2.5 ohms. Then the percentage increase in $100(R_R - R_C)/R_C$ = $100(2.5 - 2.0)/2.0$ = 25 percent. If the ambient temperature is 21°C (about 70°F), it is evident that a vertical line from the A scale will not intersect a T_C = 21°C line (just to the left of the +20 line), so choose the B scale. A line drawn from the 25-percent point on the B scale to intersect a line just to the left of T_C = 20°C, then across to the B vertical scale as shown by the dotted line gives a temperature rise reading of 63.75°C.

NOTE: The bridge must be a dc type. If it is ac, as many are, the reactance of the coil will mask the true dc reading. If a bridge is not available, then a low-reading ohmmeter can be usedbut it will be much less accurate.

RELATIVE POLARITIES

In Chapter 15, we discussed various winding arrangements for which it is necessary to know the relative polarities, or "phasing," of the windings.

Consider first the case of two secondary windings, as in Fig. 16-4A. Connect the windings in series and put a voltmeter across each secondary winding in turn, and then across both together as shown. If the voltmeter reads the sum of the voltages, the relative secondary polarities are as indicated by the dots, so that $V_o = V_2 + V_3$. If the reading

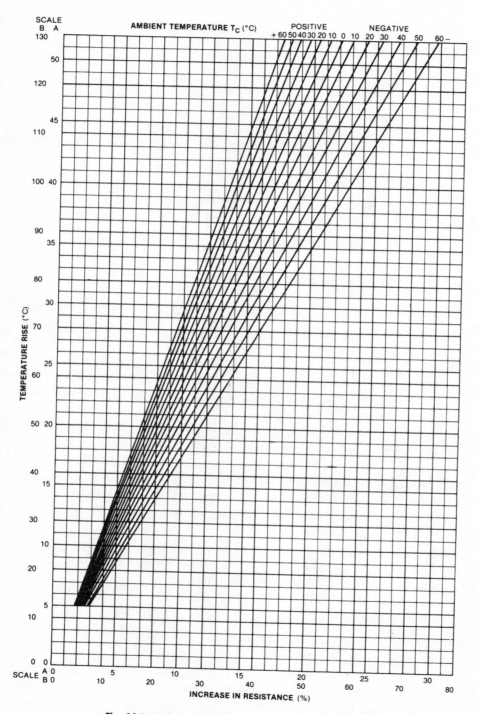

Fig. 16-3. Finding temperature rise by resistance measurement.

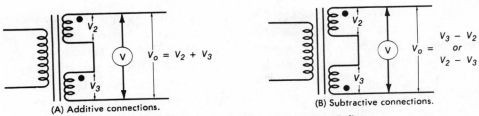

Fig. 16-4. Finding relative polarities of secondary windings.

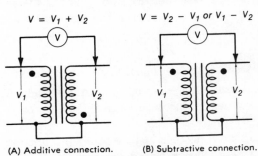

Fig. 16-5. Finding relative polarities of primary and secondary windings.

is the difference between the two, then the relative polarities are as shown in Fig. 16-4B and $V_o = V_3 - V_2$ or $V_o = V_2 - V_3$.

In the case of the polarities of the primary and secondary (needed for booster and choke circuits), make the connection shown in Fig. 16-5A. If the voltmeter reads the sum of voltages, then the polarities are as indicated. If the reading is the difference, the polarities are as shown in Fig. 16-5B.

Similarly, the polarities of two transformers can be established. If $V_o = V_2 + V_3$, then the winding polarities are as indicated by the dots in Fig. 16-6A. That is to say, they are additive. If the voltage is the difference, the polarities are as shown in Fig. 16-6B.

In these tests, it is essential that the input voltage be appropriate to the transformer under test. For power transformers, use the designed voltage or less. If the voltage is not known, apply a *very* low voltage from another transformer secondary winding and only for the shortest possible time.

Once the polarities have been determined, mark the leads with tape for future reference.

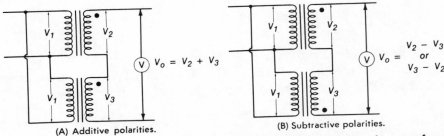

Fig. 16-6. Finding relative polarities of two transformers.

All of these techniques to determine relative polarity can be avoided by simply inspecting the windings. If the transformer is one you wound yourself, then you know the direction in which the windings were wound relative to each other. For instance, it is likely that the windings were wound one after the other without removing the coil from the mandrel; thus they are all in the same direction. In this case, the start of the primary winding has the same polarity as the end of the secondary and can be so marked. If in the unlikely event that the windings are reversed relative to each other, then the start of the primary has the same polarity as the start of the secondary.

Better still, if you remember to mark the windings when you first put them on you will be ahead of the game.

MEASURING INDUCTANCE

In commercial practice, bridge circuits of various kinds are used to measure inductance. Instruments that measure inductance (especially iron-core inductance), however, are not commonly part of the experimenter's equipment.

But there are alternative methods that are sufficiently accurate for most purposes. One such method, and possibly the simplest, is shown in Fig. 16-7. A high-input-impedance voltmeter is required, an accurately known resistor R_S, and a source of ac power, such as a transformer with an appropriate voltage output.

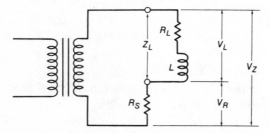

Fig. 16-7. Setup to determine inductance L.

For best accuracy, the resistor should be roughly of the same order as the impedance of the inductor. The impedance Z_L of an inductor is given by

$$Z_L = \sqrt{X_L{}^2 + R_L{}^2} \qquad (16\text{-}1)$$

where

X_L is the resistance of the inductor,
R_L is the resistance of the inductor.

The inductive reactance, in turn, is $X_L = 2\pi f L$.

Suppose that L is expected to be roughly 50 henrys and R_L measures roughly 450 ohms. Then using Equation 16-1 and making f, say, 60 hertz, we have

$$Z_L = \sqrt{(2 \times 3.14 \times 60 \times 50)^2 + 450^2}$$
$$= 18,846 \text{ ohms}$$

Clearly, in this case, R_L can be ignored and the resistance made on the same order as X_L. But we have a 12,000-ohm resistor (R_S) on hand, so we elect to use it. The ac voltage required from the source depends on what voltage is needed across L to simulate operational conditions.

The ac voltmeter goes across R_S, across the inductance L, and across both, to give the three readings V_R, V_L, and V_Z. The object now is to calculate the reactance X_L. This can be done by drawing an impedance triangle as in Fig. 16-8. Here all the lines are drawn to a specific scale starting with the known value R_S. Any scale can be used: the larger the better. For example, if R_S is 12,000 ohms, ¼ inch could represent 500 ohms, or 2000 ohms to the inch; then the line representing R_S would be 6 inches long. Mark off this distance on paper as in Fig. 16-8.

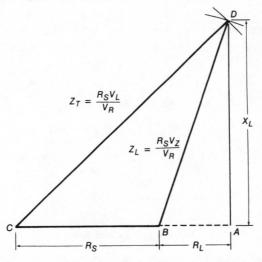

Fig. 16-8. Impedance triangle used to determine X_L and hence L.

Now calculate $R_S V_L / V_R$ ohms and convert the result to inches using the same scale as before. Draw an arc of this radius centered on one end of the R_S line. Similarly, calculate the value $R_S V_Z / V_R$ ohms; convert it to inches and draw another arc of this radius, with its center on the other end of the R_S line, to intersect the first arc.

From this intersection, drop a perpendicular to the base line. The length of this perpendicular represents the reactance X_L of the coil; measure it and convert it to ohms. Because $X_L = 2\pi f l$, the inductance can now be found from the formula $L = X_L/2\pi f = X_L/6.28f$.

Incidentally, the distance AB represents quite accurately the resistance of the inductor. The term $R_S V_L / V_R$ is the total impedance Z_T consisting of R_L, X_L, and R_S in series, while $R_S V_Z / V_R$ is the impedance of the inductor Z_L consisting of R_L and X_L only.

Inspection of Fig. 16-8 shows that if R_L can be accurately measured to start with, the problem can be solved without drawing the triangle. Use

$$X_L = \sqrt{Z_T^2 - (R_L + R_S)^2}$$

because $Z_T^2 = X_L^2 + (R_L + R_S)^2$.

Inductance Measurement With DC

This scheme can also be used if dc current is to flow in the windings. The dc can be introduced into the circuit as shown in Fig. 16-9. The dc current is read on meter A and adjusted by inserting the requisite amount of voltage. The practicality of this, however, is governed by the amount of dc needed and the series resistance of the circuit. The higher the current and resistance, the higher is the voltage needed.

A practical point when using dc is to be sure that it is blocked from the ac voltmeter by a capacitor in the instrument or one temporarily placed in series with one of the leads.

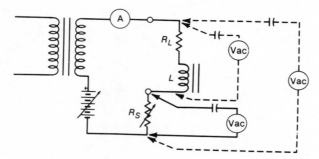

Fig. 16-9. Measuring **L** with dc current.

ADJUSTING THE GAP

If the inductor being measured is carrying dc and the gap has not yet been adjusted, this is the time to do it. With the voltmeter connected across R_S, make the gap larger and smaller by inserting different thicknesses of paper. When the voltage is lowest, the inductance is highest and the gap is set to its optimum value. Clamp the core and recheck.

The gap can also be adjusted using the circuit of Fig. 16-10. Here the rectifier circuit provides the dc current which is measured by meter A and adjusted by R_S. The voltmeter is connected first across the inductor to check the ac voltage. If it is too high, connect a smoothing capacitor C as shown in dotted line to reduce the ac content of the voltage. Trial and error determines the value of capacitor needed.

Check the dc current and adjust R_S. Recheck the voltage across the inductor and readjust C if necessary. When the currents and voltages are right, put the voltmeter across R_S and then adjust the gap in the choke for the *lowest* reading on the voltmeter. This represents the highest inductance and the optimum gap.

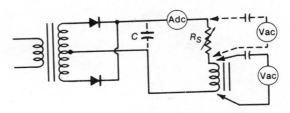

Fig. 16-10. Circuit for adjusting the gap.

LEAKAGE INDUCTANCE

Leakage inductance is an important factor in audio-frequency transformers, but also in other designs. It can be measured by short circuiting the secondary winding and measuring the primary inductance. It is not necessary to pass dc through the winding, even if dc is normally present, because leakage inductance is independent of saturation effects in the core. Also, leakage inductance is independent of frequency, so any convenient frequency will do; 60 hertz is fine. The values of leakage inductance are generally quite small. A fair degree of accuracy is obtainable on the inexpensive type of *RC* bridge popular with hobbyists, which has provision for measuring inductance against an external standard. Of course, a known inductor is needed to act as a standard against which to make the measurement.

RATIOS

All the ratios of a transformer—the turns ratio, impedance ratio, and current ratio—can be easily deduced from the voltage ratio by using the basic ratio equations discussed in Chapter 3. The voltage ratio is, of course, easy to obtain by measuring the voltage on each winding. But be sure that the applied voltage is no more than that for which the transformer is designed. Also, measure the voltages with the secondary windings open (that is to say, under no-load conditions).

If it is uncertain what the input voltage should be, apply a very low voltage.

NUMBERS OF TURNS

It is sometimes required to find the number of turns on a winding. This can be done if there is enough space to thread a few turns of thin wire around it (Fig. 16-11). Apply the voltage to the coil and measure the voltage across the temporary winding. Divide the number of turns by the voltage to get the turns-per-volt figure for the transformer. The number of turns on any winding is then the turns per volt times the volts across the winding.

As a matter of interest, this measurement can be done by putting turns around an outside limb of the core. In this case, however, the turns-per-volt figure is obtained by dividing the number of turns by twice the measured voltage. The reason for this is that each side limb of the core carries only half of the total flux, while the center limb carries all the flux.

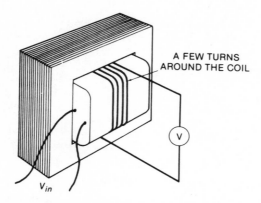

Fig. 16-11. Determining turns-per-volt figure.

TESTING THE INSULATION

In most experimenter transformers, the voltages are relatively low and insulation requirements are dictated more by mechanical than electrical considerations. For example, the wire tables recommend thicker interlayer insulation than would be necessary if only electrical stresses were considered. Also, the voltages expected to occur between the windings and core, and between winding and winding, are generally low, and in some types of design, such as amplifier input transformers, internal stresses might be on the order of fractions of a volt so that insulation is again based on mechnical stress.

Even in most run-of-the-mill power transformers, operating from 110- to 120-volt supplies, mechanical rather than electrical stress is likely to be the overriding consideration. Nevertheless, it is a wise precaution to take account not only of working conditions but also of deterioration of insulation with time and of circuit conditions that could result in unusual electrical stresses within the transformer and to plan wide safety margins.

In Chapter 6, rules were given for determining insulation requirements that in effect put the minimum stress to be catered to at 1000 volts in low-voltage transformers (1000 volts plus twice the working voltage). In specific cases, the designer might elect to design for less stress than this, but he or she should aim for a safe figure and then flash-test the finished transformer to ensure that no breakdown occurs.

If these precautions are taken, you need a test-voltage source that is capable of delivering from a low voltage up to, say, 1400 volts. This is a fair amount of voltage, but keep in mind that *peak* voltage is what counts in testing insulation. Therefore in rms terms, the 1400 volts peak is really only 1000 volts rms (1000 × 1.414 = 1400).

A flash tester that delivers from 70 to 1400 volts peak (50 to 1000 volts rms) is shown in Fig. 16-12.

In this circuit, appropriate tapping points are selected to give the required voltage. The output clips attach to the test points in the transformer—one to the core and one to a winding, or each to a separate winding. If the neon lamp lights when switch S is

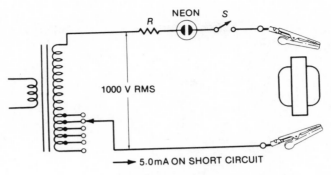

Fig. 16-12. Flash tester.

made, there is insulation breakdown between the test points. The purpose of resistor *R* is to limit the current to a maximum of about 5 mA in the event of breakdown.

If one wishes to design and make the flash-tester transformer, there is no special problem. Because this transformer delivers no power—only volts—it can be made quite small. But the following practical points should be noted. A large core area keeps the turns per volt small and reduces the winding chore. Fine-wire gauges smaller than about No. 42 are not easy to manage on a manual winder because of probable frequent breakage. Regulation, efficiency, and so on need not be considered since there is no load current. So the core window need not be filled with copper. In other words, although the transformer *can* be made small, there would be advantages in using a large core and a wire gauge that is not too small.

This being the case, one can consider the following alternative in which there is no need to make a transformer at all. You get what is needed almost for nothing and the entire tester can be made for a buck or two—if you don't mind it being relatively large. This inexpensive and practically effortless solution is to acquire two power transformers from junked TV sets today, which people will almost pay you to take away. These transformers will almost certainly have a center-tapped secondary winding on the order of 250—0—250 or even 350—0—350 volts. Two such transformers of, say, 250—0—250 volts with the secondaries connected in series give a selection of peak voltages as in Fig. 16-13A. A finer adjustment of voltage can be obtained by rewinding *half* of one transformer secondary with tapping points to give the same result as in Fig. 16-13B.

Remember that these voltages are lethal. Switch *S* off before adjusting the leads. If you wish, insulated test probes can be purchased and used instead of clips, but clips have the great advantage of providing hands-off testing, which might be safer—if you remember to use the switch.

CHECKING SALVAGED WIRE

If you propose to make use of salvaged wire—either for reasons of economy or procurement difficulties—watch out for damaged enamel, which can result in shorted turns and the consequent demise of your transformer. If you propose to use a lot of

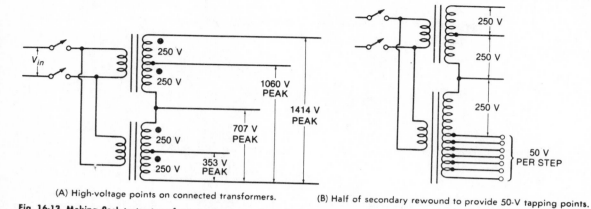

(A) High-voltage points on connected transformers.

(B) Half of secondary rewound to provide 50-V tapping points.

Fig. 16-13. Making flash-tester transformer using two salvaged tv power transformers.

salvaged wire, consider it worthwhile to construct a wire-checking device, such as that in Fig. 16-14.

This insulation checker consists of a driven metal spool that pulls the wire through a small metal bath containing mercury. The wire rides on a plastic or wooden pulley that keeps the wire submerged in the mercury. The end of the wire in the driven spool is bared and makes firm electrical contact with the spool. A wiper bears on the spool spindle and connects the wire to an external detector circuit consisting of a power source and an indicator lamp. The electrical impulse counter is a convenient accessory but is not absolutely essential.

If the wire insulation is damaged, even minutely, the circuit is completed through the mercury, the lamp lights, and the fault is registered on the counter. The number

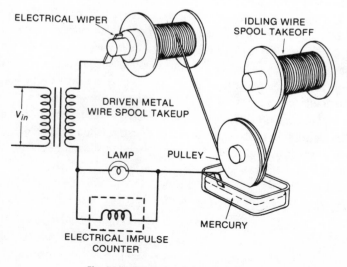

Fig. 16-14. Insulation defect detector.

of defects per given length of wire is thus counted. Generally speaking, it is not essential that the wire be perfect in order to use it. So long as the probability of two defects coming together in adjacent turns is not great, you might want to use it. Or if the defects are few, a touch of varnish on the spots should suffice.

DETECTING SHORTED TURNS

Because even new wire is possibly not perfect, it is not a bad idea to check for shorted turns anyway, before assembling the core to the coil. It is a particularly good idea if salvaged wire has been used. This check can be done using the circuit in Fig. 16-15.

Here a feedback oscillator works at a frequency of about 400 to 1000 hertz. (The precise frequency is not important.) The coils are wound on an open-ended iron core constructed from I-type laminations. The circuit is adjusted by resistor R so that it is just into oscillation. If necessary, the spacing between L_1 and L_2 can be adjusted. The coil to be tested is placed over the core. If there is a shorted turn in the coil, it will absorb energy from the circuit and stop the oscillations. This condition is detected in the headphones.

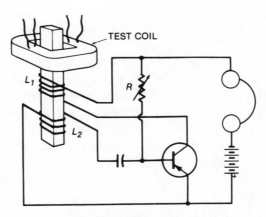

Fig. 16-15. Shorted-turn tester.

MEASURING METER RESISTANCE

As promised in Chapter 13, here is how to measure the resistance of a meter. Put together the circuit shown in Fig. 16-16. The chances are that everything needed is in your parts stock. The resistance values depend on the battery voltage and the meter range, as will be discussed shortly.

The method of measurement is as follows. With the switch S open as shown, R_1 is adjusted until the meter reads full scale. The switch is then closed and R_2 is adjusted until the meter reads precisely half-scale. The resistance of R_2 is then equal to the resistance of the meter. Resistance R_2 is then measured by the best means available to obtain the meter resistance.

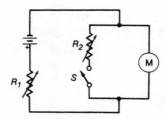

Fig. 16-16. Circuit for measuring meter resistance.

The accuracy of the final result depends, of course, on the accuracy of the method used to measure R_2, the linearity and accuracy of the meter being measured, and the care with which the measurement is done. Assuming that the measurement of R_2 can be done with good accuracy—on a bridge, for example—the end result is usually good enough for most practical purposes.

Selection of Component Values

Suppose that meter M in Fig. 16-16 is a 1.0-mA meter. Then the maximum current that can be permitted in the circuit with S open is, of course, 1.0 mA. If the battery voltage is 1.5 volts, then R_1, when adjusted, must be $R_1 = V/I_M = 1.5/0.001 = 1500$ ohms. Therefore, choose a value for R_1 which at maximum is somewhat greater than 1500, say, 2000 or 3000 ohms, and be sure that R_1 is set at maximum before starting the measurement procedure. Failure to do this results in the meter pointer being slammed into the end stop by excessive current. Remember that the object of this exercise is to avoid just such a calamity in the first place.

Find a suitable maximum value for R_2 by trial and error. For a 1.0-mA meter, start with a resistance with a maximum value of perhaps 150 or 200 ohms.

With S open, adjust R_1 for full-scale deflection. Close the switch and adjust R_2 for half-scale deflection; for best results, the half-scale reading should occur when the resistance of R_2 is in the top half of its travel. If R_2 is very close to the bottom of its movement when the half-scale reading is obtained, it would be a good idea to select a resistance with a *lower* maximum value and repeat the measurement.

SEVENTEEN

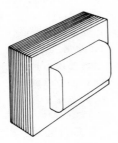

In Search
of Supertran

In today's burgeoning electrotechnology, which seems to be outstripping the fantasies of yesterday's science fiction, is the dowdy old low-frequency power transformer overdue for updating? Are discoveries that promise unprecedented advances in transformer design even now poised on the scientific brink? In that nebulous by-and-by known as the "foreseeable" future, will there be mighty midgets delivering incredibly vast power and dot-sized units tucked almost invisibly into diminutive circuits on silicon chips? Will someone create *supertran*, the super power transformer?

These are the dreams. This chapter tries to assess the realities. But first consider a brief review of the past that often acts as a fingerpost to the future.

THE PAST

The practicability of the transformer was first demonstrated by Elihu Thomson in 1879. Then, as now, it consisted simply of insulated copper wire wrapped around a ferromagnetic core. It was a device so simple, so fundamental, that the main thrust of development could be applied only to the core because there was little to be done with the windings. To be sure, there have been improvements to winding configurations and materials, all of them important, but it was from the core that the main dividends of research were (and are) expected to flow.

The core was initially of solid soft iron with a permeability of probably little more than 2,000. By 1883, Lucien Gaulard and John Dixon Gibbs had replaced the solid iron core with bunched lengths of iron wire to reduce eddy current loss. William Stanley of Westinghouse later formed the wires into a complete magnetic circuit. Then he made a laminated core with strips cut from iron photographic plates. Finally, stacked H and I pieces punched out of sheet iron was developed with the I pieces being fitted over the open ends of the Hs to complete the magnetic circuit. This was the forerunner of the modern stamping and was probably made of common carbon steel.

Between 1910 and 1915, a 4 percent silicon-steel alloy appeared with a permeability of around 3600 and an electrical resistivity some five times greater than for iron, thus markedly improving eddy current loss still further.

With refinements in processing and alloy content, the magnetic characteristics of carbon steel and silicon-steel steadily improved over the years. Permeability edged towards 9,000 and losses fell. The maximum flux density also improved; this is important, because as discussed in Chapter 3, high maximum flux density is needed to reduce transformer size for given power capability and frequency. Around the year 1950, the advent of grain-oriented silicon-steel provided permeability of 50,000 and more, low loss, and maximum flux density of 20,000 gauss compared to carbon steel's 18,000 gauss and straight silicon-steel's 16,500 or so was, therefore, a landmark advance. The surprisingly high maximum flux density of common carbon steel, it should be noted, is offset in many applications by its high loss.

About 1965, cobalt-vanadium-iron arrived with the highest known flux density to date of 23,000 gauss. Unfortunately, the price per pound is 30 to 70 times that of the older alloys. This high cost still leaves the field clear for common carbon steel, silicon-steel, and grain-oriented silicon-steel, which are to this day the preferred materials for general low frequency power transformer design. Cobalt-vanadium-iron is reserved for special cost-is-no-object applications.

The four alloys just discussed are now known as the *high flux alloys*.

NEW ALLOYS

The foregoing doesn't mean there were no other developments in magnetic metals. On the contrary, many new metals were, in fact, produced. In the period from 1920 to the present time, alloys in various permutations and percentages of nickel, iron, cobalt, vanadium, silicon, copper, and many other elements were tested. These, combined with sophisticated new heat treatments in gaseous environments and in magnetic fields, yielded an array of metals with dazzling magnetic characteristics suited to myriad of low flux applications outside the power transformer design field, for example: inductive loading of telephone and telegraph cables, high-speed relays, high-frequency generators, magnetic screening, TV scanning transformers, high-frequency power units and scores more. Ferrites, too, were born with their countless high-frequency and pulse applications; all magnificent in their respective fields, but not in the running for supertran because they don't have the high maximum flux density needed.

IS NEW PROGRESS OVERDUE?

Consider the distressing facts that this senescent device, the low frequency power transformer, is too large and cumbersome by far relative to today's ever-shrinking circuitry. It wastes hundreds of millions of dollars per year on electrical power in core losses and I^2R losses in the windings. It accounts for billions of dollars of capital costs in the power generating and distributing networks of the world and in the equipment of the home, commerce, research, industry, armed services, and the rest. Considering all this, is it not time for a dramatic surge forward in transformer development in order to keep pace with the needs of advancing technology in other fields—time for our electro-wizards to pull some new wonder out of the bag?

To be sure, interesting things have been happening, but are they leading to supertran? Let's look at them.

MAGNETIC GLASS

One strange item is currently being ladled from the bubbling cauldron of science—*glass* said to be suitable for transformer cores. It's so suitable, in fact, that the developers, Allied Chemical Company of New Jersey, claim that if all the residential distribution transformers in the U.S. power networks had glass cores, half of the estimated $200 million presently wasted in heating the core might be saved—a cool $100 million per year. And that doesn't count losses in other kinds of transformers throughout the continent.

Understand the term glass here in the context of the following abbreviated definition: "Glass is an inorganic product of fusion which has cooled to a rigid condition without crystallizing."

Metals, as normally produced, are crystalline in structure. By first melting the metal then rapidly quenching it, it can be brought through the transition stage from liquid to solid without crystallizing. One way of doing this is to direct a jet of the molten metal onto a chilled rotating wheel, as shown schematically in Fig. 17-1. An extremely thin ribbon (or sometimes a filament, depending on the set up) is formed. It has to be thin to achieve the super-fast cooling needed to avoid crystallization. The product is not transparent or brittle like conventional silicate glass.

Only certain metals, among them, fortunately, the most useful of the magnetics, can be transformed in this way. In this glass form, they are among the most easily magnetized and have low loss; they also have a high maximum flux density—though not as high as any of the current high-flux alloys. This latter limitation seems to exclude it at present from the supertran class as defined herein. But it is a relatively new material, still in its formative stages, and could prove to be a powerful stride in the right direction. The serious designer should investigate its possibilities and keep in touch with its progress.

GAS PRESSURE

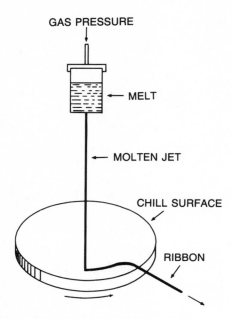

← MELT

← MOLTEN JET

CHILL SURFACE

RIBBON

Fig. 17-1. Schematic of chill block casting process for glassy metal.

SUPERCONDUCTIVITY

Is core development the only answer for supertran? Maybe in the light of recent research, the windings (which you may recall were deliberately bypassed at the beginning of this chapter) are on their way back into the picture.

In gales of excitement, dwellers in science's ivory towers are currently speculating that new discoveries in superconductivity could be the beginning of the most startling scientific revolution in history; an upheaval they say could radically change all electrotechnology and the working concepts of every electro-based device and system in existence.

Superconductivity was first reported by H. Kamerlingh Onnes in 1911 when he discovered that zero resistance (not just near zero, but actual zero ohms) occurred in certain metals at cryogenic temperatures close to absolute zero or zero degrees Kelvin (−459 degrees F) (see Table 17-1 for comparative temperatures on Kelvin, Fahrenheit, and Celsius scales).

This means that a current once established in a closed ring of superconductive material will flow *forever* without I^2R loss or need of any kind of generator to sustain it as long as the temperature is maintained below the critical level, a severe limitation to practical applications.

In 1986, however, an IBM laboratory in Zurich achieved zero resistance in a ceramic oxide conductor at 35 degrees K (−396 degrees F), a temperature not exactly suited to brazen simians, but still significantly warmer than absolute zero.

Table 17-1. Temperature Conversions between Kelvin, Fahrenheit, and Celsius

Given Kelvin:	$F = \frac{9}{5}(K - 273.16) + 32$
	$C = K - 273.16$
Given Fahrenheit:	$K = \frac{5}{9}(F - 32) + 273.16$
	$C = \frac{5}{9}(F - 32)$
Given Celsius:	$K = C + 273.16$
	$F = \frac{9}{5}C + 32$

Shortly after, reports began to roll in from laboratories all over the world of other ceramic material, that exhibited zero resistance at successively warmer temperatures: at 77 degrees K (-321 degrees F), then at 93 degrees K (-292 degrees F), achieved by Paul Chu at the University of Houston. Also reported was "evidence" of zero resistance at a relatively balmy 240 degrees K (-28 degrees F), and even higher at a chicken-roasting 500 degrees K or 440 degrees F. The word "evidence" here really means maybe, maybe not; that is to say that while a positive indication might have been obtained in a specific material, the result could not be consistently repeated. Therefore, Chu's record of 93 degrees K still stands for the moment.

Incidentally, how do you detect superconductivity in a substance? One method used in earlier experiments is to place a ring of the specimen material in a magnetic field supplied by, say, a bar magnet. Bring the temperature of the specimen to the level at which you hope to find superconductivity. Withdraw the magnet. The action of withdrawing the magnet induces a current flow in the ring that, in turn, creates a magnetic field around it. Test for persistance of the magnetic field. If you find it, you have "evidence" of superconductivity, or zero resistance, within the conductor. Laboratories today have equipment that automatically graphs resistance against temperature to give a curve similar to that shown in Fig. 17-2A. Here, superconductivity is shown by the sudden downturn in the curve.

An important characteristic of superconducting materials, discovered in 1933 by W. Meissner, is that they are perfectly diamagnetic. This means that in a magnetic field, they actually squeeze the magnetic flux out of the conductor, leaving zero field inside it. In this state, the superconductor will repel a magnet regardless of which pole is presented to it. A dramatic laboratory demonstration of superconductivity uses this repulsion to magically levitate a small magnet so that it floats freely in space above the superconductor, Fig. 17-2B.

SUPERTRAN AT LAST?

How might transformer design be affected by this phenomenon?

Let's take a superficial, and, admittedly, tongue-in-cheek, look at superconductivity's enchanted kingdom. For comparison, see Chapter 6, which deals with the effects of resistance R in conventional windings.

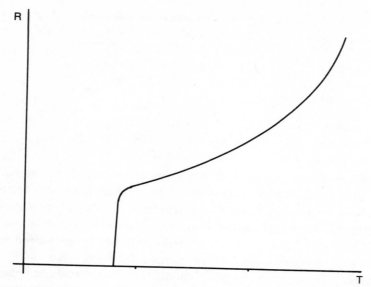

Fig. 17-2A. As temperature is reduced, sudden downturn in curve indicates onset of superconductivity.

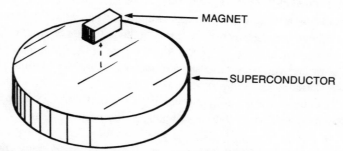

Fig. 17-2B. Magnet floats above a "slug" of superconducting material.

If the windings of a transformer were made "ohmless", the I²R power loss and IR voltage drop would, of course, be zero. Then regardless of how great the current flow, the conductors chosen could be of the smallest possible gauge without heating up. In turn, the cross-sectional area of the core could be reduced to the tiniest size possible by increasing many-fold the number of turns of the exceedingly fine wire in accordance with the simple rearrangement of Equation 3-14, Chapter 3:

$$a = \frac{V \times 10}{4FfNB}$$

It is obvious here that increasing turns N decreases cross-sectional area a. The other core dimensions can also then be reduced to the smallest proportions that can accommodate the fine wire, thus reducing the core size and the watts-per-pound loss to insignificant proportions. More wonderful still, in the topsy-turvy world of superconductivity, given the ability to manufacture and handle "zero resistance" wire in gauges small enough, then all transformers, regardless of the required power capability—whether hundreds of KVA or just fractions of a watt—*could be made more or less the same small size and virtually lossless*! Ergo—ultra supertran!

WHERE IS THE CATCH?

Speculation like this with its vague and unsubstantial proofs is easy. The hard part is to solve the problems that stand in the way of making and using a real "zero resistance" transformer.

Consider, for instance, some of the peculiarities of superconductive materials. If the current in the conductor exceeds a critical value, the property of zero resistance simply vanishes. Similarly, if the magnetic field in which the material is working is too strong, the superconductivity again vanishes. Today's superconductors are susceptible to those conditions; additionally, they are fragile and non-ductile and therefore are not easily coil wound, especially in the superfine gauges envisaged here.

The biggest problem could prove to be that of intergrating the thin superconductors into conventional circuits. The relatively large conventional conductors carry massive current, and the presence of a superconductor could create intolerable resistance and destructive heat at the point of contact.

There is also the matter of maintaining the ambient temperature around the transformer at below the critical zero resistance level. Even with materials that are "ohmless" at relatively warm temperatures, this still might not be easy in many applications.

These are not trivial problems, nor are they necessarily the only ones likely to arise in superconductivity's development.

IN THE CRYSTAL BALL

Although the crystal globes of some science visionaries are aglow with pictures of superconductivity's future marvels, the pictures are, after all, just prophetic images of what is still a laboratory novelty. There is nothing here to suggest the imminent transition of the phenomenon to a practical workaday technology such as transformer design. On the contrary, our more cautious savants guess that significant progress in that direction could take 20, 30, or 40 years—if ever. As for arriving at supertran via improved cores—a process that has already taken 109 years to bring the power transformer to its present unremarkable state—the industry believes that alloys with maximum flux densities better than 30,000 gauss are unlikely in the foreseeable future.

Dispensers of research funds might—and probably will—argue that on the evidence, supertran is a doomed concept, or at best, a too-distant dream, and pull tight the drawstrings of their moneybags, choking off the dream right there.

This is not to say, however, that solutions will not be found. There are plenty of individual adventurers and groups around who are willing to throw their all into the pot at the whiff of a good idea. For all we know, the legendary dark horse, some new discovery, might already be galloping up on the outside in a bid to win the supertran stakes. This might be, perhaps, some revolutionary concept that might entirely supersede the traditional transformer principle, even as the transistor, for example, superseded the vacuum tube.

If at last the problem cannot be solved, why not simply change the problem by concentrating on systems that don't need transformers, in the same way, say, that electrical conductors and related apparatus are currently being replaced by fiberoptics in telephone communications.

In the meantime, supertran, or a viable substitute thereof, potentially one of the worlds richest prizes, still awaits the winning.

Go to it—and good luck.

──────── **Epilog: New Theory of Superconductivity Cools the Debate** ────────

The notion that room-temperature superconductivity is the "open sesame" to an Aladdin's cave of glittering electrical treasures might regrettably be in eclipse. In September, 1988, two months after this chapter was written, scientists of the California Institute of Technology (Caltech) presented a new theory of superconductivity, involving the quantum mechanics of bonds between atoms, to a meeting of the American Chemical Society in Los Angeles. This theory plausibly explains for the first time the main features of the phenomenon and predicts that room-temperature superconductivity *is not possible* with today's materials. Nevertheless, the theory could assist in the development of superconductors for higher temperatures than are presently possible, assert Caltech's scientists.

Asked by the *New York Times* to comment on the theory, Dr. Paul Chu (mentioned earlier in this chapter) pointed out that he had not yet studied the new theory but cautiously replied, "If it can really predict the conditions of superconductivity, it is very important."

And so the pendulum swings.

References

MAGNETISM AND ELECTROMAGNETISM

Ferromagnetism, Richard M. Bozorth, D. Van Nostrand Company, 1951

Telecom Principles, R. N. Renton, A.M.I.E.E., New Era Publishing Co., 1950

Encyclopedia Britannica, Encyclopedia Britannica Inc., 1963

TRANSFORMER DESIGN

Magnetic Circuits and Transformers, E. E. Staff M.I.T, John Wiley & Sons, 1943

Electronic Transformers, Harold H. Nordenburg, Reinhold Publishing Corp., 1964

Transistor Circuit Design, Texas Inst., McGraw-Hill, 1963

Electronic Transformers & Circuits, Reuben Lee, John Wiley & Sons, 1962

Transformers, E. E. Wild, Blackie and Son, Ltd., 1946

Radio Engineers Handbook, Frederick Emmons Terman, Sc.D., McGraw-Hill, 1943

American Electricians' Handbook, McGraw-Hill, 1961

CORE MATERIALS

Magnetic Materials, Peter Muller-Munk, Allegeny-Ludlum Steel Corp., 1947

Materials for High Flux Devices, G. B. Finke, Magnetic Metals Co.

High Permeability and Low Remanence Tape Wound Transformer Cores, Guenter B. Finke, Magnetic Metals Co.

Tape Wound Core Design Manual, Magnetic Metals Co.

Encyclopedia Britannica, Encyclopedia Britannica Inc., 1963

Metallic Glasses a New Technology, J.J. Gilman, Allied Chemical Co.

Technology of Metallic Glasses, Lance A. Davis, Nicholas J. DeCristofaro, Carl H. Smith, Allied Chemical Co., 1980

A New High Flux Magnetic Material for High Frequency Application, David Nathasingh, Carl H. Smith, Allied Chemical Co.

SUPERCONDUCTIVITY

Superconductivity, D. Shoenburg, 1952

Encyclopedia Britannica, Encyclopedia Britannica, Inc., 1963

''Superconductivity,'' Arthur Fisher, *Popular Science*, April 1988

Race for the Superconductor, TV Transcript, WGBH BOSTON, 1987

Index

Other Bestsellers From TAB

TAB BOOKS Inc.
Help Us Help You!

So that we can better fill your reading needs, please take a moment to complete and return this card. We appreciate your comments and suggestions.

1. I am interested in books on the following subjects:

- ☐ automotive
- ☐ aviation
- ☐ business
- ☐ computer, hobby
- ☐ computer, professional
- ☐ engineering (specify): _____
- ☐ other (specify) _____
- ☐ other (specify) _____

- ☐ electronics, hobby
- ☐ electronics, professional
- ☐ finance
- ☐ how to, do-it-yourself

2. I own/use a computer:

- ☐ IBM _____
- ☐ Apple _____
- ☐ Commodore _____
- ☐ Other (specify) _____

- ☐ Macintosh _____
- ☐ ATARI _____
- ☐ AMIGA _____

3. This card came from TAB book (specify title and/or number):

_____

4. I purchase books:

- ☐ from general bookstores
- ☐ from technical bookstores
- ☐ from college bookstores
- ☐ other (specify) _____

- ☐ through the mail
- ☐ by telephone
- ☐ by electronic mail

Comments _____

_____

Name _____

Address _____

City _____

State _____ Zip _____